IEEE Recommended Practice for Protection and Coordination of Industrial and Commercial Power Systems

IEEE Recommended Practice for Protection and Coordination of Industrial and Commercial Power Systems

Sponsor

**Industrial and Commercial Power Systems Committee
of the
IEEE Industry Applications Society**

Library of Congress Catalog Number 75-27282

Foreword

At the October 1965 IEEE Industrial and Commercial Power Systems Conference in Buffalo a paper was presented to the attendees by Tom Higgins and Norman Peach. That paper, resulting from a suggestion by A. C. Friel (deceased), proposed a new IEEE publication covering protection and coordination for industrial and commercial power systems. The first sentence of that paper states "The proposed publication is intended for the plant electrical engineer with broad responsibilities for the plant electric system. He is familiar with the general principles of industrial system design, but has not had an opportunity to specialize in system protection." The Power System Protection Subcommittee of the Industrial and Commercial Power Systems Committee (I and CPS) accepted the responsibility of preparing the new book.

During the next seven years, thirteen of the fifteen chapter subjects were presented formally at sessions of IEEE conferences in which the I and CPS Committee participated. The last three years were occupied in reviewing and updating the papers and obtaining the necessary approvals. Copies of the manuscript were forwarded to each of the 74 I and CPS Committee members and approved through the usual IEEE Standards Procedures.

The content of this standards document is evident from an inspection of the title of each chapter. The value of proper and adequate protection in electric power distribution systems has been emphasized by the provisions of the Occupational Safety and Health Act of 1969. Good coordination helps to ensure the availability of electric energy, a requirement of steadily increasing economic importance.

The I and CPS Committee will maintain this document current with the state of developing technology, and revisions of this book will be issued at reasonable intervals in order to maintain all chapters current with the rapidly changing techniques of circuit protection. Suggestions for improvements are welcomed by the I and CPS Committee. These should be directed to the IEEE Standards Board, 345 E. 47th Street, New York, NY 10017.

This standards document, IEEE Std 242-1975 Recommended Practice for Protection and Coordination of Industrial and Commercial Power Systems will be known as the IEEE Buff Book. It takes its place alongside the following existing companion I and CPS Committee sponsored documents which have already proven their worth.

Recommended Practice for Electric power Distribution for Industrial Plants (IEEE Red Book), IEEE Std 141-1969.

Recommended Practice for Grounding of Industrial and Commercial Power Systems (IEEE Green Book), IEEE Std 142-1972.

Recommended Practice for Electric Power Systems in Commercial Buildings (IEEE Gray Book), IEEE Std 241-1974.

Recommended Practice for Emergency and Standby Power Systems (IEEE Orange Book), IEEE Std 446-1974.

Acknowledgement

Appreciation is expressed to members of the Power System Relaying Committee of the IEEE Power Engineering Society for their review and valuable suggestions in the preparation of this standards document.

Protection and Coordination of
Industrial and Commercial Power Systems

Prepared by

Power System Protection Subcommittee of the
Industrial and Commercial Power Systems Committee

Working Group Members and Contributors

D. Dalasta, Norman Peach, *Cochairmen*
R. H. Kaufman, R. L. Smith, *General Reviewers*
W. A. Weddendorf, *Past Subcommittee Chairman*

Chapter 1 — First Principles: Thomas D. Higgins, *Chairman*, Norman Peach.

Chapter 2 — Calculation of Short-Circuit Currents: Norman Peach, *Chairman*, Russell O. Ohlson, B.E. DeRoy.

Chapter 3 — Selection and Application of Overcurrent Relays: Kao Chen, *Chairman*.

Chapter 4 — Instrument Transformers: Warren H. Cook, *Chairman*, James W. Thomas, E.F. Hussa, Jr.

Chapter 5 — Fuse Characteristics: Charles W. James and Harold Fahnoe, *Cochairmen*, Kenneth W. Swain.

Chapter 6 — Low-Voltage Circuit Breakers: Paul J. Reifschneider, *Chairman*, Ward Laubach.

Chapter 7 — Coordination: William A. Weddendorf and Robert L. Smith, *Cochairmen*, Bruce G. Bailey, A.A. Regotti, Edgar R. Burgin (deceased).

Chapter 8 — Ground-Fault Protection: D. Dalasta, *Chairman*, Bruce G. Bailey, Norman Peach, Russell O. Ohlson, Roy H. Comstock, George R. Elder, B.W. Whittington.

Chapter 9 — Motor Protection: D. Vern Fawcett, *Chairman*.

Chapter 10 — Transformer Protection: P.E. Breniman, *Chairman*, N. Depool, T.W. Olsen.

Chapter 11 — Conductor Protection: Timothy T. Ho and Keith Grimm, *Cochairmen*, Russell S. Davis, Kent Stiner.

Chapter 12 — Bus and Switchgear Protection: William P. Burt, *Chairman*, Samuel P. Axe.

Chapter 13 — Service Supply Line Protection: D. Dalasta, *Chairman*, J. Harry Beckman, Thomas D. Higgins, Eugene W. Hendron.

Chapter 14 — Generator Protection: Eugene M. Smith, *Chairman*, Roger T. Greenwood.

Chapter 15 — Maintenance Testing and Calibration: Henry S. Orth, *Chairman*, Wesley C. Dorothy, J.C. Wilson.

Editorial Consultant: Ingeborg Stochmal. Production: Michael Weinstein.
Artwork: Dorothy Brosterman. Jacket Design: T. Dale Goodwin.

low-voltage system (electric power). An electric system having a maximum rms ac voltage of 1000 volts or less.

medium-voltage system (electric power for industrial and commercial systems only). An electric system having a maximum rms ac voltage above 1000 volts to 72 500 volts.

high-voltage system (electric power for industrial and commercial systems only). An electric system having a maximum rms ac voltage above 72 500 volts to 242 000 volts.

extra-high-voltage system (electric power). An electric system having a maximum rms ac voltage above 242 000 volts to 800 000 volts.

Voltage Class	NOMINAL SYSTEM VOLTAGE (volts)			Maximum System Voltage	Reference
	Two-Wire	Three-Wire	Four-Wire		
Low Voltage		Single-Phase Systems			IEEE Std 100-1972 (ANSI C42.100-1972) defines low-voltage system as "less than 750 V." It is now being proposed as "maximum rms ac voltage of 1000 V."
	(120)	120/240		127 or 127/254	
		Three-Phase Systems			
			208Y/120	220	
	(240)		240/120	254	
	480		480Y/277	508	
	(600)			635	
Medium Voltage	(2400)			2 540	These voltages are listed in ANSI C84.1-1970 but are not identified by voltage class.
	4160			4 400	
	(4800)			5 080	
	(6900)			7 260	
		12 470Y/7200		13 200	
		13 200Y/7620		13 970	Note that additional voltages in this class are listed in ANSI C84.1-1970 but are not included because they are primarily oriented to electric-utility system practice.
	13 800	(13 800Y/7970)		14 520	
	(23 000)			24 340	
		24 940Y/14 400		26 400	
	(34 500)	34 500Y/19 920		36 510	
	(46 000)			48 300	
	69 000			72 500	
High Voltage	115 000			121 000	ANSI C84.1-1970 identifies these as "higher voltage three-phase systems" (as well as 46 kV and 69 kV).
	138 000			145 000	
	(161 000)			169 000	
	230 000			242 000	
Extra High Voltage	345 000			362 000	ANSI C92.2-1974
	500 000			550 000	
	735 000—765 000			800 000	

System Voltages shown without brackets are preferred.
See paragraph 2.2 on page 8 of ANSI C84.1-1970.

NOTES:

(1) Voltage class designations were approved for use in documents published by the IEEE Industrial and Commercial Power Systems Committee at its Committee meeting on May 16, 1973 (and on June 6, 1974 and January 28, 1975).

(2) Limits for the 600-volt Nominal System Voltage established by ANSI C84.1a-1973.

(3) The maximum and minimum voltage ranges for system voltages from 120 V to 230 kV are given in ANSI C84.1-1970, "Voltage Ratings for Electric Power Systems and Equipment (60 Hz)."

(4) It is appropriate to observe that the voltage class of Medium Voltage has had long acceptance in industrial practice. In contrast, the term Medium Voltage does not as easily apply to electric-utility practice because of the traditional use of overlapping voltage areas of transmission and distribution (as defined in the IEEE Std. 100-1972, "Standard Dictionary of Electrical and Electronics Terms").

(5) The preferred maximum voltage for three-phase system voltages from 345 to 735-765 kV are given in ANSI C92.2-1974, "Preferred Voltage Ratings for Alternating-Current Electrical Systems and Equipment Operating at Voltages Above 230 Kilovolts Nominal." When ANSI C92.2-1974 superseded ANSI C92.2-1967 the specific identification of the Nominal System Voltages for the EHV systems were withdrawn. They are now only identified as "Typical 'Nominal' Voltages." (The limits for the EHV systems which appear in ANSI C84.1-1970, page 9, are superseded by ANSI C92.2-1974.)

Contents

FIGURES

TABLES

1. First Principles

1.1 **Introduction.** All power systems, whether they be utility, industrial, commercial, or residental, have the common purpose of providing electric energy to the utilization equipment as safely and as reliably as is economically feasible. The relative importance of economic, reliability, and safety considerations may vary somewhat with the type of system, but all three elements must be taken into consideration in any good system design, and certain minimum safety and reliability requirements must be satisfied.

The design of protection and coordination for utility systems lies beyond the scope of this publication. Although some extremely large industrial systems surpass many utility systems in complexity and scope, the protection and coordination of such systems also lies beyond the scope of this publication. Nevertheless, much of the material presented may be applicable to these systems. The protection and coordination of residential systems may fall within the area of interest of many of the same builders and architects who are concerned with the design of small to medium industrial and commercial systems, but in most instances, the National Electrical Code (NEC) (NFPA No 70, 1975, and ANSI C1-1975) and applicable local codes adequately cover safety requirements for such systems. No specific attempt will be made to cover these systems, although, as in the case of large industrial and utility systems, much of the material presented may be applicable.

This publication will be limited to coverage of system protection and coordination as it pertains to system design treated in IEEE Std 141-1969, Electric Power Distribution for Industrial Plants, and IEEE Std 241-1974, Electric Power Systems in Commercial Buildings.

1.2 **Need for System Protection and Coordination.** This publication deals with one of the most important and least appreciated and understood aspects of electric power system design, the proper selection, application, and coordination of that group of components which constitute system protection for industrial plants and commercial buildings.

If the designer needed to consider only normal operation, his task would be relatively easy. He could assume that there would be no equipment failures, no operating mistakes, and no "acts of God" such as floods, fires, wind storms, or lightning strokes. He would only have to design an installation capable of producing or receiving and delivering sufficient electrical energy for the utilization equipment to satisfy the initial load requirements with some allowance for reasonable load growth. A design based solely on normal operational requirements would, in practice, be totally inadequate and would inevitably result in intolerable equipment outages. Any sound design of an electric power system must be predicated on the assumptions that equipment will fail, that people will make mistakes, and that "acts of God" will occur.

The function of system protection and coordination is to minimize damage to the system and its components and to limit the extent and duration of service interruption whenever equipment failure, human error, or "acts of God" occur on any portion of the system. Economic considerations and the choice of system components will determine the degree of system protection and coordination which can be feasibly designed into a system. Failure to design into a system provisions for protection and coordination sufficient to satisfy at least the minimum system safety and reliability requirements will inevitably result in unsatisfactory performance. Modifying an inadequate existing system to attain greater safety and reliability will prove more expensive and in most cases less satisfactory than designing these features into the system at the outset.

1.3 Philosophy of System Protection and Coordination.

This publication presents in a step-by-step, simplified yet comprehensive, form the principles of system protection and the proper application and coordination of those components of system protection which may be required to protect industrial and commercial power systems against any abnormalities which could reasonably be expected to occur in the course of system operation.

The designer has available to him several methods to minimize the effects of such system abnormalities on the system itself or on the utilization equipment which it supplies. He can design into the electric system features which will

(1) Quickly isolate the affected portion of the system while maintaining normal service for the rest of the system and minimizing damage to the affected portion

(2) Minimize the magnitude of the available short-circuit current to minimize potential damage to the system, its components, and the utilization equipment it supplies

(3) Provide alternate circuits, automatic throwovers, and automatic reclosing devices where applicable to minimize the duration and extent of supply and utilization equipment outages

The system protection aspect of electric power system design encompasses all of these methods. This publication deals almost exclusively with the quick isolation of the affected portion of the system. The use of the other methods will be dictated by economic and reliability considerations governing the choice of basic system design and selection of system components. The function of system protection may be defined as "the detection and prompt isolation of the affected portion of the system whenever a short circuit or other abnormality occurs which might cause damage to, or adversely affect, the operation of any portion of the system or the load which it supplies."

1.3.1 *Initial Planning of System Protection.* The system protection aspect is one of the most essential features of electric system design and must be considered concurrently with other essential features. There is a tendency to consider the subject of system protection only after all other features have been determined and basic system design has irrevocably been fixed. Such an approach often results in a poorly designed system which can be adequately protected only at a disproportionately high cost. The end result is usually poor protection and unsatisfactory coordination of protective devices. Since the basic purpose of electric power system design is to supply electric energy to the utilization equipment as safely and reliably as is economically feasible, such an approach will seldom result in sound system design. The question of system protection is so basic to both the safety and the reliability of electric supply and can have such profound influence on the economics of system design that examining system protection needs only after all other design features have been fixed is completely unrealistic. Any competent designer must examine the question of system protection at each stage of the design planning and develop a fully integrated system protection plan which is capable of being properly coordinated, and is flexible enough to accommodate system expansion.

1.3.2 *Need for Simplicity.* With the exception of some extremely large industrial plants, industrial or commercial installations do not have the personnel, equipment, or know-how to adequately service and maintain system protection for a complex system. In laying out small to medium industrial systems and commercial systems of all sizes, the designer should endeavor to keep the final design as simple as is compatible with safety, reliability, and economic considerations. Designing for additional reliability and flexibility leads to additional complexity in the system and hence in the protection scheme. In the absence of a clear comprehension of the problems associated with system protection and coordination and the equipment and trained personnel required to properly service and maintain the protective scheme, any additional complexity may prove self-defeating by actually lowering the level of system reliability.

Although the cost of system protection is usually small compared to the cost of the entire system, this cost can be minimized by designing a simple system. The cost of servicing system protection and the need for extensive testing can be minimized and the probability of obtaining and maintaining good protection coordination can be maximized through simplification of the system design. By endeavoring to keep the system protection and coordination scheme as simple as possible the designer will, by eliminating some contemplated flexibility and reliability features which would be difficult to service and maintain, actually materially lower the overall cost of the design.

1.4 Abnormalities to Be Protected Against. Much of the material in the ensuing chapters of this publication will be geared to the problems of industrial plants which are generally more complex than commercial-building power systems. However, electric requirements for commercial buildings have grown to the point where many of these systems approach or surpass in complexity those of typical industrial plants. Potential losses to some commercial enterprises, arising from unnecessary interruptions of electric service, already exceed potential losses to the majority of industrial operations. Although the examples in the text are substantially applicable to electric power distribution for industrial plants, the same basic principles and the same basic approach are equally applicable to electric systems for commercial buildings.

Electric power systems in today's industrial plants and large commercial establishments handle enormous quantities of energy which, if not properly confined, can cause catastrophic damage and seriously endanger personnel. A review of the trend in electric energy usage in such establishments indicates that such energy usage has been doubling every seven to eight years and shows little signs of leveling off. Many industrial processes and commercial operations demand a high degree of continuity of electric power supply because of the great costs of production down time, whether such production be basically

the output of industrial products as in the case of industrial processes or the output of services as in the case of most commercial installations. Trends toward highly automated operations have not only increased the number of processes and commercial operations requiring a high degree of service reliability, but have also reduced the tolerable limits of voltage variation. Protection for electric systems should be designed with the following objectives in mind:

(1) Prevent injury to personnel
(2) Prevent or minimize damage to equipment
(3) Minimize interruption of power
(4) Minimize the effect of the disturbance on the uninterrupted portion of the system, both in its extent and duration
(5) Minimize the effect on the utility system

The first eight chapters deal with a step-by-step approach to the general problem of system protection and coordination and are devoted to the detection and prompt isolation of the affected portion of the system whenever a short circuit occurs. Other sources of disturbances such as lightning, load surges, loss of synchronism, etc, usually have little overall effect on system coordination and can be handled on an individual basis for the specific equipment to be protected. Many of these items will be covered in those chapters dealing with specific equipments such as transformers, motors, generators, etc.

Treatment of the overall problem of system protection and coordination of electric power systems will be restricted to the selection, application, and coordination of devices and equipments whose primary function is the isolation and removal of short circuits from the system. Short circuits may be phase to ground, phase to phase, phase to phase to ground, three phase, or three phase to ground. Short circuits may range in magnitude from extremely low current faults having high impedance paths to extremely high current faults having very low impedance paths. However, all short circuits produce abnormal current flow in one or more phase conductors or in the neutral or grounding circuit. Such disturbances can be detected and safely isolated.

1.5 Protective Equipment Used to Isolate Short Circuits. The isolation of short circuits requires the application of protective equipment which will sense an abnormal current flow and remove the affected portion from the system. The sensing device and interrupting device may be completely separate, interconnected only through external control wiring, or they may be the same device or separate devices mechanically coupled to function as a single device.

1.5.1 *Overcurrent Relays.* Overcurrent relays are sensing devices only and must be used in conjunction with some type of interrupting device to interrupt a short circuit and isolate the affected portion of the system. These relays may be either directional or nondirectional in their action. They may be instantaneous or time delay in response. Various time—current characteristics, such as inverse time, very inverse time, extremely inverse time, and definite minimum time, are available over a wide range of current settings. For specific applications, various types of differential overcurrent relays are available. Overcurrent relays and their selection, application, and settings are covered in detail in Chapter 3. Such relays generally are used in conjunction with instrument transformers which are covered in Chapter 4.

1.5.2 *Fuses.* Fuses are the oldest and simplest of all protective devices. The fuse is both the sensing and the interrupting device. They are installed in series with the circuit and operate by the melting of a fusible link in response to the current flow through them on an inverse time—current basis. They are one-shot devices since their fusible elements are destroyed in the process of interrupting the current flow. Fuses may have only the ability to interrupt short-circuit current up to their maximum rating or the ability to limit the magnitude of short-circuit current by interrupting the current flow before it reaches its maximum value. Fuses, their characteristics, applications, and limitations are described in detail in Chapter 5.

1.5.3 *Circuit Breakers.* Circuit breakers are interrupting devices only and must be used in conjunction with sensing devices to fulfill the detection function. In the case of medium voltage (1—72.5 kV) circuit breakers, the sensing devices are separate protective relays or combinations of relays covered in Chapter 3. In the case of low-voltage (under 1000 V) circuit breakers, sensing devices may be external protective relays or com-

binations of relays. In most low-voltage applications, either molded-case circuit breakers or other low-voltage circuit breakers having series sensing devices built into the equipment are used. Low-voltage circuit breakers, their application, and characteristics are covered in Chapter 6.

1.5.4. *Fused Interrupter Switches.* Fused interrupter switches are a hybrid form of circuit protection which function exactly the same as fuses under short-circuit conditions but which may, under certain circumstances, function as circuit interrupters. Application of these devices to function as automatic circuit interrupters to interrupt currents higher than simple overloads requires a detailed knowledge of these devices, the risks involved, and the consequences of device failure. All such applications require specialized engineering and economic evaluations so that they will be treated and coordinated as fuses and no attempt will be made to incorporate the interrupter in the coordination scheme.

1.6 Basic Protection. The designer of an electric power system must first determine the load requirements, including the sizes and types of load, the location of principal loads, and any special requirements. He should also determine the available short-circuit current at the point of delivery, the time—current curves and settings of the nearest utility company protective device, and any contract restrictions on ratings and settings of protective devices in the user's system.

The designer can then proceed with a preliminary system design. Chapter 2 covers the fundamentals of short-circuit behavior and the calculation of short-circuit duty requirements which will permit evaluation of the preliminary design for compatibility with available ratings of circuit breakers and fuses. At this point some modification of the design may be necessary because of economic considerations or equipment availability. Preliminary system design should be evaluated from the standpoint of system coordination as covered in Chapter 7. If the protection provided in the preliminary design cannot be coordinated with utility company settings and contractual restrictions on protective device settings, the design must be modified to provide proper coordination.

1.7 Special Protection. In addition to developing

a basic protection design, the designer may also need to develop protective schemes for specific pieces of equipment or for specific portions of the system. Such specialized protection must be coordinated with the basic system protection as covered in Chapter 7. Specialized protection applications include

Chapter 9 Motor Protection
Chapter 10 Transformer Protection
Chapter 11 Conductor Protection
Chapter 12 Bus and Switchgear Protection
Chapter 13 Service Supply Line Protection
Chapter 14 Generator Protection

1.8 Ground-Fault Protection. Ground-Fault protection is an essential part of system protection, and is given detailed coverage in Chapter 8 for two basic reasons. First, although many of the devices used to attain ground-fault protection are similar to those covered in Chapters 3 through 6, the need for this protection and the potential problems of proper application are frequently not properly appreciated. Hence, separate treatment of this important element of system protection is used to highlight its importance. Second, proper coordination of ground-fault protection seldom has any significant effect on overall system coordination, although its effect must be taken into consideration in the same general manner covered in Chapter 7.

1.9 Field Followup. Proper application of the principles covered in the first fourteen chapters of this publication will result in the installation of system protection capable of coordinated selective isolation of system faults. But this capability will be to no avail if the proper field followup is not planned and executed. This followup consists of three elements: proper installation, including testing and calibration of all protective devices; proper operation of the system and its components; and a proper preventive maintenance program, including periodic retesting and recalibration of all protective devices. A separate chapter, Chapter 15, has been included to cover testing and maintenance.

1.10 Standards References. The following standards publications were used as references in preparing this chapter.

IEEE Std 141-1969, Electric Power Distribution for Industrial Plants
IEEE Std 142-1972, Grounding of Industrial and Commercial Power Systems (ANSI C114.1-1973)
IEEE Std 241-1974, Electric Power Systems in Commercial Buildings
NFPA No 70, National Electrical Code (1975), (ANSI C1-1975)

2. Calculation of Short-Circuit Currents

2.1 General Discussion. Certain simplifying assumptions are customarily made when calculating fault current. An important assumption is that the fault is "bolted," that is, it has zero impedance. This assumption not only simplifies calculation, but also applies a safety factor since the calculated values are a maximum, and equipment selected on this basis is rarely stressed beyond its full rating. Furthermore a three-phase fault is customarily assumed because this type of fault generally results in the maximum short-circuit current available in a circuit. In high-voltage systems the three-phase fault is frequently the only one calculated.

The actual fault current is usually less than the bolted three-phase value. Bolted line-to-line currents are about 87 percent of the three-phase value, while bolted line-to-ground currents can range from about 25 or 60 percent to 125 percent of the three-phase value, depending on system parameters. (Line-to-ground currents of more than the three-phase value rarely occur in industrial and commercial systems.) Actual faults, especially line-to-ground faults, usually involve arcing. Ground-fault currents, particularly in low-voltage systems, are often less than normal load currents, yet can be extremely destructive. These considerations have led to various techniques specifically directed to ground-fault protection (see Chapter 8).

2.1.1 *Sources of Fault Current.* The basic sources of fault current are (1) the utility supply system, (2) generators, (3) synchronous motors, and (4) induction motors.

A typical modern electric utility represents a large and complex interconnection of generating plants. The individual generators, in a typical system, are not affected by a maximum short circuit in an industrial plant as such. Transmission and distribution lines and transformers introduce impedance between the utility generators and the industrial customer. Were it not for this impedance, the utility system would be an infinite source of fault current. Before performing a calculation, accurate values of present and projected available short-circuit currents and the X/R ratio, or the $R + j X$ source impedance at the delivery point, must be obtained from the supplying utility.

In-plant generators react to system short circuits in a characteristic way. Fault current from a generator decreases exponentially from a relatively high initial value to a lower steady-state value some time after the initiation of the fault. Since a generator continues to be driven by its prime mover, and to have its field energized from its separate exciter, the steady-state value of fault current will persist unless interrupted by some circuit interrupter.

For purposes of fault-current calculation, the variable reactances of a generator can be represented by three reactance values.

X_d'', the direct-axis subtransient reactance, determines the current magnitude during the first cycle after the fault occurs. (It is the time-zero effective reactance of the armature circuit including the effect of all transient-induced currents in the magnetic field structure as well as the excitation winding.)

X_d', the direct-axis transient reactance, deter-

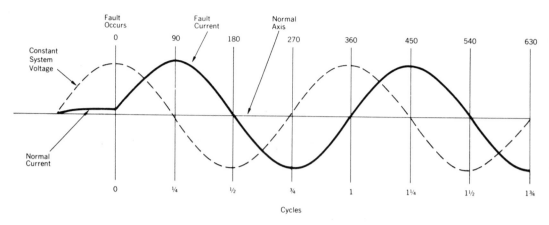

Fig 1
A Symmetrical Short-Circuit Current Wave

mines the current magnitude in the 1/2 to 2 s range. (It is the time-zero effective reactance of the armature circuit with only the current in the excitation winding itself allowed to flow.)

X_d, the direct-axis synchronous reactance, determines the current flow after a steady-state condition is reached. (It occurs when all transient currents in the excitation winding and field structure are reduced to zero.)

As most fault protective devices, such as circuit breakers or fuses, operate before steady-state conditions are reached, the generator synchronous reactance is seldom used in calculating fault currents for the application of these devices.

Synchronous motors supply current to a fault in much the same manner as do synchronous generators. The drop in system voltage due to a fault causes the synchronous motor to receive less power from the system for driving its load. The inertia of the motor and its load acts as a prime mover, and with field excitation maintained, the motor acts as a generator to supply fault current. This fault current diminishes because the motor will slow down as the kinetic energy is dissipated, thus reducing the voltage generated and because of motor field excitation decay.

The same designation as described for a generator is used to express the variable reactance of a synchronous motor. However, numerical values of the three reactances X_d'', X_d', and X_d will often be different for motors than for generators.

The fault-current contribution of an induction motor results from generator action produced by inertia driving the motor after the fault occurs. In contrast to the synchronous motor, the field flux of the induction motor is produced by induction from the stator rather than from a direct-current field winding. Since this flux decays on removal of source voltage resulting from a fault, the contribution of an induction motor drops off rapidly, ultimately disappearing completely upon loss of voltage. As field excitation is not maintained, there is no steady-state value of fault current as for synchronous machines. As a consequence, induction motors are assigned only a reactance which is about equivalent to the synchronous-machine subtransient reactance X_d''. This value will be about equal to the locked-rotor reactance, and hence the fault-current contribution will be about equal to the full-voltage starting current of the machine for a fault at the motor terminals.

Wound-rotor induction motors normally operate with their rotor rings short-circuited and will contribute fault current in the same manner as a squirrel-cage induction motor. Occasionally large wound-rotor motors are operated with some external resistance maintained in their rotor circuits, and then they have sufficiently low short-circuit time constants that their fault contribution is insignificant. A specific investigation should be made before neglecting the contribution from a wound-rotor motor.

Capacitor discharge current, because of its very short time constant of less than one cycle, can be neglected in most cases. However, there are application conditions in industrial and commercial power systems in which very high transitory short-circuit currents can be developed when a short-circuit occurs close to a bank of energized capacitors. These transitory currents, generally of much higher frenquency than normal operating frequency, may exceed in magnitude the power-frequency short-circuit currents and persist long enough to impose severe duty on the circuit parts which carry this current.

2.1.2 *Short-Circuit-Current Behavior.* When a short circuit occurs, a new circuit is established with lower impedance, and the current consequently increases. In the case of a bolted short circuit the impedance is drastically reduced, and the current increases to a very high value in a fraction of a cycle. Fig 1 represents a symmetrical short-circuit current, that is, a short-circuit current that has the same axis as the normal current which flowed before the fault occurred. To produce a symmetrical short-circuit current under the usual condition that the short-circuit power factor be essentially zero, the fault must occur exactly when the normal voltage is maximum. In Fig 1 the system voltage is assumed to remain constant although the current changes.

The total short-circuit current is made up of components from any sources connected to the circuit (Fig 2). The contributions from rotating machinery decrease at various rates, so that the symmetrical current is initially at a maximum, then decreases until a steady-state value is reached. This decrease is known as the alternating-current decrement of the short-circuit current. Fig 2 shows a decreasing symmetrical short-circuit current.

Most short-circuit currents are not symmetrical, but are offset from the normal-current axis for several cycles. If the power factor is essentially zero until a steady-state value is reached and the short circuit occurs at the zero point on the voltage wave, the current starts to build up from zero, but cannot follow the normal-current axis because the current must lag behind the voltage by 90°. Although the current is symmetrical with respect to a new axis, it is asymmetrical with respect to the original axis. Fig 3 illustrates the maximum possible asymmetry. The magnitude of current offset for a typical fault will be between the two extremes of complete symmetry and complete asymmetry because the odds are

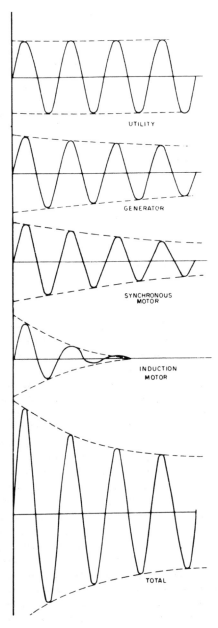

Fig 2
Decreasing Symmetrical Short-Circuit
Current

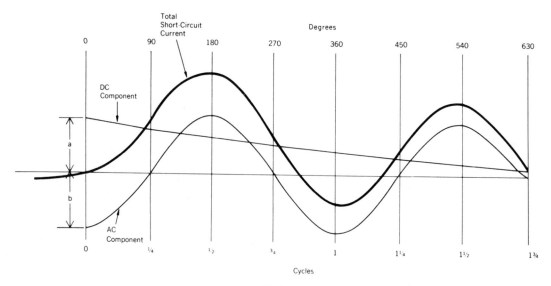

Fig 3
Analysis of Asymmetrical Current

gainst the fault occurring at either the voltage peak or the voltage zero.

When the resistance of the system to the point of fault is not negligible (not zero power factor), the point of the voltage wave to produce maximum or minimum asymmetry will be different from that for the circuit containing negligible resistance. Maximum asymmetry occurs if the fault starts at voltage zero at a time angle equal to $90° + \theta$ (measured in degrees from the zero point of the voltage wave), where $\tan \theta$ equals the X/R ratio of the circuit.

An analysis of the asymmetrical current wave is made in Fig 3. The offset of the asymmetrical current wave from a symmetrical wave having equal peak-to-peak displacement is a positive value of current that may be considered as a direct current. The asymmetrical current, therefore, may be thought of as the sum of an alternating-current component b and a direct-current component a. At the instant of fault initiation (0 cycles in Fig 3), b is negative and $a + b = 0$. At 1/4 cycle the symmetrical alternating-current component is zero, and the total current is equal to the direct-current component. At about 1/2 cycle the total current is maximum, being the sum of the maximum positive alternating-current

component and the direct-current component.

The direct-current component decreases eventually to zero as the stored energy it represents is expended in the form of I^2R losses in the resistance of the system. The initial rate of decay of the direct-current component is inversely proportional to the X/R ratio of the system from the source to the fault. The lower the X/R ratio, the more rapid is the decay. This decay is called the "direct-current decrement." The total short-circuit current is thus affected by both an alternating- and a direct-current decrement before reaching its steady-state value.

2.1.3 *Short-Circuit-Current Concepts.* From the foregoing it can be seen that the short-circuit current behaves differently in the first few cycles than it does later (if allowed to persist). Former practice was to determine an asymmetrical value of short-circuit current by applying simple multipliers to the calculated symmetrical value of short-circuit current. The trend in recent years, now nearly complete, is to rate protective equipment on a basic symmetrical value. Asymmetry is then accounted for by various application formulas depending on the class of equipment. Some of these application formulas, embodied in national standards, are quite complicated [1].

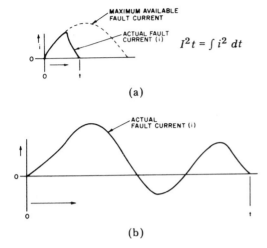

$$I^2 t = \int i^2 \, dt$$

(a)

(b)

Fig 4
Finding $I^2 t$ in Fault-Current Waves
(a) Current-Limiting Fuse
(b) 1.5 Cycle Circuit Breaker

Short-circuit-current calculation concepts are also examined from another point of view. While the symmetrical rating concept was gaining in acceptance, the operating times of interrupting devices were being reduced. The interrupting period of protective devices became crowded into the first few cycles and in the case of current-limiting fuses, into the first half-cycle. To an increasing number of engineers, peak symmetrical values of short-circuit current have not appeared fully adequate to define short-circuit conditions.

The concept of $I^2 t$ has been introduced to supplement the symmetrical current concept because it represents the actual thermal and magnetic stresses imposed on equipment carrying short-circuit current in the first few cycles [2]. The quantity $I^2 t$ represents $\int i^2 \, dt$, the time integral of the current squared for the time under consideration (Fig 4). An $I^2 t$ rating is being applied increasingly to electrical apparatus, and conceivably in the future some protective equipment may be coordinated on an $I^2 t$ basis rather than a maximum current basis.

2.2 Calculation Preliminaries
2.2.1 *Purpose of Calculation*. The maximum magnitude of short-circuit current must be known in order to coordinate protective devices as well as to select adequate interrupting ratings.

For coordination, minimum as well as maximum values may be required. Furthermore it is often necessary to know the maximum let-through current to verify the withstand capability of circuit elements in series with the fault.

The fault current varies with time after the fault. A protector which does not interrupt until several cycles after initiation of the fault usually allows the fault current to decay from its maximum asymmetrical value. However, the protector and all series devices must withstand the maximum current as well as the total energy. A protective device which interrupts in a fraction of a cycle (before maximum fault current is attained) reduces the withstand requirements of series devices.

Short-circuit currents may be calculated at the following recommended times.

(1) *The First Cycle.* First-cycle maximum symmetrical values are always required. They are often the only values needed for low-voltage systems and for fuses in general.

(2) *From 1.5 to 4 Cycles.* Maximim values are required for high-voltage circuit-breaker application, but they are not further considered in this chapter (see IEEE Std 141-1969, Electric Power Distribution for Industrial Plants).

(3) *About 30 Cycles.* These reduced fault currents are needed for estimating the performance of time-delay relays and fuses. Often minimum values must be calculated to determine whether sufficient current is available to open the protective device within a satisfactory time.

2.2.2 *One-Line Diagram.* A one-line diagram should be prepared as the first step in making a short-circuit-current study. This diagram should show all the sources of short-circuit current and all significant circuit elements. All the reactance and resistance values needed for the calculation, or set of calculations, should be included in the diagram. Much of the reactance and resistance data needed can be obtained from Tables 1—4 or from the manufacturer. This should be done well in advance of the study.

2.3 Calculating Short-Circuit Currents on Low-Voltage Systems. Low-voltage distribution systems are usually simpler than high-voltage systems, and the more common low-voltage protective devices are generally easier to apply. For many applications a satisfactory value of short-

Table 1

Resistance R and Reactance X for Typical Three-Phase Primary System and Transformers, in Ohms, Line-to-Neutral

Maximum Three-Phase Short-Circuit	208 V		240 V		480 V		600 V	
A. Primary System								
kVA	R_S (ohms)	X_S (ohms)	R_S (ohms)	X_S (ohms)	R_S (ohms)	X_S (ohms)	R_S (ohms)	X_S (ohms)
15 000	0.000 576	0.002 820	0.000 765	0.003 781	0.003 058	0.015 095	0.004 823	0.023 596
25 000	0.000 348	0.001 692	0.000 459	0.002 266	0.001 835	0.009 063	0.002 880	0.014 158
50 000	0.000 168	0.000 840	0.000 236	0.001 140	0.000 917	0.004 531	0.001 457	0.007 079
100 000	0.000 084	0.000 420	0.000 111	0.000 570	0.000 445	0.002 280	0.000 729	0.003 539
150 000	0.000 060	0.000 276	0.000 070	0.000 375	0.000 306	0.001 501	0.000 486	0.002 359
250 000	0.000 036	0.000 168	0.000 042	0.000 222	0.000 195	0.000 917	0.000 278	0.001 423
500 000	0.000 012	0.000 084	0.000 028	0.000 111	0.000 083	0.000 445	0.000 139	0.000 694
1 000 000	0.000 012	0.000 036	0.000 014	0.000 056	0.000 056	0.000 222	0.000 069	
Q^*	8.64/Q	42.36/Q	11.54/Q	56.71/Q	46.15/Q	226.57/Q	72.18/Q	340.1/Q
B. Transformers								
kVA at %Z	R_T (ohms)	X_T (ohms)	R_T (ohms)	X_T (ohms)	R_T (ohms)	X_T (ohms)	R_T (ohms)	X_T (ohms)
300 at 5	0.001 440	0.007 044	0.001 918	0.009 452	0.007 645	0.037 808	0.012 075	0.058 990
450 at 5	0.000 960	0.004 704	0.001 278	0.006 255	0.005 087	0.025 215	0.008 050	0.039 211
500 at 5	0.000 864	0.004 224	0.001 154	0.005 699	0.004 587	0.022 685	0.007 252	0.035 394
600 at 5	0.000 720	0.003 528	0.000 959	0.004 721	0.003 809	0.018 904	0.006 038	0.029 495
750 at 5.75	0.000 660	0.003 240	0.000 890	0.004 351	0.003 503	0.017 375	0.005 552	0.027 066
1000 at 5.75†	0.000 504	0.002 424	0.000 667	0.003 267	0.002 641	0.013 038	0.004 164	0.020 299
1500 at 5.75	0.000 336	0.001 620	0.000 445	0.002 182	0.001 751	0.008 701	0.002 776	0.013 533
2000 at 5.75	0.000 252	0.001 212	0.000 333	0.001 640	0.001 334	0.006 505	0.002 082	0.010 132
3000 at 5.75	0.000 168	0.000 816	0.000 222	0.001 098	0.000 890	0.004 337	0.001 338	0.006 766
4000 at 5.75	0.000 120	0.000 612	0.000 167	0.000 820	0.000 667	0.003 253	0.001 041	0.005 101
5000 at 5.75	0.000 096	0.000 480	0.000 139	0.000 653	0.000 528	0.002 613	0.000 833	0.004 095
6000 at 5.75	0.000 084	0.000 408	0.000 111	0.000 542	0.000 445	0.002 168	0.000 694	0.003 401
M at N‡	0.08N/M	0.442N/M	0.115N/M	0.567N/M	0.46N/M	2.27N/M	0.72N/M	3.4N/M

From IEEE JH 2112-1 [3].

NOTE: The table is based on $X/R = 5$. Do not use for other X/R values.

* Q is any given value of primary system kVA.

† For a 1000-kVA 8-percent impedance transformer, the reactance value would be 8/5.75 times the values in this table. Network transformers have 5 or 7 percent instead of 5.75 percent.

‡ M is any given transformer kVA and N the corresponding %Z.

circuit current can be obtained by reference to a table. Where actual calculations must be made, the basic Ohm's law method described in this chapter can be used.

NOTE: By reducing all system impedances to a single per-unit base and combining impedances, as will be shown in Section 2.4.3, the same results as obtained in this section can be found with the parallel inductance method. This latter method is simpler than the method illustrated in this section when the system is large and contains many motors.

2.3.1 *Service Entrance.* The service entrance is ordinarily the first point at which to determine the available short-circuit current. The available short-circuit current (present and future) from the utility source must be obtained from the utility company. For three-phase systems it should be expressed on a three-phase symmetrical basis. In the case of single-phase systems it should be expressed on a single-phase symmetrical basis as related to the type of service supplied. In cases where the supply transformer is not accessible, and where there is no significant impedance ahead of the main service protection, select a main protective device with an interrupting rating not less than the maximum available short-circuit current.

For the case just described the feeder protector may be required to interrupt a higher available short-circuit current than the main protector because the feeder protectors may have to pass a motor contribution of fault current while the

Table 2
Typical Resistance and Reactance Values for Building Wire and Cable, in Ohms per 100 ft,
Line-to-Neutral, at Normal Operating Temperature

Wire Size (MCM)	Temperature (°C)	R_{dc}	R_{ac}	Magnetic or Interlocked Armor Cable Conduit			Nonmagnetic Conduit		
				X	Z	R/Z	X	Z	R/Z
Single-Conductor Cable in Conduit									
8	60	0.07275	0.07275	0.00585	0.0730	0.99	0.00468	0.0729	1.00
4	60	0.02928	0.02928	0.00525	0.0297	0.98	0.00420	0.0296	0.99
2	75	0.01947	0.01964	0.00491	0.0202	0.97	0.00392	0.0200	0.98
1	75	0.01530	0.01554	0.00515	0.0163	0.95	0.00412	0.0161	0.96
0	75	0.01218	0.01241	0.00510	0.0134	0.93	0.00408	0.0131	0.95
000	75	0.00768	0.00798	0.00480	0.0093	0.86	0.00384	0.0088	0.90
0000	75	0.00608	0.00639	0.00464	0.0079	0.81	0.00371	0.0074	0.86
250	75	0.00516	0.00546	0.00461	0.0071	0.76	0.00368	0.0066	0.83
350	75	0.00368	0.00397	0.00456	0.0060	0.66	0.00365	0.0054	0.73
500	75	0.00257	0.00291	0.00432	0.0052	0.56	0.00346	0.0045	0.64
750	75	0.00172	0.00208	0.00417	0.0047	0.44	0.00334	0.0039	0.53
1000	75	0.00129	0.00170	0.00416	0.0045	0.38	0.00333	0.0037	0.45
1500	75	0.00086	0.00137	0.00408	0.0043	0.32	0.00326	0.0035	0.39
Two- or Three-Conductor Cable in Conduit									
8	60	0.07275	0.07275	0.00541	0.0729	1.00	0.00389	0.0728	1.00
4	60	0.02928	0.02928	0.00404	0.0296	0.99	0.00349	0.0295	0.99
2	75	0.01947	0.01964	0.00378	0.0200	0.98	0.00326	0.0199	0.98
1	75	0.01530	0.01554	0.00397	0.0161	0.96	0.00342	0.0159	0.98
0	75	0.01218	0.01241	0.00393	0.0130	0.95	0.00339	0.0129	0.96
000	75	0.00768	0.00798	0.00370	0.0088	0.91	0.00319	0.0086	0.93
0000	75	0.00608	0.00639	0.00358	0.00731	0.87	0.00308	0.0071	0.90
250	75	0.00516	0.00546	0.00355	0.00651	0.84	0.00306	0.00626	0.87
350	75	0.00368	0.00397	0.00352	0.00531	0.75	0.00303	0.00499	0.79
500	75	0.00257	0.00291	0.00333	0.00442	0.66	0.00287	0.00409	0.71
750	75	0.00172	0.00208	0.00321	0.00383	0.54	0.00278	0.00347	0.60
1000	75	0.00129	0.00170	0.00320	0.00362	0.47	0.00277	0.00325	0.52
1500	75	0.00086	0.00137	0.00315	0.00342	0.40	0.00271	0.00304	0.45

For aluminum cables of the same physical size, multiply the resistance by 1.64. (See NEC, article 9, table 8.)

This table is taken from IEEE JH 2112-1, Protection Fundamentals for Low-Voltage Electrical Distribution Systems in Commercial Buildings. The letter symbol used in the table for kilocircular mils (MCM) has been deprecated and replaced by kcmil.

main protective device does not (Fig 5). For a fault on feeder A (Fig 5), protective device A must interrupt the utility's contribution to fault current I_U and also the motor contribution I_M from the motors on feeder B. Similarly, a fault on feeder B receives the motor contribution from feeder A. Each feeder protector must be checked for the motor contribution from all other feeders to make sure that the total available fault current does not exceed the interrupting rating of the protective device involved.

The short-circuit current contributed to a fault by induction motors varies from motor to motor. It is seldom possible to determine the motor contribution I_M precisely, but experience has demonstrated that the following equation gives satisfactory values for most low-voltage systems:

I_M = 4 times the sum of rated current of motors connected to a faulted bus

This may be found by summing the full-load currents in amperes of the maximum number of motors that may be running at any one time, and which feed into the fault without an intervening transformer. Where motors are connected to the faulted circuit through a transformer, the transformer impedance and transformation ratio must be considered. Where an actual rated current tabulation is impractical, certain assumptions may sometimes be made. For a building where lighting is obviously a large part of the load, 50 percent of the total load at all voltages may be assumed to be motor load, if the office building has air-conditioning. A building without air-con-

Table 3
Typical Busway Resistance R, Resistance X, and Impedance Z, in Ohms per 100 ft, Line-to-Neutral

Busway Type	Rating (amperes)	R (ohms)	X (ohms)	Z (ohms)	X/R
Plug-in, copper bars	225	0.0836	0.0800	0.1157	
	400	0.0437	0.0232	0.0495	
	600	0.0350	0.0179	0.0393	
	800	0.0218	0.0136	0.0257	
	1000	0.0145	0.0135	0.0198	
Plug-in, aluminum bars	225	0.1090	0.0720	0.1313	
	400	0.0550	0.0222	0.0592	
	600	0.0304	0.0121	0.0327	
	800	0.0243	0.0154	0.0288	
Low-impedance feeder	800	0.0219	0.0085	0.0235	
	1000	0.0190	0.0050	0.0196	
	1350	0.0126	0.0044	0.0134	
	1600	0.0116	0.0035	0.0121	
	2000	0.0075	0.0031	0.0081	
	2500	0.0057	0.0025	0.0062	
	3000	0.0055	0.0017	0.0058	
	4000	0.0037	0.0016	0.0040	
Current-limiting	1000	0.013	0.063	0.064	4.85
	1350	0.012	0.061	0.062	5.08
	1600	0.009	0.056	0.057	6.22
	2000	0.007	0.052	0.052	7.45
	2500	0.006	0.049	0.049	8.15
	3000	0.005	0.046	0.046	9.20
	4000	0.004	0.042	0.042	10.50

From IEEE JH 2112-1 [3].

Table 4
Typical Impedance Values for Single-Phase Transformers in Percent Z

kVA	25	37.5	50	75	100	167	250	333	500
Distribution-Pole Top									
Primary	13 800Y/7970 V	4160Y/2400 V	7200Y/4160 V						
Secondary	240/120 V	240/120 V	240/120 V						
%IR	1.5	1.3	1.2	1.2	1.2	1.2	1.1	1.0	0.9
%IX	1.3	1.6	1.6	1.7	1.7	2.7	4.7	4.7	4.7
Underground									
Primary	13 800Y/7970 V								
Secondary	240/120 V								
%IR	1.0	0.7	0.8	0.9	0.8	0.9			
%IX	1.2	1.3	1.1	1.4	1.5	1.4			
Dry Type									
Primary	4160Y/2400 V								
Secondary	240/120 V								
%IR	3.0	3.0	2.2	2.2	2.0	1.5			
%IX	2.9	3.0	2.8	2.5	3.8	3.5			
Primary	13 800Y/7970 V	7200Y/4160 V							
Secondary	240/120 V	240/120 V							
%IR	3.6	3.4	2.5	2.0	2.1	1.6			
%IX	2.7	2.9	2.5	2.5	3.6	3.3			

From IEEE JH 2112-1 [3].

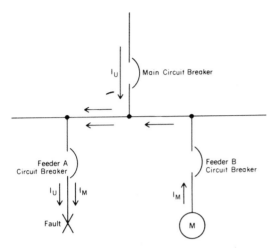

Fig 5
Current through Feeder Protectors
Includes Sum of Current through Main
Protector and Motor Contribution

ditioning or elevators, such as a warehouse, may have essentially zero motor load. In a manufacturing plant the motor load might be close to 100 percent of the connected load.

For services having high values of available short-circuit current, it is common practice to use some type of current-limiting fuse. Current practice dictates that feeder protective devices cannot be selected with an interrupting rating based on the let-through current of the current-limiting fuse ahead of them. Feeder protective devices must have an interrupting rating based on the short-circuit current available at the line-side terminals of the current-limiting fuse. (An exception is the case where significant busway or cable impedance is introduced between the current-limiting fuse and the feeder circuit breaker.)

2.3.2 *Fault at a Point in the System.* To calculate the available short-circuit current at a given point in the low-voltage circuit, the following fundamental equations, derived from Ohm's law, may be used:

$$I = V_{LN}/Z \qquad \text{(Eq 1)}$$

$$Z = V_{LN}/I \qquad \text{(Eq 2)}$$

$$Z = \sqrt{R^2 + X^2} \qquad \text{(Eq 3)}$$

$$I = \frac{V_{LN}}{\sqrt{R^2 + X^2}} \qquad \text{(Eq 4)}$$

$$X = \frac{Z(X/R)}{\sqrt{(X/R)^2 + 1}} \qquad \text{(Eq 5)}$$

$$R = \frac{X}{X/R} \qquad \text{(Eq 6)}$$

where
I = Three-phase first-cycle short-circuit symmetrical rms current
V_{LN} = Line-to-neutral voltage
R = Total resistance, in ohms, on a line-to-neutral basis
X = Total first-cycle reactance, in ohms, on a line-to-neutral basis
Z = Total first-cycle impedance, in ohms, on a line-to-neutral basis

At the service entrance, if only I and V_{LN} are known, one may assume that $X = 0.98\ V_{LN}/I\ \Omega$ and $R = 0.20\ V_{LN}/I\ \Omega$ for the source reactance and source resistance, respectively. This assumption is based on general experience that the short-circuit current power factor at many low-voltage service entrances is about 0.20 (or $X/R = 5$). For spot networks with large transformers the X/R ratio could exceed 10 since larger transformers have higher X/R ratios. The assumption is consistant with other assumptions made in low-voltage short-circuit-current calculations. The values of X and R thus obtained for the circuit up to the service entrance can be combined with X and R values for circuit elements on the load side of the service entrance when calculating available short-circuit currents at other points in the system.

For example, assume that the short-circuit current available from the supplying utility is 125 000 A, and the line-to-neutral voltage is 277 V. It is desired to reduce the short-circuit current at the feeder circuit breakers to 75 000 A by introducing additional impedance Z_A between the main service equipment and the feeder bus. For an initial calculation assume that the impedance of the circuit up to the load-side terminals of the service entrance protective device Z_U is equal to the reactance X_U of the same circuit. These values are obtained as follows:

I_U = 125 000 A
$Z_U = V/I_U = 277/125\ 000 = 0.0022\ \Omega$
$X_U = 0.98V/I_U = 0.98(277/125\ 000)$
$\qquad = 0.0022\ \Omega$
$R_U = 0.20V/I_U = 0.20(277/125\ 000)$
$\qquad = 0.0004\ \Omega$

The circuit up to the feeder bus must have a total impedance Z of

$$Z = V/I = 277/75\,000 = 0.0037 \ \Omega/\text{phase}$$

(Motor contribution is considered negligible in this case.)

The additional impedance required is

$$Z_A \cong Z - Z_U = 0.0037 - 0.0022$$

$$= 0.0015 \ \Omega/\text{phase}$$

The desired additional impedance may be obtained by inserting a reactor in the circuit, or a length of suitable busway. A check calculation may now be made using the actual resistance and reactance values of the inserted element (typical busway with $X/R = 5$):

$$R = R_U + R_A$$

$$R_A = 0.0003 \ \Omega$$

$$X = X_U + X_A$$

$$X_A = 0.0015 \ \Omega$$

$$I = \frac{V_{LN}}{\sqrt{R^2 + X^2}} = \frac{277}{\sqrt{(0.0007)^2 + (0.0037)^2}}$$

$$= 73\,560 \ \text{A}$$

This value is the available symmetrical short-circuit current at the feeder bus.

2.3.3 *Fault at Transformer Load Terminals.* Where the plant is supplied from a three-phase transformer, and where data on the transformer are available, fault calculations should be made taking the transformer into consideration. The short-circuit current I_T available at the load terminals of the transformer is given by the formula

$$I_T = \frac{V_{LN}}{\sqrt{(R_S + R_T)^2 + (X_S + X_T)^2}} \quad (\text{Eq } 7)$$

where

V_{LN} = Line-to-neutral voltage at transformer low-voltage terminals

R_S = Resistance of source and up to transformer, in ohms

X_S = Reactance of source and up to transformer, in ohms

R_T = Resistance of transformer, in ohms

X_T = Reactance of transformer, in ohms

Assume a 480Y/277V system supplied by a 500 kVA three-phase transformer with 5 percent impedance on a primary system with a short-circuit current capacity of 25 000 kVA (Fig 6). Obtaining values for the appropriate 480 V transformer and the proper supply system from Table 1, we have

$$I_T = \frac{277}{\sqrt{(0.001835 + 0.004587)^2 + (0.009063 + 0.022685)^2}}$$

$$= \frac{277}{\sqrt{(0.006422)^2 + (0.031748)^2}}$$

$$= \frac{277}{\sqrt{0.0004 + 0.001007}} = \frac{277}{\sqrt{0.001047}}$$

$$= \frac{277}{0.03236} = 8560 \ \text{A}$$

To the fault current I_T available at the transformer secondary terminals and derived from the utility source, the contribution of motors and generators, if any, must be added. The available current at fault F_1 in Fig 6 will be the sum of the transformer contribution (8560 A) and the motor contribution. Motor contribution I_M, in this case, comes from the 100 hp motor with a rated full-load current of 125 A. According to the previously described rule,

$$I_M = 4 \text{ times the rated current of the}$$
$$\text{motor connected to the faulted bus}$$

The fault current at F_1 is therefore

$$I_{F1} = I_T + I_M = 8560 + (4 \times 125) = 8560 + 500$$
$$= 9060 \ \text{A}$$

For a fault at F_1 the main protective device sees I_T (8560 A) while the motor feeder protective device sees I_M (500 A). The impedance of the secondary bus and of the protective devices is considered negligible in these calculations.

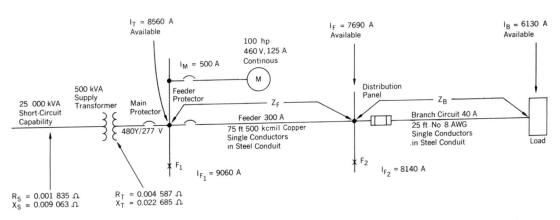

Fig 6
Example of Short-Circuit-Current Calculation in a Low-Voltage System

2.3.4 *Faults on Feeders and Branch Circuits.*
The short-circuit current at the end of feeders and branch circuits is substantially reduced by the impedance of the cables and busways running from the service entrance out to these points. The same basic formulas apply, only now the resistance factor becomes more important. Resistance and reactance values for common sizes of wire and cable are given in Table 2. For busway, data are available in the manufacturers' catalogs; Table 3 gives typical values.

Returning to Fig 6, let us determine the short-circuit current available at the distribution panel, temporarily ignoring the motor contribution. For 75 ft of a 300 A feeder, single-conductor cable in steel conduit, the ampacity and voltage drop of a 500 kcmil copper cable is adequate (see IEEE Std 141-1969, chapter 2). From Table 2 find the resistance and reactance of this cable:

$$R_F = 75 \times 0.00291/100 = 0.002183 \ \Omega$$
$$X_F = 75 \times 0.00432/100 = 0.003240 \ \Omega$$

The resistance and reactance up to the cable is

$$R_S + R_T = 0.00642 \ \Omega$$
$$X_S + X_T = 0.03175 \ \Omega$$

The total resistance and reactance out to the distribution panel is

$$R = R_S + R_T + R_F = 0.00642 + 0.00218$$
$$= 0.00860 \ \Omega$$
$$X = X_S + X_T + X_F = 0.03175 + 0.00324$$
$$= 0.03499 \ \Omega$$

The available current at the distribution panel (excluding motor contribution) is

$$I_F = \frac{277}{\sqrt{R^2 + X^2}} = \frac{277}{\sqrt{(0.00860)^2 + (0.03499)^2}}$$
$$= \frac{277}{0.0360} = 7690 \ A$$

The available short-circuit current has thus been reduced from 8560 A at the feeder bus to 7690 A at the distribution panel.

A fault F_2 on the distribution panel will be fed the available 7690 A from the supply source and also a fault-current contribution from the 100 hp motor (Fig 6). This motor contribution will also be reduced by the impedance of the feeder Z_F, approximately in proportion to the reduction in available current from the transformer,

$$I_{M2} = (7690/8560) \ 500 = 450 \ A$$

The total current feeding into fault F_2 and imposed on the feeder protective device is

$$I_{F2} = 7690 + 450 = 8140 \ A$$

This total value of fault current will be imposed on the feeder protector. A more rigorous approach would be to represent the motor as an impedance and solve for the total current I_{F2}.

Again referring to Fig 6 find the available short-circuit current at the load at the end of a 25 ft, 40 A branch circuit. No 8 AWG conductor in steel conduit is adequate for this circuit. From Table 2,

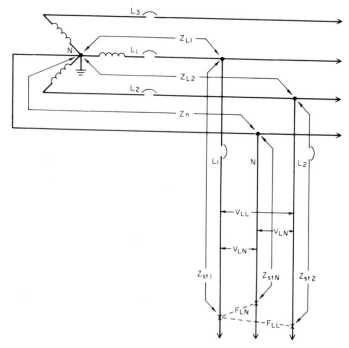

Fig 7
Single-Phase Tap from a Three-Phase System

$R_B = 25 \times 0.07275/100 = 0.018188 \ \Omega$

$X_B = 25 \times 0.00585/100 = 0.001463 \ \Omega$

The total resistance and reactance out to the load is

$R = R_S + R_T + R_F + R_B$

$= 0.00860 + 0.018188 = 0.02679 \ \Omega$

$X = X_S + X_T + X_F + X_B$

$= 0.03499 + 0.001463 = 0.03645 \ \Omega$

$I_B = \dfrac{277}{\sqrt{R^2 + X^2}} = \dfrac{277}{\sqrt{(0.02679)^2 + (0.03645)^2}}$

$= \dfrac{277}{0.0452} \approx 6130 \text{ A}$

The available short-circuit current at the end of the branch circuit, originally supplied by the transformer, is 6130 A compared with 7690 A at the distribution panel. The motor contribution to a fault on the load is correspondingly reduced: (6130/7690) 450 = 360 A. The total current feeding a fault at the load end of the branch circuit is, therefore, 6130 + 360 = 6490 A.

2.3.5 *Fault at a Local Transformer.* Consider a three-phase transformer used to obtain 208Y/120 V from a 480 V feeder.

(1) Determine the line-to-neutral resistance and reactance values of the higher voltage system out to the transformer, using Tables 1—4.

(2) Put all values on a common voltage base (208 V in this case) by multiplying by the transformer ratio squared $(480/208)^2$.

(3) Add the resistance and reactance values of the transformer from Table 1, such as the values for 208Y/120 V.

(4) Add the resistance and reactance values (Table 2) of the lower voltage system, such as the 208Y/120 V system, out to the point of short circuit.

(5) Use the familiar formula

$$I = \frac{V_{LN}}{\sqrt{R^2 + X^2}} \qquad \text{(Eq 4)}$$

2.3.6 *Single-Phase Taps from a Three-Phase System.* Single-phase taps may be taken from a three-phase system as shown in Fig 7, where two

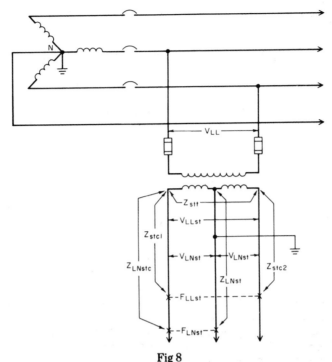

Fig 8
Single-Phase Transformer Connected across Single-Phase Tap from a Three-Phase Line

phase conductors and the three-phase neutral are extended from the three-phase four-wire system. A line-to-line fault on the single-phase system may be calculated using the same parameters as for three-phase fault calculations. The voltage driving the fault current (line-to-line voltage) equals $\sqrt{3}\ V_{LN}$, and the impedance to the fault is twice the line-to-neutral impedance. Thus,

$$I_{st} = \frac{\sqrt{3}\ V_{LN}}{2\ Z_{LN}} = \frac{0.87\ V_{LN}}{Z_{L1-N} + Z_{st}} \quad \text{(Eq 8)}$$

where

I_{st} = Line-to-line short-circuit current of single-phase tap

V_{LN} = Line-to-neutral voltage of three-phase system

Z = Line-to-neutral total impedance

Z_{L1-N} = Line-to-neutral impedance of three-phase system to point of single-phase tap

Z_{st} = Line-to-neutral impedance of single-phase tap from point of tap to point of fault

In the case of a line-to-neutral fault on a single-phase tap from a three-phase system (Fig 7), first the short-circuit current available between the line and neutral at the point of the tap from the three-phase circuit must be determined. This calculation involves complications which put it beyond this simplified method.

This available current I_{pst}, when obtained, may be used to find the impedance up to the tap, $Z_{pst} = V_{LN}/I_{pst}$, from which the short-circuit current at the single-phase fault may be found:

$$I_{stN} = \frac{V_{LN}}{Z_{pst} + Z_{st1} + Z_{stN}} \quad \text{(Eq 9)}$$

where

I_{pst} = Fault current available at point of tap

I_{stN} = Fault current at single-phase line-to-neutral fault

Z_{pst} = Impedance of three-phase circuit at point of tap

Z_{st1} = Impedance of single-phase line conductor from point of tap to single-phase fault (neutral return)

Z_{stN} = Impedance of neutral conductor from point of tap to single-phase fault (line return)

Table 5
Symmetrical RMS Short-Circuit Currents at Secondary Terminals of Single-Phase 240/120 V Transformers Connected Line to Line on the Primary

Transformer Rating (kVA)	Symmetrical RMS Current	
	L—L (amperes)	L—N (amperes)
25	5230	7593
37.5	7536	11 375
50	10 333	15 744
75	14 842	22 783
100	19 712	30 295
167	32 579	50 222

From IEEE JH 2112-1 [3].

2.3.7 *Single-Phase Tap through a Single-Phase Transformer.* In the case where a single-phase tap is made from a three-phase line through a single-phase transformer (Fig 8), the effect of the transformer must be taken into account. Because of the complications involved, it is best to determine the short-circuit current available at the secondary terminals of the transformer I_{stt} from a table such as Table 5 for line-to-line primaries. The effective impedance of the circuit up to the transformer secondary terminals may then be found by dividing the line-to-line voltage at the secondary by the available current, $Z_{stt} = V_{LLst}/I_{stt}$. The short-circuit current of a line-to-line fault on the single-phase circuit is then found by the formula

$$I_{stLL} = \frac{V_{LLst}}{Z_{stt} + Z_{stc}}$$

where

I_{stLL} = Short-circuit current of line-to-line single-phase fault

V_{LLst} = Line-to-line voltage of single-phase circuit beyond transformer secondary

Z_{stt} = Impedance of circuit up to transformer secondary terminals

Z_{stc} = Impedance of single-phase conductors, outgoing and return, from transformer secondary terminals to point of the fault. $Z_{stc} = Z_{stc1} + Z_{stc2}$, as shown in Fig 8

In the case of line-to-neutral faults on the single-phase circuit supplied by the single-phase transformer (Fig 8) the available short-circuit current between line and neutral I_{sttN}, at the transformer secondary terminals, is obtained from Table 5. The effective impedance Z_{sttN} of the circuit up to the line-to-neutral secondary terminals of the transformer may then be found, $Z_{sttN} = V_{LNst}/I_{sttN}$. With this value the short-circuit current of a line-to-neutral fault on the single-phase circuit beyond the transformer secondary may be found:

$$I_{stLN} = \frac{V_{LNst}}{Z_{sttN} + Z_{stc1} + Z_{LNst}} \qquad (Eq\ 9)$$

where

I_{stLN} = Short-circuit current of line-to-neutral single-phase fault

V_{LNst} = Line-to-neutral voltage of single-phase circuit beyond the transformer secondary

Z_{sttN} = Line-to-neutral impedance of circuit up to transformer secondary terminals

Z_{stc1} = Impedance of single-phase line conductor from transformer secondary terminals up to point of fault

Z_{LNst} = Impedance of neutral conductor of single-phase circuit from transformer secondary to point of fault

NOTE: Because of the pronounced effect of resistance in most low-voltage circuits, impedance calculations similar to those shown in Section 2.4.5, Fig 16, are usually required.

2.3.8 *Short-Circuit-Current Tables.* A simple method of finding the available short-circuit current at any point in a low-voltage circuit is to use tables based on the type of circuit in question. Not all systems are applicable. Tables 6 through 9 are based on the following prerequisites:

(1) The system must be served from a single transformer.

(2) The transformer must have a nominal impedance of 4 1/2 percent for ratings up to and including 500 kVA, and 5 1/2 percent for ratings above 500 kVA.

(3) The motor contribution which is included must be based on 100 percent motor load.

These tables can be used for systems not served by installations of single transformers by choosing a value in column 0 equal to or greater than the available short-circuit current at the service

Table 6
Available Three-Phase Symmetrical RMS Fault Current at 208 V, in Amperes

Transformer Rating (kVA)	Conductor Size per Phase	Distance from Transformer to Point of Fault (feet)								
		0	5	10	20	50	100	200	500	1000
	No. 4	11 500	10 700	10 000	8 500	5400	3200	1750	720	350
	No. 0	11 500	11 120	10 750	10 050	8070	5850	3600	1620	860
150	250 MCM	11 500	11 300	11 050	10 550	9250	7600	5550	3000	1600
	2–250 MCM	11 500	11 400	11 250	11 050	10 300	9240	7600	4820	3000
	No. 4	17 220	15 700	13 950	12 000	6100	3400	1800	750	400
	No. 0	17 220	16 450	15 600	14 100	10 400	6750	3600	1700	900
225	250 MCM	17 220	16 700	16 200	15 200	12 600	9750	6500	3200	1700
	2–250 MCM	17 220	17 000	16 700	16 200	14 700	12 700	9600	5600	3250
	2–500 MCM	17 200	17 100	16 900	16 500	15 300	13 700	11 300	7200	4500
	No. 4	23 000	20 400	17 100	12 600	6500	3500	1800	750	400
	No. 0	23 000	21 600	20 200	17 500	13 950	7500	4000	1750	900
300	250 MCM	23 000	22 100	21 200	19 500	15 300	12 200	7300	3350	1750
	2–250 MCM	23 000	22 500	22 000	21 200	18 500	15 300	11 300	6000	3300
	2–500 MCM	23 000	22 750	22 450	21 700	19 550	16 800	13 300	7900	4550
	No. 4	38 200	30 800	24 000	15 400	6900	3500	1800	800	400
	No. 0	38 200	34 400	30 400	24 000	14 200	8000	4000	1800	1000
500	250 MCM	38 200	36 000	33 800	29 400	20 100	13 600	8000	3400	1800
	2–250 MCM	38 200	36 900	35 700	33 300	27 000	20 100	13 200	6400	3500
	2–500 MCM	38 200	37 400	36 500	34 600	29 400	23 800	17 000	9000	5000
	No. 4	47 200	35 800	26 000	16 000	6900	3400	1700	800	400
	No. 0	47 200	41 900	36 300	27 300	14 800	8000	4100	1800	950
750	250 MCM	47 200	43 600	40 000	34 300	23 000	14 000	8000	3200	1700
	2–250 MCM	47 200	45 100	43 300	40 000	31 900	22 800	14 400	6900	3500
	2–500 MCM	47 200	45 900	44 300	41 700	34 600	27 000	18 300	9200	5000
	No. 4	62 700	43 000	29 100	17 000	7800	3700	1800	700	400
	No. 0	62 700	53 500	44 300	31 200	16 000	8500	4400	1800	950
1000	250 MCM	62 700	56 600	51 000	42 000	26 000	15 900	8800	3400	1870
	2–250 MCM	62 700	59 900	56 300	50 400	37 800	25 900	15 500	6900	3500
	2–500 MCM	62 700	61 800	58 200	54 700	42 400	31 500	21 000	10 000	5300
	No. 4	92 400	53 000	33 000	18 100	7800	3900	2000	800	600
	No. 0	92 400	73 500	57 000	36 500	17 800	9200	4600	2000	1000
1500	250 MCM	92 400	80 000	69 500	52 000	30 000	17 400	9200	3800	2000
	2–250 MCM	92 400	85 700	79 500	68 500	46 000	30 000	17 600	7000	3800
	2–500 MCM	92 400	88 000	83 000	74 000	57 000	38 000	23 800	11 000	6000
	No. 4	121 800	58 000	33 800	18 200	7200	3800	1800	600	—
	No. 0	121 800	88 000	63 700	38 000	17 000	8800	4200	1800	800
2000	250 MCM	121 800	100 200	83 800	60 000	31 000	17 000	8500	3200	1800
	2–250 MCM	121 800	110 800	100 500	83 000	50 000	30 000	17 000	6800	3500
	2–500 MCM	121 800	114 200	106 000	91 000	62 000	40 000	23 900	10 000	5000

This table is taken from IEEE JH 2112-1, Protection Fundamentals for Low-Voltage Electrical Distribution Systems in Commercial Buildings. The letter symbol used in the table for kilocircular mils (MCM) has been deprecated and replaced by kcmil.

The fault currents listed are maximum available symmetrical rms values based on liquid-filled transformers, with nominal impedances of 4.5 percent for ratings up to and including 500 kVA and 5.5 percent for ratings above 500 kVA, and include motor contribution based on 100-percent motor load. Modern transformers have impedances down to 3.0 percent below 500 kVA and approximately 5.75 percent above 500 kVA.

Table 7
Available Three-Phase Symmetrical RMS Fault Current at 240 V, in Amperes

Transformer Rating (kVA)	Conductor Size per Phase	Distance from Transformer to Point of Fault (feet)								
		0	5	10	20	50	100	200	500	1000
150	No. 4	9980	9520	9000	8000	5580	3440	1900	800	400
	No. 0	9980	9700	9450	9000	7600	5850	3900	1800	950
	250 MCM	9980	9820	9660	9350	8500	7220	5550	3200	1900
	2–250 MCM	9980	9900	9800	9650	9200	8400	7200	4900	3200
225	No. 4	14 940	13 800	12 800	10 600	6500	3800	2000	800	450
	No. 0	14 940	14 500	14 000	12 900	10 100	7100	4300	2000	1000
	250 MCM	14 940	14 600	14 300	13 600	11 800	9500	6800	3500	1800
	2–250 MCM	14 940	14 700	14 500	14 300	13 200	11 700	9400	6000	3500
	2–500 MCM	14 940	14 800	14 700	14 500	13 600	12 500	10 600	7500	5000
300	No. 4	19 970	18 000	16 000	12 700	7000	4000	2000	800	400
	No. 0	19 970	19 100	18 100	16 200	11 800	7800	4500	2000	1000
	250 MCM	19 970	19 300	18 700	17 500	14 500	11 200	7500	3600	2000
	2–250 MCM	19 970	19 500	19 300	18 700	17 000	14 500	11 200	6400	3600
	2–500 MCM	19 970	19 600	19 400	19 000	17 600	15 600	13 000	8200	5200
500	No. 4	33 100	28 000	22 900	15 900	7800	4200	2200	900	500
	No. 0	33 100	30 800	28 000	23 100	14 800	9000	4900	2000	1000
	250 MCM	33 100	31 500	30 000	27 000	20 300	14 200	8800	4000	2000
	2–250 MCM	33 100	32 300	31 400	29 800	25 300	20 100	14 000	7000	3900
	2–500 MCM	33 100	32 600	32 000	30 700	22 200	22 500	17 000	9600	5500
750	No. 4	40 900	33 000	26 000	17 000	8000	4000	2000	900	500
	No. 0	40 900	37 400	33 900	27 000	15 900	9200	5000	2000	1000
	250 MCM	40 900	38 300	36 000	32 000	23 000	15 000	8900	3900	2050
	2–250 MCM	40 900	39 800	38 500	36 000	30 000	22 900	15 000	7300	4000
	2–500 MCM	40 900	40 100	39 100	37 100	32 000	26 100	19 000	10 100	5600
1000	No. 4	54 400	41 000	29 500	18 000	8200	4200	2100	950	400
	No. 0	54 400	48 800	42 200	32 100	17 900	9900	5000	2050	1000
	250 MCM	54 400	50 100	46 300	39 900	27 000	17 000	9500	4000	2050
	2–250 MCM	54 400	52 100	50 000	46 000	36 800	26 900	17 000	8000	4050
	2–500 MCM	54 400	52 800	51 000	48 000	40 300	31 800	22 000	11 200	6000
1500	No. 4	80 100	53 200	35 500	20 500	9900	4800	2500	1200	900
	No. 0	80 100	66 500	55 000	40 000	20 000	10 500	5800	2800	1800
	250 MCM	80 100	72 000	64 500	52 000	32 000	19 500	10 100	4500	3000
	2–250 MCM	80 100	76 000	72 000	64 000	47 000	32 000	19 500	8500	4800
	2–500 MCM	80 100	77 500	74 000	68 000	53 500	40 000	25 500	12 000	6500
2000	No. 4	105 600	60 500	38 000	21 000	8800	4300	2200	800	—
	No. 0	105 600	83 000	64 000	42 000	20 000	10 300	5500	2500	1200
	250 MCM	105 600	90 500	79 000	60 000	34 500	19 800	10 200	4500	2400
	2–250 MCM	105 600	97 500	91 000	78 000	54 000	34 000	19 000	8500	4600
	2–500 MCM	105 600	100 000	94 500	84 000	62 500	43 500	2700	12 000	6200

This table is taken from IEEE JH 2112-1, Protection Fundamentals for Low-Voltage Electrical Distribution Systems in Commercial Buildings. The letter symbol used in the table for kilocircular mils (MCM) has been deprecated and replaced by kcmil.

The fault currents listed are maximum available symmetrical rms values based on liquid-filled transformers, with nominal impedances of 4.5 percent for ratings up to and including 500 kVA and 5.5 percent for ratings above 500 kVA, and include motor contribution based on 100-percent motor load. Modern transformers have impedances down to 3.0 percent below 500 kVA and approximately 5.75 percent above 500 kVA.

Table 8
Available Three-Phase Symmetrical RMS Fault Current at 480 V, in Amperes

Trans-former Rating (kVA)	Conductor Size per Phase	Distance from Transformer to Point of Fault (feet)								
		0	5	10	20	50	100	200	500	1000
150	No. 4	4990	4930	4880	4770	4420	3800	2800	1480	790
	No. 0	4990	4940	4920	4880	4700	4400	3850	2650	1680
	250 MCM	4990	4960	4930	4910	4800	4600	4250	3350	2500
	2-250 MCM	4990	4970	4940	4920	4900	4800	4600	4050	3350
225	No. 4	7470	7380	7240	7000	6140	4880	3300	1600	840
	No. 0	7470	7400	7320	7200	6800	6200	5100	3180	1860
	250 MCM	7470	7420	7360	7300	7040	6640	5900	4400	3000
	2-250 MCM	7470	7440	7400	7350	7220	7000	6600	5580	4300
	2-500 MCM	7470	7460	7450	7400	7300	7100	6800	6000	5000
300	No. 4	9985	9800	9600	9100	7600	5600	3560	1620	840
	No. 0	9985	9840	9750	9520	8800	7650	5900	3400	1920
	250 MCM	9985	9880	9800	9660	9240	8500	7300	5000	3240
	2-250 MCM	9985	9920	9825	9790	9580	9200	8450	6800	5020
	2-500 MCM	9985	9950	9850	9800	9660	9400	8820	7500	5880
500	No. 4	16 550	16 000	15 400	14 000	10 250	6800	3800	1600	800
	No. 0	16 550	16 200	15 950	15 250	13 250	10 500	7400	3500	1900
	250 MCM	16 550	16 300	16 050	15 700	14 500	12 700	10 000	5900	3500
	2-250 MCM	16 550	16 350	16 250	16 100	15 450	14 400	12 500	9000	6000
	2-500 MCM	16 550	16 400	16 350	16 300	15 700	14 800	13 400	10 500	7500
750	No. 4	20 450	19 700	18 700	16 800	11 700	7500	4000	1600	800
	No. 0	20 450	20 000	19 500	18 700	16 000	12 400	8100	3800	2000
	250 MCM	20 450	20 200	19 800	19 250	17 500	15 000	11 500	6600	3800
	2-250 MCM	20 450	20 250	20 200	19 700	19 000	17 500	15 000	10 500	6600
	2-500 MCM	20 450	20 400	20 250	19 900	19 300	18 200	16 300	12 000	8400
1000	No. 4	27 200	26 000	24 200	21 000	13 400	7900	4400	1800	800
	No. 0	27 200	26 700	25 900	24 300	20 000	14 400	9000	4100	2200
	250 MCM	27 200	26 900	26 400	25 300	22 400	18 600	13 600	7200	4000
	2-250 MCM	27 200	27 000	26 700	26 200	24 500	22 200	18 500	12 100	7200
	2-500 MCM	27 200	27 100	26 800	26 500	25 300	23 300	20 300	14 500	9500
1500	No. 4	40 050	37 000	33 100	26 000	14 400	8200	4000	1400	600
	No. 0	40 050	38 800	36 800	33 200	24 500	16 000	9200	4000	2000
	250 MCM	40 050	39 100	37 800	35 600	29 900	23 000	15 200	7500	4000
	2-250 MCM	40 050	39 600	39 000	37 900	34 100	29 000	22 500	13 000	7400
	2-500 MCM	40 050	39 700	39 200	38 200	35 500	31 600	25 900	16 400	10 100
2000	No. 4	52 800	47 400	40 700	30 000	15 100	8200	4200	1900	1000
	No. 0	52 800	50 200	47 000	41 200	28 000	17 000	9700	4200	2400
	250 MCM	52 800	51 000	49 000	45 400	36 200	26 500	16 500	8000	4200
	2-250 MCM	52 800	51 800	50 900	48 900	43 100	36 000	26 700	14 000	8000
	2-500 MCM	52 800	52 100	51 300	49 900	45 100	39 200	30 800	18 500	11 000

This table is taken from IEEE JH 2112-1, Protection Fundamentals for Low-Voltage Electrical Distribution Systems in Commercial Buildings. The letter symbol used in the table for kilocircular mils (MCM) has been deprecated and replaced by kcmil.

The fault currents listed are maximum available symmetrical rms values based on liquid-filled transformers, with nominal impedances of 4.5 percent for ratings up to and including 500 kVA and 5.5 percent for ratings above 500 kVA, and include motor contribution based on 100-percent motor load. Modern transformers have impedances down to 3.0 percent below 500 kVA and approximately 5.75 percent above 500 kVA.

Table 9
Avaliable Three-Phase Symmetrical RMS Fault Current at 600 V, in Amperes

Trans-former Rating (kVA)	Conductor Size per Phase	Distance from Transformer to Point of Fault (feet)								
		0	5	10	20	50	100	200	500	1000
150	No. 4	3990	3950	3910	3850	3670	3340	2710	1640	960
	No. 0	3990	3960	3930	3880	3820	3670	3360	2600	1850
	250 MCM	3990	3970	3950	3910	3860	3780	3580	3080	2430
	2–250 MCM	3990	3980	3970	3940	3910	3860	3760	3480	3100
225	No. 4	5980	5920	5870	5740	5300	4610	3500	1880	1010
	No. 0	5980	5940	5900	5850	5640	5300	4700	3280	2100
	250 MCM	5980	5950	5920	5890	5760	5550	5150	4180	3090
	2–250 MCM	5980	5960	5940	5930	5860	5750	5540	4920	4140
	2–500 MCM	5980	5970	5960	5950	5900	5820	5650	5180	4620
300	No. 4	7990	7880	7800	7560	6800	5560	3900	2000	1050
	No. 0	7990	7920	7880	7740	7380	6800	5800	3740	2300
	250 MCM	7990	7940	7910	7800	7600	7200	6540	5000	3500
	2–250 MCM	7990	7960	7940	7850	7760	7580	7200	6200	5000
	2–500 MCM	7990	7980	7960	7900	7840	7700	7400	6600	5600
500	No. 4	13 230	13 000	12 700	12 000	9980	7350	4600	2000	1000
	No. 0	13 230	13 100	12 960	12 600	11 600	10 180	7700	4200	2400
	250 MCM	13 230	13 130	13 100	12 920	12 300	11 300	9650	6400	4200
	2–250 MCM	13 230	13 170	13 130	13 060	12 720	12 180	11 200	9000	6580
	2–500 MCM	13 230	13 200	13 170	13 120	12 880	12 500	11 700	9800	7650
750	No. 4	16 360	16 100	15 750	14 800	11 800	8200	5000	2200	1050
	No. 0	16 360	16 200	16 000	15 550	14 200	12 000	8700	4800	2550
	250 MCM	16 360	16 250	16 100	15 800	14 950	13 400	11 200	7100	4300
	2–250 MCM	16 360	16 350	16 150	16 000	15 600	14 800	13 300	10 200	7300
	2–500 MCM	16 360	16 350	16 200	16 050	15 800	15 200	14 000	11 400	8700
1000	No. 4	21 750	21 100	20 250	18 500	13 800	9000	5000	2200	1200
	No. 0	21 750	21 500	21 000	20 250	17 800	14 400	9800	4800	2550
	250 MCM	21 750	21 570	21 200	20 750	19 300	16 900	13 400	8000	4700
	2–250 MCM	21 750	21 650	21 500	21 250	20 500	19 200	16 800	12 000	8200
	2–500 MCM	21 750	21 730	21 600	21 400	20 750	19 700	17 900	13 800	10 000
1500	No. 4	32 050	30 550	28 700	25 250	16 300	9600	5300	2300	1000
	No. 0	32 050	31 250	29 500	28 800	23 800	17 500	10 800	4800	2500
	250 MCM	32 050	31 500	30 800	29 800	26 600	22 250	16 300	8800	4800
	2–250 MCM	32 050	31 800	31 500	31 000	29 200	26 600	22 800	14 300	8800
	2–500 MCM	32 050	31 900	31 600	31 200	29 800	27 600	29 000	17 200	11 500
2000	No. 4	42 200	39 700	36 300	30 000	17 400	10 000	5100	2100	1200
	No. 0	42 200	40 900	39 500	36 000	27 800	19 000	11 500	5000	2600
	250 MCM	42 200	41 300	40 050	38 100	32 900	26 000	18 000	9100	5000
	2–250 MCM	42 200	41 700	41 000	40 000	36 900	32 200	25 900	15 800	9200
	2–500 MCM	42 200	42 000	41 300	40 600	38 100	34 200	28 800	19 600	12 500

This table is taken from IEEE JH 2112-1, Protection Fundamentals for Low-Voltage Electrical Distribution Systems in Commercial Buildings. The letter symbol used in the table for kilocircular mils (MCM) has been deprecated and replaced by kcmil.

The fault currents listed are maximum available symmetrical rms values based on liquid-filled transformers, with nominal impedances of 4.5 percent for ratings up to and including 500 kVA and 5.5 percent for ratings above 500 kVA, and include motor contribution based on 100-percent motor load. Modern transformers have impedances down to 3.0 percent below 500 kVA and approximately 5.75 percent above 500 kVA.

entrance. Other tables can be constructed for different circuit parameters. Where suitable tables are not available it is necessary to make calculations using the general methods previously described.

2.4 Calculating Short-Circuit Currents on High-Voltage and Multivoltage Systems. The per-unit method of calculation, while suitable for all short-circuit-current calculations, is the preferred method for high-voltage systems and for combinations of these systems with low-voltage systems. The greater complexity of these systems makes the per-unit method advantageous. When making other than first-cycle calculations one must keep in mind that the time—current coordination of protective devices also requires determination of the minimum values of short-circuit current. However, minimum values are sometimes estimated as percentages of the maximum values, so the calculation of maximum fault-current values is the first objective.

2.4.1 *Representation of Impedances.* The principal problem in calculating short-circuit currents is assigning values of impedance to each circuit element and then converting to a per-unit base form to provide an easy method of combining series and parallel elements. Impedance values not included in Tables 1—4 must be obtained from other sources, such as IEEE Std 141-1969 (chapter 4), handbooks, or equipment nameplates. In some cases it will be necessary to obtain the values directly from the manufacturer.

The determination of short-circuit current has been shown to be dependent principally upon the reactance X of the elements from the source (or sources) to the fault point. This holds true for all elements except cable, open-wire lines, and buses. When the ratio of reactance to resistance (X/R ratio) of the entire system from the sources to the fault is greater than 5, negligible error will result from neglecting resistance. Neglecting R introduces an error that always makes the calculated short-circuit current slightly larger than the actual short-circuit current. It is common practice to refer to reactances X, even when they represent impedances Z.

Per-unit impedances, expressed on a chosen base, can be combined directly, regardless of how many voltage levels are crossed from source to

fault. The per-unit system developed in this chapter is illustrated as it applies to three-phase balanced electric systems.

2.4.2 *Per-Unit Quantities on a Chosen Base.* In the per-unit system there are four base quantities: base power in kilovolt-amperes, base voltage in volts, base impedance in ohms, and base current in amperes. The relationship between base, per unit, and actual quantities is

$$\text{per-unit quantity} = \frac{\text{actual quantity}}{\text{base quantity}}$$

$$\text{actual quantity} = \text{per-unit quantity}$$

$$\times \text{ base quantity}$$

Normally a base power in kilovolt-amperes is selected. Then, for a base voltage at each level in the system, base current and base impedance are derived. Base voltages must be related to the turns ratios of the interconnecting transformers. The base power may be any convenient value, in kilovolt-amperes, often (but not necessarily) the rating of a transformer in the system, such as 10 000 kVA. Base voltages are usually line-to-line voltages in kilovolts. The formulas for deriving the bases are

$$I_b = \frac{kVA_b}{\sqrt{3} \times kV_b} \qquad (Eq\ 10)$$

$$Z_b = \frac{kV_b \times 1000}{\sqrt{3} \times I_b} = \frac{(kV_b)^2 \times 1000}{kVA_b}$$

$$= \frac{(kV)^2}{MVA_b} \qquad (Eq\ 11)$$

where
I_b = Base current, in amperes
kV_b = Base voltage, in kilovolts, line-to-line
kVA_b = Base power, in kilovolt-amperes
MVA_b = Base power, in megavolt-amperes
Z_b = Base impedance, in ohms per phase (the impedance will usually be expressed as a reactance, X_b)

Circuit element impedances are usually described in actual ohms or in percent on an equipment rating as a base. Cable impedance is generally expressed in ohms and transformer impedance as percent impedance on the self-cooled rating in kilovolt-amperes; for example, 5 percent on a 500 kVA transformer. These actual element reac-

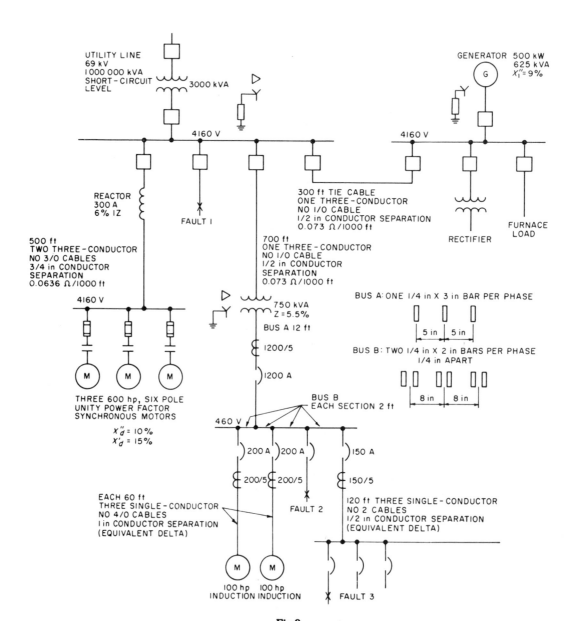

Fig 9
One-Line Diagram for Typical Industrial Distribution System

tance values (resistance is neglected) can be converted to per unit by the formulas

$$X_{pu} = \frac{X_a \times kVA_b}{1000 \times kV_b^2} \qquad \text{(Eq 12)}$$

$$X_{pu} = \frac{X_\% \times kVA_b}{10 \times kVA_e} \left(\frac{kV_e}{kV_b}\right)^2 \qquad \text{(Eq 13)}$$

$$X_{pu} = \frac{X_{pu\ e} \times kVA_b}{kVA_e} \left(\frac{kV_e}{kV_b}\right)^2 \qquad \text{(Eq 14)}$$

where

X_{pu} = Per-unit reactance on chosen kVA base power

$X_{pu\ e}$ = Per-unit reactance on elementary kVA rating

X_a = Actual element reactance, in ohms

$X_\%$ = Percent element reactance based on element kVA rating

kVA_e = Actual element power rating, in kilovolt-amperes

kV_e = Actual element voltage rating, in kilovolts

2.4.3 *Impedance Diagram.* A one-line diagram should be prepared showing all sources of short-circuit current and all significant circuit elements as illustrated in Fig 9. The next step is to convert the one-line diagram into an impedance diagram or diagrams. The impedance diagram can use either reactance or resistance values, or both. For a reactance diagram, simply replace all elements of the one-line diagram with their calculated per-unit reactances. For calculating a system X/R ratio (as described later) it will be necessary to prepare a resistance diagram. This is constructed by replacing all circuit elements with their corresponding per-unit resistance values.

It is then necessary to combine reactances and resistances to the point of fault into a single equivalent reactance or resistance. Occasionally in complex systems it becomes necessary to combine three branch elements which form a wye or delta configuration. Fig 10 shows how to combine reactances (or resistances).

2.4.4 *Short-Circuit-Current Calculation.* The short-circuit current to be calculated will depend upon the particular application desired. This chapter will be concerned with (1) symmetrically rated fuses and low-voltage circuit breakers, (2) asymmetrically rated fuses, and (3) relaying devices. The short-circuit currents involved are (1)

maximum first cycle, (2) maximum approximately 30 cycles, and (3) minimum approximately 30 cycles. (For calculation of interrupting duties and withstand duties of high-voltage circuit breakers refer to IEEE Std 141-1969.)

The impedance diagrams for the three applications differ in impedance values assigned to rotating machines and connections.

(1) *Maximum First-Cycle Duties for Symmetrically Rated Fuses and Low-Voltage Circuit Breakers.* Subtransient reactance should be used to represent all rotating machines in the equivalent network. Modified reactances for groups of low-voltage induction and synchronous motors fed from a low-voltage distribution transformer are used as shown in IEEE Std 141-1969. If the total motor horsepower rating is equal to the transformer self-cooled rating in kilovolt-amperes, a per-unit reactance of 0.25 on the transformer self-cooled rating may be used to represent the group of motors as a single impedance. This value must be converted to the system base being used.

The operating prefault voltage, in volts or kilovolts, must be determined. This should be the highest typical operating voltage at the short-circuit location. Most industrials operate typically with a bus voltage equal to or less than rated (1.0 pu of base) under maximum load conditions. Maximum load corresponds to maximum rotating machines connected and hence maximum short-circuit-current duty. It is usual to calculate using V = 1.0 pu. For three-phase bolted fault currents, the circuit is reduced to a single value of reactance X_{pu} for each short-circuit location.

The symmetrical short-circuit-current duty is calculated by dividing V_{pu} by X_{pu} or Z_{pu} and multiplying by I_b:

$$I = \frac{V_{pu}}{X_{pu}} \times I_b \qquad \text{(Eq 15)}$$

where I is the three-phase short-circuit symmetrical rms current.

This short-circuit current is now directly applicable for fuses or low-voltage circuit breakers whose short-circuit-current ratings or capabilities are expressed in symmetrical rms currents.

(2) *First-Cycle Duties for Asymmetrically Rated Fuses.* When equipment ratings or capabilities are expressed as total rms currents, the symmetrical short-circuit-current calculation is made in

FOR COMBINATION OF BRANCHES IN SERIES

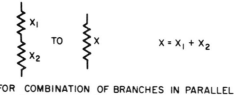

$$X = X_1 + X_2$$

FOR COMBINATION OF BRANCHES IN PARALLEL

$$X = \frac{X_1 X_2}{X_1 + X_2}$$

FOR TRANSFORMING WYE TO DELTA

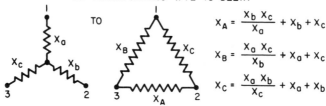

$$X_A = \frac{X_b X_c}{X_a} + X_b + X_c$$

$$X_B = \frac{X_a X_c}{X_b} + X_a + X_c$$

$$X_C = \frac{X_a X_b}{X_c} + X_a + X_b$$

FOR TRANSFORMING DELTA TO WYE

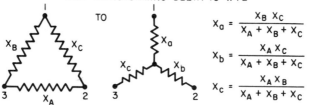

$$X_a = \frac{X_B X_C}{X_A + X_B + X_C}$$

$$X_b = \frac{X_A X_C}{X_A + X_B + X_C}$$

$$X_c = \frac{X_A X_B}{X_A + X_B + X_C}$$

Fig 10
How Reactances Are Combined

the same manner as for application (1). The result is then multiplied by a factor found in the applicable fuse standard to arrive at the asymmetrical first-cycle total rms current.

(3) *Relaying Devices.* The first-cycle short-circuit total rms current must be determined for the maximum setting of instantaneous relays. This value is multiplied by 1.6 to find the first-cycle total short-circuit-current duty per unit current, which is in turn multiplied by the base current to obtain the duty in amperes:

$$I_{1C} = 1.6 \frac{V_{pu}}{X_{pu}} \times I_b$$

where I_{1C} is the first-cycle short-circuit asymmetrical current.

The symmetrical current V_{pu}/X_{pu} in per unit can be used with reasonable accuracy for relaying currents during approximately five cycles.

For application of time-delay relays much beyond six cycles, the equivalent system network representation will include only generators and static equipment between them and the fault point. The generators should be represented by their transient impedance. All motor contributions are omitted. Only the generators that contribute fault current through the relay under consideration to the fault point must be considered for that relay application. The direct-current component will have decayed to near zero and need not be considered. The application short-circuit symmetrical rms current is V_{pu} divided by X_{pu}, where X_{pu} is derived from the equivalent reactance network consisting of generators and static equipment (cables, transformers, etc) in the relay protective fault path.

2.4.5 *Example of Per-Unit Short-Circuit-Current Calculation.* Fig 9 shows a hypothetical power distribution system typical of those to be found in industrial plants. The circuit components and arrangement have been chosen to illustrate the fundamentals and procedures involved in making fault-current calculations.

A complete fault-current study would involve the calculation of fault currents at all locations in the system. This example will illustrate the calculations for only a few significant locations.

The base power will be chosen as 10 000 kVA. Eq 11 gives the per-unit impedance base.

(1) At 4160 V, if resistance is neglected,

$$X_b = \frac{(4.16)^2 \times 1000}{10\ 000} = 1.729$$

(2) At 480 V, if resistance is neglected

$$X_b = \frac{(0.480)^2 \times 1000}{10\ 000} = 0.02404$$

These multipliers will be used in the example to simplify the conversion from ohmic to per-unit values.

The first step is to calculate the per-unit reactance value for each significant circuit element that will contribute to or limit the fault current.

(1) *Utility Supply Equivalent Reactance.* From Eq 14,

$$X_{pu} = \frac{1.0 \times 10\ 000}{1\ 000\ 000} = 0.01\ pu$$

(2) *3000 kVA Transformer.* The per-unit reactance is 0.07 on the transformer rating kilovolt-ampere base. From Eq 14,

$$X_{pu} = \frac{0.07 \times 10\ 000}{3000} = 0.233\ pu$$

(3) *625 kVA Generator.* Given $X_d'' = 9$ percent reactance, the per-unit reactance is $9/100 = 0.09$ on the generator-rating kilovolt-ampere base. From Eq 14,

$$X_{pu} = \frac{0.09 \times 10\ 000}{625} = 1.44\ pu$$

(4) *300 ft Tie Cable.* $X_a = 0.103\ \Omega/1000$ ft, 60 Hz reactance spacing factor, $X_d = -0.073\ \Omega/1000$ ft for 1/2 inch spacing. The total reactance X_{tot} is

$$X_{tot} = X_a + X_d$$
$$= 0.103 - 0.073 = 0.03\ \Omega/1000\ ft$$

For 300 ft,

$$X_{tot} = \frac{0.030 \times 300}{1000} = 0.009\ \Omega$$

$$X_{pu} = \frac{0.009}{X_b} = \frac{0.009}{1.729} = 0.005184\ pu$$

(5) *500 ft Feeder Cable.* $X_a = 0.0981\ \Omega/1000$ ft, $X_d = -0.0636\ \Omega/1000$ ft for 3/4 inch spacing. The total reactance is

$$X_{tot} = X_a + X_d$$
$$= 0.0981 - 0.0636 = 0.0345\ \Omega/1000\ ft$$

For 500 ft,

$$X_{tot} = \frac{0.0345 \times 500}{1000} = 0.01725\ \Omega$$

For two parallel conductors per phase,

$$X_{tot} = \frac{1}{2} \times 0.01725 = 0.008625\ \Omega$$

$$X_{pu} = \frac{0.008625}{X_b} = \frac{0.008625}{1.729} = 0.00486\ pu$$

(6) *700 ft Feeder Cable.* From the calculation for the 300 ft tie cable,

$$X_{tot} = 0.030\ \Omega/1000\ ft$$

For 700 ft,

$$X_{tot} = \frac{0.030 \times 700}{1000} = 0.021\ \Omega$$

$$X_{pu} = \frac{0.021}{X_b} = \frac{0.021}{1.729} = 0.0121\ pu$$

(7) *Rectifier and Furnace Loads.* These loads will neither contribute to nor limit the fault current in the system, and so are neglected for the purpose of calculating fault currents.

(8) *Current-Limiting Reactor.* The per-unit reactance on the reactor circuit kilovolt-ampere rating base = 0.06 and the reactor circuit kilovolt-ampere rating is $\sqrt{3} \times 300 \times 4.16 = 2160$ kVA. From Eq 14,

$$X_{pu} = \frac{0.06 \times 10\ 000}{2160} = 0.277\ pu$$

(9) *750 kVA Transformer.* The per-unit reactance on the transformer kilovolt-ampere base =

0.055. From Eq 14,

$$X_{pu} = \frac{0.055 \times 10\ 000}{750} = 0.733\ pu$$

(10) *600 hp Synchronous Motors.* The per-unit reactance $X_d'' = 0.10$ and the per-unit reactance $X_d' = 0.15$ on the motor kilovolt-ampere base. The motor kilovolt-ampere base is $0.8 \times 600 = 400\ kVA$.

$$X_{pu} = X_d'' = \frac{0.10 \times 10\ 000}{480} = 2.08\ pu$$

$$X_{pu} = X'_d = \frac{0.15 \times 10\ 000}{480} = 3.13\ pu$$

(11) *Bus A.* For an equivalent delta spacing of $\sqrt[3]{5 \times 5 \times 10} = 6.3$ inches, the reactance is 0.000049 Ω/ft. For 12 ft, the reactance is $12 \times 0.000049 = 0.000588\ \Omega$.

$$X_{pu} = \frac{0.000588}{X_b} = \frac{0.000588}{0.02404} = 0.0255\ pu$$

(12) *Bus B.* For an equivalent delta spacing of $\sqrt[3]{8 \times 8 \times 16} = 10.08$ inches, the reactance is 0.0000645 Ω/ft. For 2 ft, the reactance is $2 \times 0.0000645 = 0.000129\ \Omega$.

$$X_{pu} = \frac{0.000129}{X_b} = \frac{0.000129}{0.02404} = 0.0056\ pu$$

(13) *60 ft Motor Feeder Cables.* $X_a = 0.0953$ Ω/1000 ft and $X_d = -0.0572\ \Omega$/1000 ft for 1 inch spacing. The total reactance is

$$X_{tot} = X_a + X_d$$
$$= 0.0953 - 0.0572 = 0.0381\ \Omega/1000\ ft$$

For 60 ft,

$$X_{tot} = \frac{0.0381 \times 60}{1000} = 0.00229\ \Omega$$

$$X_{pu} = \frac{0.00229}{X_b} = \frac{0.00229}{0.02404} = 0.0995\ pu$$

(14) *100 hp Motors.* The per-unit reactance X_d'' = 0.20 on the motor kilovolt-ampere base. The motor kilovolt-ampere base is approximately the horsepower rating, 100 kVA.

$$X_{pu} = \frac{0.20 \times 10\ 000}{100} = 20.0\ pu$$

(15) *120 ft Feeder Cable.* $X_a = 0.108\ \Omega$/1000 ft and $X_d = -0.0729\ \Omega$/1000 ft for ½ inch spacing. The total reactance is

$$X_{tot} = 0.108 - 0.0729 = 0.035\ \Omega/1000\ ft$$

For 120 ft,

$$X_{tot} = \frac{0.035 \times 120}{1000} = 0.0042\ \Omega$$

$$X_{pu} = \frac{0.0042}{X_b} = \frac{0.0042}{0.02404} = 0.182\ pu$$

(16) *Air Circuit Breakers.* The reactance of a 1200 A circuit breaker is 0.00007 Ω; the reactance of 150 and 200 A circuit breakers is 0.001 Ω. The per-unit reactances are as follows:
    (a) 1200 A circuit breaker,

$$X_{pu} = \frac{0.00007}{0.02404} = 0.00304\ pu$$

    (b) 150 and 200 A circuit breakers,

$$X_{pu} = \frac{0.001}{0.02404} = 0.0434\ pu$$

(17) *Current Transformers.* The reactance of a 1200/5 current transformer is 0.00007 Ω and that of 150/5 and 200/5 current transformers is 0.0022 Ω. The per-unit reactances are as follows:
    (a) 1200/5 current transformer,

$$X_{pu} = \frac{0.00007}{0.02404} = 0.00304\ pu$$

    (b) 150/5 and 200/5 current transformers,

$$X_{pu} = \frac{0.0022}{0.02404} = 0.0954\ pu$$

The next step is to draw a reactance diagram showing the calculated per-unit reactance values in one-line form. Fig 11 shows the values arranged similar to the circuits of Fig 9.

Many of the reactances can be combined with ease. Series reactances can be added up and represented as a single reactance. Parallel reactances can also be combined into a single value. All rotating machine sources including plant motors and generators plus utility generators are represented by their per-unit reactances connected to an equivalent source bus. Thus the utility supply

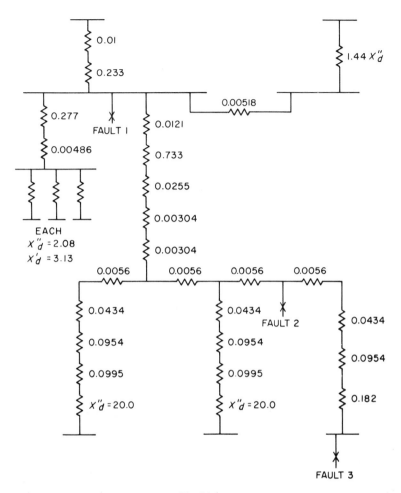

Fig 11
Impedance Diagram Constructed from One-Line Diagram

equivalent reactance is in parallel with the motor and generator reactances.

The reactance diagram should be simplified as much as possible, retaining the points at which the fault current is to be calculated. Fig 12 illustrates a simplification of Fig 11 by combining series and parallel reactance values. The dashed lines indicate buses of "equal potential" so far as the fault-current calculations are concerned.

Further simplification of the reactance diagram can be made only for a specific fault location. For a fault at location 1, for example, it is no longer necessary to retain fault locations 2 and 3, and further simplifications of the reactance diagram can be made.

(1) *Fault 1, 4160 V Fault on Feeder from Utility Tie Substation.* The simplification of the reactance diagram into a single equivalent reactance is shown in Fig 13. From Eqs 10 and 15 the symmetrical rms fault current is

$$I = \frac{10\ 000}{\sqrt{3} \times 4.16 \times 0.169} = 8220 \text{ A}$$

A multiplication factor of 1.6 must be applied to account for the effect of the direct-current component of initial (first-cycle) fault current. The assymmetrical rms fault current is then

$$I = 1.6 \times 8220 = 13\ 140 \text{ A}$$

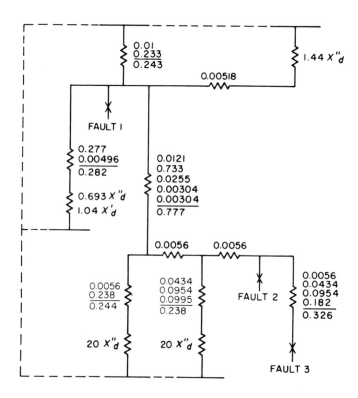

Fig 12
Impedance Diagram Simplified by Combining Impedances

For other times and applications use multiplying factors from appropriate standards such as IEEE Std 20-1973, Low-Voltage AC Power Circuit Breakers Used in Enclosers (ANSI C37. 13-1973) and ANSI C37.14-1969, Low-Voltage DC Power Circuit Breakers and Anode Circuit Breakers.

(2) *Fault 2, 480 V Fault on Feeder from Bus B.* Fig 12 must be simplified in another manner for the calculation of fault currents at location 2. In systems of 600 V and less, only the first-cycle fault current value is of interest since protective devices on those systems usually initiate operation within this time. Fig 14 shows the process of reducing the reactance diagram of Fig 12 to a single equivalent reactance.

From Eq 13 the symmetrical first-cycle fault current is

$$I = \frac{10\ 000}{\sqrt{3} \times 0.480 \times 0.881} = 13\ 620\ A$$

(3) *Fault 3, 480 V Fault on Feeder from Distribution Panelboard.* Fig 15 shows the simplified reactance diagram for fault location 3. The value 0.881 is the single equivalent per-unit reactance for fault location 2, and 0.326 is the per-unit reactance between locations 2 and 3 (see Fig 12). The total reactance from source to fault is then

$$0.881 + 0.326 = 1.207\ pu$$

A check of Table 2 indicates that the resistance of the 120 ft feeder cable is high in comparison to the system equivalent reactance from source to fault. The per-unit value of resistance is

$$\frac{0.181 \times 120}{1000} \times \frac{1}{1.729} = 0.943\ pu$$

This per-unit resistance is

$$\frac{0.943 \times 100}{1.207} = 78\ percent$$

of the system reactance from source to fault. It now becomes necessary to include the effect of

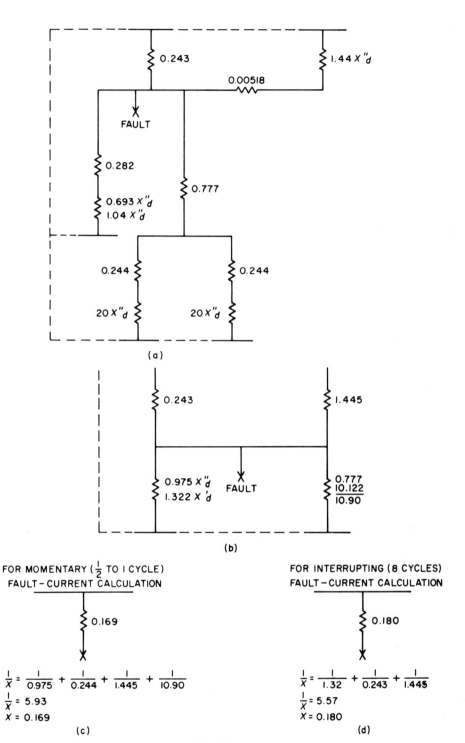

Fig 13
Impedance Calculations for Fault 1

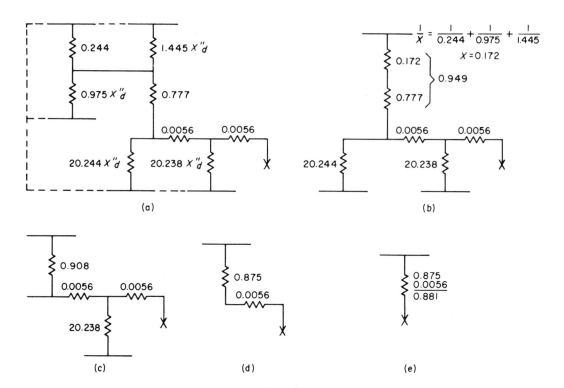

(a)

(b)

(c)

(d)

(e)

Fig 14
Impedance Calculations for Fault 2

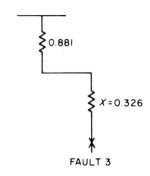

FAULT 3

Fig 15
Impedance Calculation for Fault 3

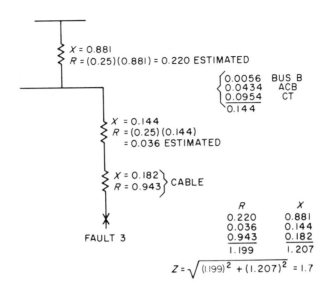

Fig 16
Reactance and Resistance Calculation for Fault 3

the resistance of the feeder cable in order to arrive at a realistic value of fault current. Fig 16 shows the simplified reactance—resistance circuit.

Account should also be taken of the resistance of the system up to the feeder. Assume a per-unit resistance equal to 25 percent of the corresponding per-unit reactance from source to feeder. This is at best an estimate, but it will hold fairly true for most low-voltage systems not having any generators or a large proportion of motors connected to it.

Fig 16 shows the calculation of the single equivalent impedance Z for fault 3. From Eqs 10 and 15 the symmetrical rms fault current is

$$I = \frac{10\ 000}{\sqrt{3} \times 0.480 \times 1.7} = 7070 \text{ A}$$

The value of symmetrical rms fault current if determined by reactance alone would be

$$I = \frac{10\ 000}{\sqrt{3} \times 0.480 \times 1.207} = 9960 \text{ A}$$

This would represent an error of 41 percent if resistance were not considered in the calculation of fault 3.

2.5 Standards References. The following standards publications were used as references in preparing this chapter.

ANSI C37.14-1969, Low-Voltage DC Power Circuit Breakers and Anode Circuit Breakers

IEEE Std 20-1973, Low-Voltage AC Power Circuit Breakers Used in Enclosures (ANSI C37.13-1973)

IEEE Std 141-1969, Electric Power Distribution for Industrial Plants

2.6 References and Bibliography

2.6.1 *References*

[1] HUENING, W.C., JR. Interpretation of New American National Standards for Power Circuit Breaker Applications. *IEEE Transactions on Industry and General Applications*, vol IGA-5, Sept/Oct 1969, pp 501-523.

[2] KAUFMANN, R.H. The Magic of I^2t. *IEEE Transactions on Industry and General Applications*, vol IGA-2, Sept/Oct 1966, pp 384-392.

[3] IEEE Committee Report. Protection Fundamentals for Low-Voltage Electrical Distribution Systems in Commercial Buildings. IEEE JH2112-1, 1974.

2.6.2 Bibliography

[4] BEEMAN, D.L., Ed. *Industrial Power Systems Handbook*. New York: McGraw-Hill, 1955, chap 2.

[5] CLARKE, E. *Circuit Analysis of AC Power Systems; vol 1, Symmetrical and Related Components*. New York: Wiley, 1943.

[6] *Electrical Transmission and Distribution Reference Book*. East Pittsburgh, PA; Westinghouse Electric Corporation, 1964.

[7] KERCHNER, R.M., and CORCORAN, G.F. *Alternating-Current Components*, 4th ed. New York: Wiley, 1960, chap 12.

[8] OHLSON, R.O. Procedure for Determining Maximum Short-Circuit Values in Electrical Distribution Systems. *IEEE Transactions on Industry and General Applications*, vol IGA-3, Mar/Apr 1967, pp 97-120.

[9] REED, M.B. *Alternating Current Circuit Theory*, 2nd ed. New York: Harper and Row, 1956.

[10] STEVENSON, W.D., JR. *Elements of Power Systems Analysis*. New York: McGraw-Hill, 1962.

[11] WAGNER, C.F., and EVANS, R.D. *Symmetrical Components*. New York: McGraw-Hill, 1933.

3. Selection and Application of Overcurrent Relays[1]

3.1 General Discussion of Protective System. As stated in Chapter 1, industrial and commercial power systems should be designed in such a way that protective equipment can operate to isolate faults quickly and limit the extent and duration of service interruptions. Protective devices such as fuses, circuit breakers, and relays are the watchmen of a power system functioning to detect and dispose of trouble expeditiously.

In a typical industrial system various protective schemes may be used to achieve these purposes. However, the most widely used scheme is an overcurrent protective system, which means a multiplicity of coordinated individual devices: fuses, direct-acting trip coils or plungers, and inverse time type relays. The characteristics of fuses and their proper applications will be discussed in Chapter 5. The direct-acting trip devices (circuit-breaker mounted tripping system) are encountered occasionally on medium-voltage power circuit breakers (over 1000 V) and almost always on low-voltage power circuit breakers (1000 V and below). They will be fully discussed in Chapter 6. This chapter will confine its discussions to the selection and application of overcurrent relays.

Overcurrent relays are generally used on circuit breakers over 1000 V and to some extent on low voltage circuit breakers where greater accuracy is

required than can be provided by certain direct-acting trip coils on the circuit breakers.

3.2 Principle and Construction of Overcurrent Relays. The overcurrent relays used in the industry are mainly of the electromagnetic attraction and induction type. Solid-state overcurrent relays have become available in recent years. Their characteristics are generally comparable to induction relays, except that they can provide faster reset times and have inherently reduced overtravel. The simplest overcurrent relay using the electromagnetic attraction principle is the solenoid type. The basic elements of this relay are a solenoid and a movable plunger.

When the circuit is energized with current at or above the minimum pickup value, the coil raises the plunger up into the operated position which closes the contacts in the trip circuit. A calibration screw is provided to adjust the initial position of the plunger. A common range of calibration in relays of this type is in the ratio of 1:4.

3.2.1 *Induction-Type Time-Delay Overcurrent Relay.* The most commonly used time-delay relays for system protection use the induction disc principle. The same principle is used on alternating-current watthour meters, and when applied to relay construction, it provides many varieties of time characteristics, depending on minor differences in electrical and mechanical design. The principal component parts of an induction-type overcurrent relay are shown in Fig 17.

---

[1] For discussion of protective relays other than overcurrent, refer to Chapters 13 and 14 and to individual chapters on protected equipment.

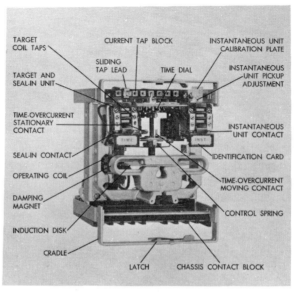

Fig 17
Induction-Disc Overcurrent Relay
with Instantaneous Attachment
(Relay Removed from Drawout Case)

The elements of an induction disc type relay are shown in Fig 18. The disc is mounted on a rotating shaft restrained by a spring. The moving contact is fastened to the shaft. The operating torque on the disc is produced by an electromagnet. A damping magnet provides restraint after the disc starts to move. This feature provides the desired time characteristic. The time scale indicates the initial position of the moving contact when the relay is de-energized. Its setting controls the time necessary for the relay to close its contact. A relay constructed on these principles has an inverse time characteristic. This means the relay operates slowly on small values of overcurrent, but as the overcurrent increases, the time of operation decreases. There is a limit to the speed at which the disc can travel. If the current continues to increase, the time curve of the relay will tend to reach a constant value. Different time—current curves of relays can be obtained by minor modifications of electromagnetic design.

An auxiliary seal-in relay is incorporated into the relay case to lighten the current-carrying duty of the moving contact. It also operates the target indicator.

3.2.2 *Instantaneous Trip Attachments.* The induction relay is generally provided with an auxiliary alternating-current operated instantaneous clapper-type current element. This element is provided with an adjustable calibrated range of 1:4 and is set for a current higher than that which should operate the time-delay element. The contacts of this element are either connected in parallel with the contacts of the time-delay element or they are connected to separate terminals.

3.2.3 *Solid-State Overcurrent Relay.* Some new overcurrent relay designs utilize solid-state technology. Time—current curves are obtained through the use of *RC* timing circuits. Time—current characteristic curves and tap ranges are similar to those provided in induction relays. Solid-state overcurrent relays have the same applications as induction relays and are particularly useful in certain severe environmental conditions, or where fast reset is required.

3.3 Overcurrent-Relay Types and Their Characteristic Curves. The overcurrent relays are generally available in the following current ratings:

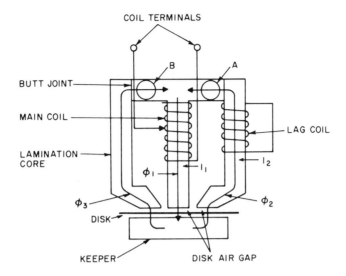

Fig 18
Elementary Induction-Type Relay

Range	Taps
0.5—2.5 (or 0.5—2)	0.5, 0.6, 0.8, 1.0, 1.2, 1.5, 2.0, 2.5
1.5—6 (or 2—6)	1.5, 2, 2.5, 3, 3.5, 4, 5, 6
4—16 (or 4—12)	4, 5, 6, 7, 8, 10, 12, 16

NOTE: Some models have extended ranges.

The relays can be specified to have either single or double circuit closing contacts for tripping either one or two circuit breakers.

The time—current characteristics for a variety of relays are shown in Figs 19—25. These characteristics give the contact closing times for the various time dial settings when the indicated multiples of tap current are applied to the relay.

Fig 26 illustrates the comparative shapes of five typical induction relay curves at the number 5 time dial.

3.4 Special Types of Overcurrent Relays. By adding different elements to the basic overcurrent relays as discussed in Section 3.2, several special types of overcurrent relays are derived.

3.4.1 *Voltage-Controlled (Restrained) Overcurrent Relay.* When a fault occurs, the system voltage collapses to a relatively low value, but when an overload occurs, the voltage drop is relatively small. If an overcurrent relay is provided with a restraining torque proportional to voltage, it will recognize the difference between a fault where the voltage drops and an overload where the voltage is maintained.

Such relays will provide sensitive fault protection without tripping on small overloads. They are used for generator circuits (see Chapter 14) for backup for external faults. The voltage-controlled overcurrent relay merely supervises the operating coil circuit at a preset drop in voltage. The voltage-restrained relay modifies the time—current curve in proportion to the drop in voltage.

3.4.2 *Directional Overcurrent Relay.* This relay consists of two units, an overcurrent element and a directional element. The contact circuits are arranged in such a way that tripping occurs only when current has proper relationship to the voltage with power flow in the tripping direction. The same function is also obtained by the use of a single element that combines a suitable time-delay and directional characteristic. The directional

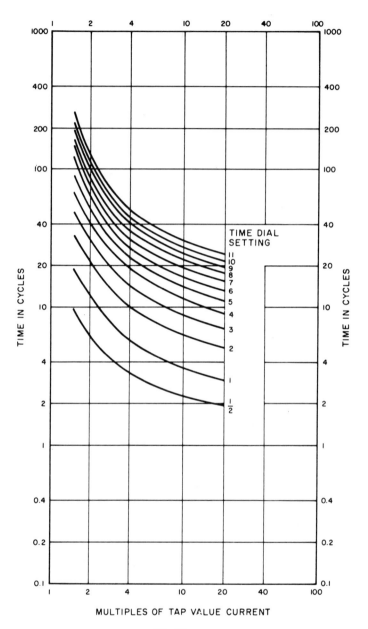

Fig 19
Typical Time—Current Curves for Short-Time Relay

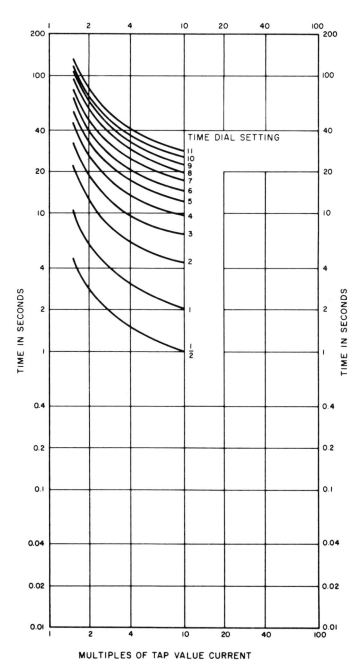

Fig 20
Typical Time—Current Curves for Long-Time Relay

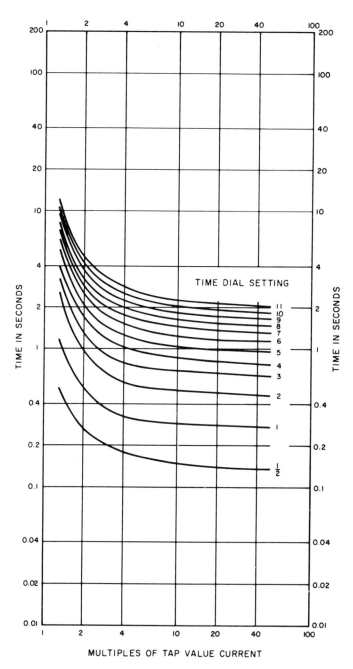

Fig 21
Typical Time—Current Curves for Definite Minimum-Time Relay

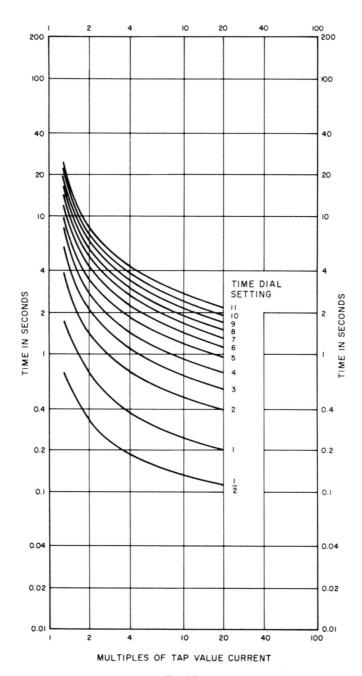

Fig 22
Typical Time—Current Curves for Moderately Inverse-Time Relay

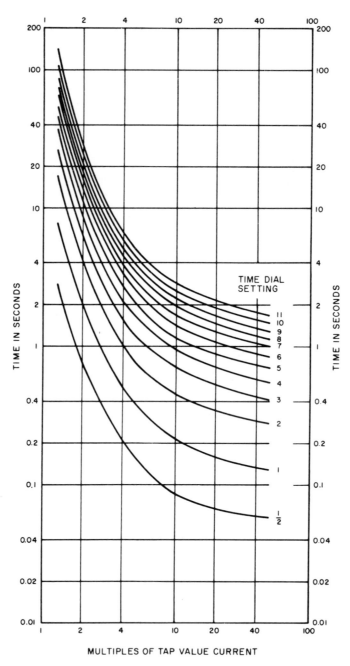

Fig 23
Typical Time—Current Curves for Inverse-Time Relay

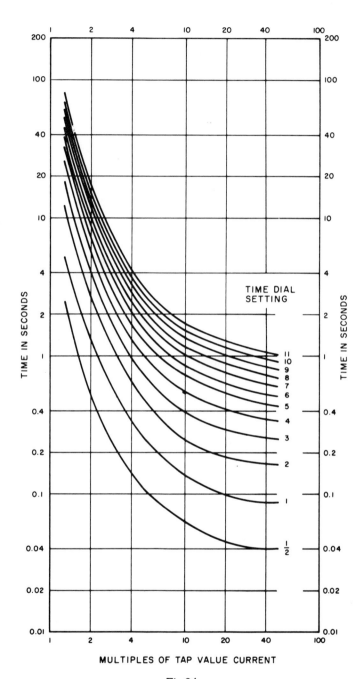

Fig 24
Typical Time—Current Curves for Very Inverse-Time Relay

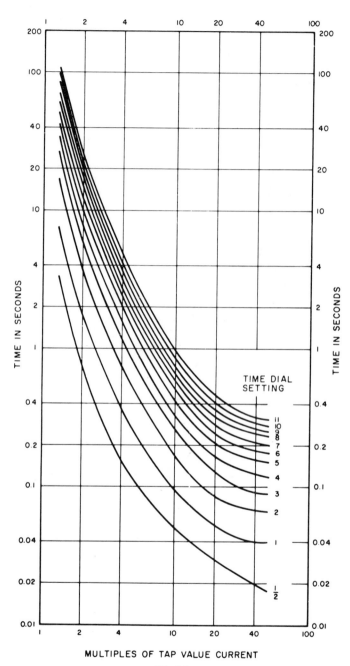

Fig 25
Typical Time—Current Curves for Extremely Inverse-Time Relay

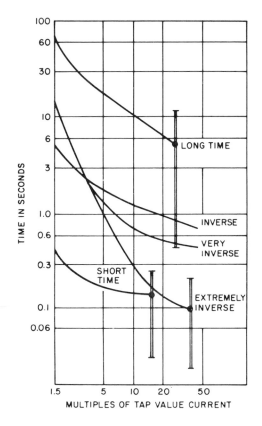

Fig 26
Comparison of Typical Curve Shapes for
Overcurrent Relays

control of an overcurrent time-delay relay is shown in Fig 27. The directional element is somewhat similar in principle to a contact-making wattmeter, held in the open position by a spiral spring when de-energized. This element is designed with high sensitivity to assure positive operation over the widest range of current and voltage relations encountered during a fault.

The actual tripping of the circuit is done by a contact on the overcurrent element. With the directional control element, the overcurrent element does not operate until the current is flowing in the proper direction and is above the pickup setting. The overcurrent element cannot operate on a fault in the nontripping direction.

3.4.3 *Directional Overcurrent Relay with Voltage Restraint.* Sometimes it is desirable to re-

strain the directional element from closing its contacts on normal load current flow in the tripping direction by using voltage to restrain the operation of the directional element. The use of voltage restraint can allow a wider choice of relay settings to fit the requirements of systems which have wide variations between minimum and maximum fault conditions.

The voltage restraint is greatly reduced when there is a phase-to-phase or a three-phase fault. Proper relay operation during a three-phase fault will depend on the sensitivity of the relay. Minimum trip current for this type of relay may be determined by reference to the manufacturer's instructions.

3.4.4 *High-Speed Directional Overcurrent Relay.* A high-speed directional element and an overcurrent element are coordinated to operate upon the occurrence of a short circuit that reverses the normal flow of current. Proper coordination of elements requires that a contact on one high-speed element open before the contact on another can close when the relay is suddenly de-energized or a quick reversal in power flow occurs.

This type of relay can be used to protect inplant generation operating in parallel with a utility system from a fault on the utility system.

3.5 Application of Overcurrent Relays

3.5.1 *Differential Protection.* Differential protection operates by comparing the current flowing into a circuit with the current flowing out of the circuit. Normally the currents are the same and the difference between them is zero. If a fault occurs in the circuit, part of the current flowing out will be deflected into the fault, and a differential current will flow. Differential protection may be applied to any section of a circuit and is used extensively to protect motors, generators and transformers against internal faults. It detects internal faults immediately and is not affected by overloads or faults outside the differentially protected section.

Fig 28 shows differential protection applied to a generator. Both ends of the windings must be available for the installation of the current transformers which have their secondary windings connected in series. Under normal conditions the current flowing in each current transformer sec-

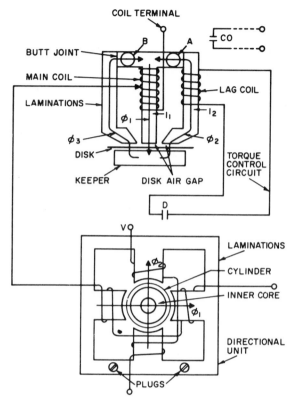

Fig 27
Directionally Controlled Overcurrent Relay

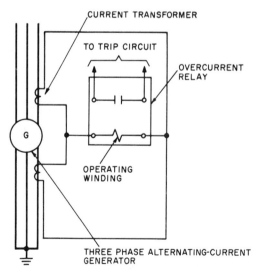

Fig 28
Overcurrent Relay Used for Differential
Protection of a Generator, One Phase Shown

ondary winding will be the same and the differential current flowing through the relay operating winding will be zero. In case of an internal fault in the generator, the secondary currents will no longer be the same and the differential current will flow through the relay operating coil and cause the trip circuit to be energized.

Overcurrent relays in a differential connection as indicated in Fig 28 afford very sensitive protection when used with ideal current transformers. However, the current transformers will not always give exactly the same secondary current for the same primary current. This difference is caused by minor variation in their manufacture and by differences in their secondary loadings. Where there is a prolonged direct-current component in the primary fault current, such as invariably occurs in medium to large generating plants, the current transformers will not saturate equally and a substantial operating current can be expected to flow. Hence if overcurrent relays are

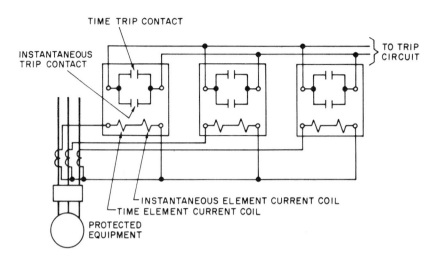

TIME TRIP CONTACT

INSTANTANEOUS
TRIP CONTACT

TO TRIP
CIRCUIT

INSTANTANEOUS ELEMENT CURRENT COIL
TIME ELEMENT CURRENT COIL

PROTECTED
EQUIPMENT

Fig 29
Overload Protection Combined with Instantaneous Short-Circuit Protection
for Motors Using Long-Time-Delay Overcurrent Relays

used, they have to be set so that they do not operate on the maximum error current which can flow in the relay during an external fault. To meet this problem without sacrificing sensitivity, the percentage differential type relay is usually used.

3.5.2 *Overload Protection of Rotating Apparatus.* In installations where some scheme of overload protection is desired, an appropriate type of overcurrent relay is a time-delay relay. An instantaneous relay would require a setting above the maximum momentary peak loads. Since the short-circuit current as determined by the synchronous reactance may be of the order of full load, the overcurrent protection of generators cannot be as satisfactory as the overcurrent protection for motors.

An effective and flexible scheme of overload protection for motors consists of the use of long-time-delay type relays with instantaneous trip attachments as shown in Fig 29. These relays have long-time-delay settings at motor starting current magnitudes and thus can be set on a low current tap, thereby affording protection against sustained moderate overloads. The time delay is sufficient to avoid relay operation on momentary peaks which the apparatus can carry safely. The instantaneous trip attachment should be set

for a current value higher than maximum peak load or maximum motor starting current, thus providing instantaneous protection against short circuits or extremely heavy overloads which may damage the apparatus before the overcurrent relay can operate. For more treatment on motor protection, refer to Chapter 9.

3.5.3 *Differential Protection of Two-Winding Wye—Delta Transformer Bank.* When differential relays are used for the transformer protection, the inherent characteristics of the power transformers introduce a number of problems which do not exist in the protection of generators and motors. Magnetizing inrush current in the transformer bank appears to the relay as an internal fault. The voltage transformation of the power transformer may not be exactly matched in the current transformer circuit so that unequal secondary currents occur on through faults.

If the secondary currents on the two sides of the transformer differ in magnitude, the relay currents can be mached by means of a current-balancing autotransformer. If the high-voltage and low-voltage line currents are not in phase due to a wye—delta connection in the tranformer, the secondary currents can be brought into phase by connecting the current transformer in delta on the wye side, and in wye on the delta side. Fig 30

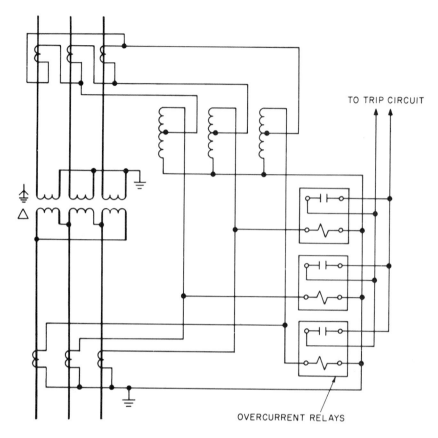

TO TRIP CIRCUIT

OVERCURRENT RELAYS

Fig 30
Differential Protection of Two-Winding Wye—Delta Transformer Bank
Using Induction-Type Overcurrent Relays

shows such an application in which standard overcurrent relays are used as the differential relays.

Since the current transformers on the two sides of a power transformer bank will have different ratios and may be of different types, the differential current caused by dissimilar ratio characteristics will be considerably greater than in the case of generator protection. This is the reason why ordinary overcurrent relays cannot be given sensitive settings, and transformer differential relays are generally used instead. For detailed discussions on differential protection of the transformers, refer to Chapter 10.

3.5.4 *Protection against Phase-to-Phase Faults Within a Zig-Zag Connected Transformer.* When a zig-zag connected grounding transformer is used to provide a source of ground current for a system where the power transformers are not grounded, overcurrent relays are sometimes used for protection against phase-to-phase faults within the transformer, while a differential relay may be used to secure protection against phase-to-ground faults. To protect against ground faults elsewhere on the system, which may fail to be automatically cleared and which will eventually overheat the grounding transformer, a thermal relay can be used. However, to provide faster protection against more severe faults of this nature, a long-time-delay overcurrent relay is usually used instead of the thermal relay.

3.5.5 *Differential Protection of Station Buses.* In most bus protectiom problems, the differential scheme is utilized to distinguish between an

IEEE Std
242-1975

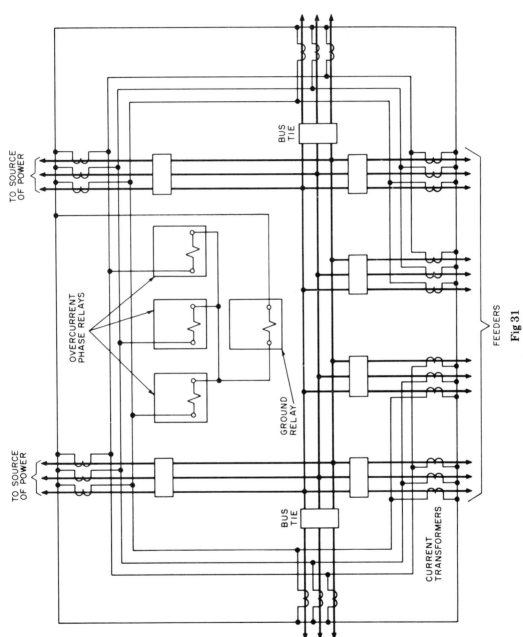

Fig 31

Phase and Ground Overcurrent Protection of Eight-Circuit Bus Using Induction-Type Overcurrent Relays

internal and external fault by comparing the magnitudes of the currents flowing in and out of the bus. The major differences between bus protection and generator or transformer protection are in the number of circuits in the protected zone and in the magnitudes of currents involved in the various circuits. In Fig 31 there are two sources of power, four feeders, and two bus tie circuits. Current transformers of the same ratio are inserted in the leads of all circuits entering and leaving the bus section and the vector sum of all the secondary currents is passed through the operating winding of an overcurrent relay for each phase. Under normal conditions the vector sum of all the currents will be zero, and there will be no current in the operating winding of the relay. Under a fault condition, current will flow into the fault (upsetting the balance) to close the relay contacts and initiate tripping of all circuits connected to the bus. In Fig 31 tripping circuits are not shown. Overcurrent relays can be used for phase fault protection, and where necessary, a ground relay can be used to measure the vector sum of the residual and ground currents. Both time-delay and instantaneous relay elements can be used. The conventional time-delay relay has been the most satisfactory.

Due to saturation in one or more of the current transformers used in the scheme, the magnitude of current in various branches of the circuit will be different. The resulting inaccuracy in the current transformer response will cause a current to flow in the relay even though the fault is external. Experience has shown that inverse time-delay-type relays can be given a lower setting than instantaneous relays would require to ensure against the possibility of incorrect operation. This can be accomplished because these relays are less sensitive to the distorted waveform of the current from saturated transformers. In some installations, neither the instantaneous nor the timed overcurrent relays can be used satisfactorily.

The current transformers must be capable of reproducing the primary current with sufficient fidelity to permit setting the relays so that they will operate on the minimum internal fault but will not operate on maximum external fault. Special bus differential relays are usually used for important buses. This relay arrangement can provide sensitive operation for internal faults and a decidedly insensitive characteristic to assure against operation on external fault. For more detailed treatment, refer to Chapter 12.

3.5.6 *Protection of Distribution Lines.* A radial system is one having a single source of power with feeders leaving the source bus, each feeder in turn being subdivided into a number of smaller feeders. The protection of such a system against short circuits may be obtained by overcurrent protective relays. The circuit breaker near the source, as well as the intermediate circuit breakers, must be equipped with time-delay relays. The time interval required between successive relays must be long enough to assure selective operation, and relays at the source bus will have the highest time settings. The most remote relays will have an instantaneous setting. For service supply line protection, refer to Chapter 13.

3.5.7 *Ground-Fault Protection.* Overcurrent protection for ground faults, using time as the means of selectivity, follows the same principle as outlined for phase protection. Settings can be made if the short-circuit magnitudes are known. The connections for a ground overcurrent relay in conjunction with phase overcurrent relays are shown in Fig 32. For a more detailed treatment, refer to Chapter 8.

3.6 Overcurrent-Relay Setting Study. (Refer to Chapter 7.)

3.6.1 *Basic Steps.* The preceding sections illustrated what operating time performance to expect from various system overcurrent protective devices. The next problem is to obtain the necessary data from which specific relay or other overcurrent device settings may be determined. The following basic steps are required.

(1) *Collection of Necessary Data.* A one-line diagram of the power system is needed, showing the type and rating of the protective devices and their associated instrument transformers; the impedances of all transformers, rotating machines, and feeder circuits; and the starting current and time requirements of motors, and their maximum load current.

(2) *Calculation of Fault Currents.* Maximum and minimum values of short-circuit current must be calculated, including transient variations that are expected to flow through each protective device whose performance is to be studied under varying operating conditions.

STATION BUS

TO TRIP CIRCUIT

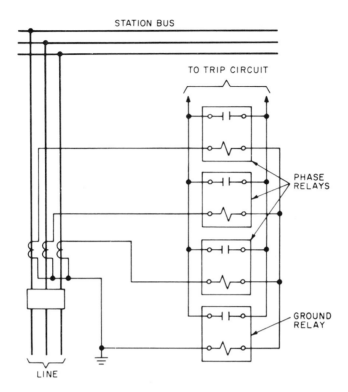

PHASE RELAYS

GROUND RELAY

LINE

Fig 32
Phase and Ground Overcurrent Protection of a Circuit
Using Induction-Type Overcurrent Relays

(3) *Equipment Performance.* Manufacturers' characteristic performance curves of the relays, trip coils and fuses, current transformers, especially bushing types, are required as well as decrement curves showing the rate of decay of the fault current supplied by generators.

(4) *Special Requirements.* Any special overcurrent protective requirements, such as those specified by NFPA No 70, National Electrical Code (1975) (ANSI C1-1975) or dictated by the load characteristics, and any special overcurrent protective device setting requirements stipulated by the utility company with which the industrial plant is to be interconnected, must be considered.

(5) *Selection and Plotting of Preliminary Settings.* Selection is based on the overload protection, and the calculated maximum and mimimum fault currents. Curves are plotted on a common voltage base and a coordination time interval is allowed.

(6) *Checking of Final Settings.* Preliminary settings are revised if required to prevent overlap. Selective operation at maximum and minimum fault current should be compatible with short-time ratings of transformers and cables. The instantaneous settings must be checked for compatibility with inrush currents of transformers and large motors.

3.6.2 *Short-Circuit-Current Calculations for a Relay Study.* The problem in a relay study is to determine the characteristics of devices that (1) will be sure to operate on the minimum value of fault current expected at certain fixed times following the initiation of the short circuit, and (2) will be selective in their sequence of operation over the range between minimum and maximum values of short-circuit current so that the protective device nearest the fault will operate first.

The magnitude of the short-circuit currents, which will determine the settings of the overcurrent protective devices, should be calculated (Chapter 2) on the basis of the fault current from any utility company ties, plus that contributed by all rotating machines directly connected to the local power system, that is, the user's own power generation and motors.

The values of fault current to which relays are responsive vary with the type of relay or relay element.

(1) Instantaneous elements or plunger-type relays are responsive to direct current as well as alternating current and operate on the first half-cycle of fault current. Therefore the initial (instantaneous) asymmetrical fault current contributed by all rotating equipment is required.

(2) High-speed induction-type relays which operate in three cycles or less are also responsive to direct-current components of current. Therefore the initial asymmetrical fault current as in (1) is also required. In reality the current magnitude will drop during the three cycles due to a decrement effect, but the influence of this reduction can be neglected for ordinary industrial systems.

(3) Time-delay relays and trip coils are too slow to be appreciably affected by the subtransient reactance values of fault current. Their maximum operating current is dependent on the initial value of symmetrical transient reactance current supplied by the utility company and local plant synchronous equipment. Such relays will usually be the first in a series of relays. Time settings of such devices at maximum current usually are more than 0.1 s.

3.6.3 *Settings of Relays as Affected by Their Function.* The maximum current settings of the relays must be high enough to carry normal load swings, and yet low enough to be sure that they will operate positively on the minimum expected short-circuit current. Normally there will be plenty of margin between the minimum short-circuit current and the relay setting dictated by the maximum permissible load.

For details of overcurrent relay settings on individual feeder circuits refer to Chapters 9, 10, and 15.

(1) *Relays on Incoming Lines and Feeders with*

Miscellaneous Load. The minimum setting of a time-delay overcurrent relay on an incoming line or feeder should be just above the expected peak load on the circuit. This will be the total of the starting current of the largest motor plus full load of the other feeders. However a system where motors restart automatically when a circuit is re-energized following an outage requires special attention.

Instantaneous relays are often omitted since they may not be able to coordinate with other instantaneous relays at the same voltage level in the branch circuits. If they are used, they should be set slightly above the total initial asymmetrical short-circuit current which all the motors could contribute to a short circuit elsewhere on the system, plus the full-load current on the nonmotor load.

(2) *Residually Connected Ground-Fault Relays.* Residual ground-fault relays are connected in the wye of the secondaries of three current transformers where they will see only the unbalanced residual current flowing during line-to-ground faults. Tap settings used vary from a low 0.5 A to a maximum of 2 to 3 A. These relays will operate satisfactorily over their entire tap range when connected to high-accuracy current transformers, but because of high burden they may not operate properly on the low relay taps when connected to the low-turns-ratio taps of some "through type" current transformers. It will usually be possible to obtain satisfactory operation by using a higher current tap on the relay or bushing current transformer, or both. For a detailed treatise on instrument transformers, refer to Chapter 4.

The use of core balance transformers provides excellent ground-fault protection and is described in detail in Chapter 8.

3.6.4 *Working Examples.* The curves of relays and other devices that are to operate selectively in series should be plotted on log—log paper using a common voltage base. The plotting can usually reveal conflicts that otherwise might escape notice.

Example 1 (Selection of time—current curves and relay trip settings for a 2400 V industrial plant distribution system.) In Fig 33 it is assumed that the power supply is sufficient to maintain a 250 MVA short-circuit level without appreciable decrement, and that there is no pump-back fault

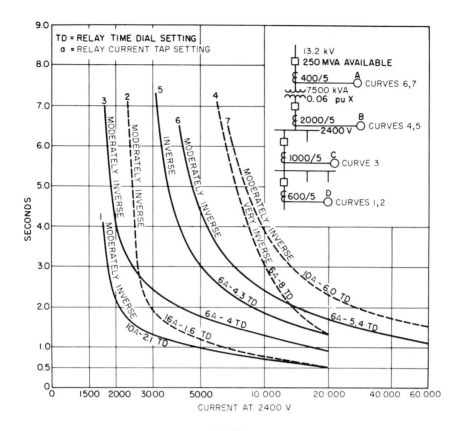

Fig 33
Selecting Time—Current Curves and Relay Tap Settings for a 2400 V
Industrial Plant Distribution System (Example 1)

NOTE: The curves are plotted on semilog coordinates for
illustrative purposes only. Normally log-log coordinates are
used. Also in actual practice tap settings above 8 A are seldom
required.

current from synchronous equipment on the 2400 V system. On this basis, the maximum symmetrical fault current would be 20 000 A on the 2400 V system and 10 250 A on the 13.8 kV system. A typical procedure would be to coordinate the three sets of 2400 V relays by having their operating times at 20 000 A 0.35 s apart, starting with 0.5 s which was the minimum required time for the end relay in the chain. Relays at A should be coordinated with those on the 2400 V system at the same value of fault current. This can be accomplished if the time—current settings selected will give 0.35 s between relays A

and B for a 20 000 A fault on the 2400 V system.

The very inverse curve (curve 4) of relay B crosses curve 6 at a high level of fault current, showing that the tripping sequence of the circuit breaker will be reversed. This would not be a serious problem since tripping either circuit breaker would shut down the entire circuit, but the identity of which circuit breaker is responsible for the tripping will be lost. If it was necessary to keep the very inverse relay B, relay A's setting would have to be increased in time, as shown by curve 7, in order to be selective with B, resulting in more damage in case of a short circuit. On the

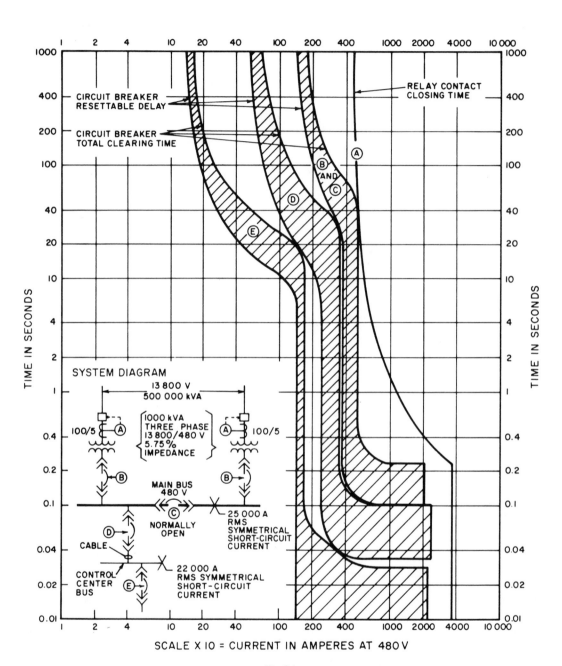

Fig 34
Selecting Primary Relays to Coordinate with Main Secondary
Circuit Breaker and Feeder Circuit Breakers (Example 2)

contrary, if an inverse relay was used for B, it would permit setting B to give performance as shown by curve 5, which would afford more sensitive and faster protection for the system. Also the very inverse relay curve for B would leave a very big "gap" between this curve and curve 3. This means poor backup protection.

Example 2 (Selection of primary relays to be selective with the main secondary protection and group feeder protection.) The tap setting was chosen to give selectivity with a secondary air circuit breaker. In Fig 34 all currents are referred to the 480 V side of the transformer, hence a relay tap setting of 8 A and time lever 1 are chosen to yield adequate time delay to allow the secondary circuit breaker to clear low-voltage faults. The time level is purposely kept low to give the best possible arcing fault protection. The 8 A tap setting corresponds to

tap setting $\times$ current transformer ratio

$$\times \frac{\text{primary voltage}}{\text{secondary voltage}}$$

$$= 8 \times \frac{100}{5} \times \frac{13\ 800}{480}$$

$$= 4600 \text{ A on the 480 V side}$$

The instantaneous trip attachment on the overcurrent relay is necesssary to give high-speed clearing of faults on the high-voltage side. The setting must be high enough to override low-voltage faults (including asymmetry). The settings can be arrived at as follows.

The short circuit current flowing through the transformer is 20 200 A. To override a low-voltage fault, the minimum instantaneous trip setting allowable is

$$1.6 \times 20\ 200 = 32\ 320 \text{ A}$$

Allowing for a 10 percent tolerance, this becomes

$$\frac{32\ 320}{0.9} = 35\ 900 \text{ A}$$

In terms of actual relay current this is

$$\frac{35\ 000}{(100/5)\ (13\ 800/480)} = 62 \text{ A}$$

An inverse-type overcurrent relay with 4–12 A range for time delay and 20–80 A for instantaneous setting should be chosen.

In Fig 34 curve A is the result of these relay settings, which coordinate exceedingly well with curves B, C, D, and E for the main secondary circuit breaker, bus tie circuit breaker, group feeder circuit breaker, and feeder circuit breaker at the control center bus.

For additional examples, refer to Chapter 7.

3.7 Standards References. The following standards publication was used as reference in preparing this chapter.

NFPA No 70, National Electrical Code (1975) (ANSI C1-1975)

4. Instrument Transformers

4.1 Current Transformers. A current transformer transforms line current into values suitable for standard protective relays and insulates the relays from line voltages. A current transformer has two windings, designated as primary and secondary, which are insulated from each other. The various types of primary windings are covered below. The secondary is wound on an iron core. The primary winding is connected in series with the circuit carrying the current to be measured, and the secondary winding is connected to the protective devices and in some instances to instruments, meters, or control devices. The secondary winding supplies a current in direct proportion and at a fixed relationship to the primary current.

4.1.1 *Types.* The three common types of construction are as follows.

(1) *Wound-Primary Type.* As the name implies, this type has more than one primary turn. The primary and secondary windings are completely insulated for their respective voltage ratings and permanently assembled on a laminated iron core. This construction permits higher accuracy at the lower ratios (Fig 35).

(2) *Through or Bar Type.* This type has the primary and secondary windings completely insulated and permanently assembled on a laminated iron core. The primary winding usually consists of a bar-type conductor passing through the core window (Fig 36).

(3) *Window or Bushing Type.* These types have a secondary winding completely insulated and permanently assembled on an iron core. The primary conductor passes through the core window and serves as the primary winding. This conductor may be a cable, bus bar, or bushing core. In applying a transformer of this type care must be taken to ensure that either the primary conductor or the transformer is insulated for full system voltage (Fig 37).

4.1.2 *Ratios.* ANSI C57.13-1968, Requirements for Instrument Transformers, designates certain ratios as standard. These ratios are shown in Tables 10 and 11. Note that the standard rated secondary current in all instances is 5 A.

4.1.3 *Application.* The general considerations for the application of current transformers follow.

(1) *Continuous-Current Rating.* The continuous-current rating should be equal to or greater than the rating of the circuit in which the current transformer is used. A secondary current of 3 or 4 A at full load is considered proper. An oversized current transformer is undesirable because the percent error is likely to be greater than a correctly rated current transformer.

(2) *Thermal Short-Time Rating.* This is the symmetrical rms primary current that the current transformer can carry for 1 s with the secondary winding short-circuited, without exceeding a specified temperature in any winding.

Fig 35
Wound-Primary-Type Current Transformer

Fig 36
Through- or Bar-Type Current Transformer

Fig 37
Window-Type Current Transformer
with Low Base

Table 10
Current Transformer Ratings
Multiratio Bushing Type

Current Ratings (amperes)		Secondary Taps
600/5	50/5	X2-X3
	100/5	X1-X2
	150/5	X1-X3
	200/5	X4-X5
	250/5	X3-X4
	300/5	X2-X4
	400/5	X1-X4
	450/5	X3-X5
	500/5	X2-X5
	600/5	X1-X5
1200/5	100/5	X2-X3
	200/5	X1-X2
	300/5	X1-X3
	400/5	X4-X5
	500/5	X3-X4
	600/5	X2-X4
	800/5	X1-X4
	900/5	X3-X5
	1000/5	X2-X5
	1200/5	X1-X5
2000/5	300/5	X3-X4
	400/5	X1-X2
	500/5	X4-X5
	800/5	X2-X3
	1100/5	X2-X4
	1200/5	X1-X3
	1500/5	X1-X4
	1600/5	X2-X5
	2000/5	X1-X5
3000/5	1500/5	X2-X3
	2000/5	X2-X4
	3000/5	X1-X4
4000/5	2000/5	X1-X2
	3000/5	X1-X3
	4000/5	X1-X4
5000/5	3000/5	X1-X2
	4000/5	X1-X3
	5000/5	X1-X4

Table 11
Current Transformer Ratings
Other than Multiratio Bushing Type

Single Ratio (amperes)	Double Ratio with Series—Parallel Primary Windings (amperes)		Double Ratio with Taps in Secondary Winding (amperes)
10/5	25 ×	50/5	25/50/5
15/5	50 ×	100/5	50/100/5
25/5	100 ×	200/5	100/200/5
40/5	200 ×	400/5	200/400/5
50/5	400 ×	800/5	300/600/5
75/5	600 ×	1200/5	400/800/5
100/5	1000 ×	2000/5	600/1200/5
200/5	2000 ×	4000/5	1000/2000/5
300/5			1500/3000/5
400/5			2000/4000/5
600/5			
800/5			
1200/5			
1500/5			
2000/5			
3000/5			
4000/5			
5000/5			
6000/5			
8000/5			
12 000/5			

(3) *Mechanical Short-Time Rating.* This is the maximum current the current transformer is capable of withstanding without damage with the secondary short-circuited. It is the rms value of the alternating-current component of a completely displaced (asymmetrical) primary current wave. Mechanical limits need only be checked for wound-type current transformers.

(4) *Voltage Rating.* Current transformers are capable of operating continuously at 10 percent above rated primary voltage. Standard voltage ratings (insulation class) for most industrial applications are 600, 2500, 5000, 8700, and 15 000 V.

(5) *Impulse and High-Potential Rating.* The values are given in Table 12.

Table 12
Current Transformer
High-Potential and Impulse Ratings

Nameplate Rating (kV)	60 Hz High-Potential Rating (kV)	Impulse (BIL) Rating (kV)
0.6	4	10
2.5	15	45
5.0	19	60
8.7	26	75
15.0	34	95

Table 13
Standard Burdens for Current Transformers

Standard Burden Designation	Characteristics		Characteristics for 60 Hz and 5 A Secondary Current		
	Resistance (ohms)	Inductance (mH)	Impedance (ohms)	Apparent Power* (VA)	Power Factor
B-0.1	0.09	0.116	0.1	2.5	0.9†
B-0.2	0.18	0.232	0.2	5.0	0.9†
B-0.5	0.45	0.580	0.5	12.5	0.9†
B-1	0.5	2.3	1.0	25	0.5‡
B-2	1.0	4.6	2.0	50	0.5‡
B-4	2.0	9.2	4.0	100	0.5‡
B-8	4.0	18.4	8.0	200	0.5‡

* At 5 A; note that VA = I^2Z, or 2 Ω at 5 A = 2 × 5^2 = 50 VA.

† Usually considered metering burdens, but data sheets may give metering accuracies at B-1.0 and B-2.0.

‡ Usually considered relaying burdens.

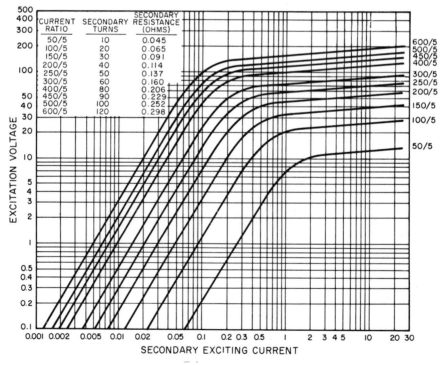

Fig 38
Typical Secondary Excitation Curves for Various Ratio Current Transformers

4.1.4 *Accuracy*. Protective-relay performance depends on the accuracy of transformation of the current transformers, not only at load currents, but also at all fault-current levels. The accuracy at high overcurrents depends on the cross section of the iron core and the number of turns in the secondary winding. The greater the cross section of the iron core, the more flux can be developed before saturation. Saturation results in an increase of ratio error. The greater the number of turns, the lower the flux required to force the secondary current through the relay.

The ANSI C57.13-1968 designates the relaying accuracy class by use of the two letters C and T and the classification number. C means that the percent ratio error can be calculated, and T means that it has been determined by test. The classification number indicates the secondary terminal voltage which the transformer will deliver to a standard burden (as listed in Table 13) at 20 times normal secondary current without exceeding a 10 percent ratio error. Furthermore, the ratio error should not exceed 10 percent at any current from 1 to 20 times rated current at any lesser burden. The standard designated secondary terminal voltages are 10, 20, 50, 100, 200, 400, and 800 V. For instance, a transformer with a relaying accuracy class of C 200 means that the percent ratio error can be calculated and that it does not exceed 10 percent at any current from 1 to 20 times the rated secondary current at a burden of 2.0 Ω. (Maximum terminal voltage = $20 \times 5 \, A \times 2 \, \Omega = 200 \, V$.)

4.1.5 *Burden*. Burden, in current transformer terminology, is the load connected to the secondary terminals and is expressed in volt-amperes at a given power factor, or as impedance with a power factor. The term "burden" is used to differentiate the current transformer load from the primary circuit load. The power factor referred to is that of the burden and not of the primary circuit. For the purpose of comparing various transformers, ANSI has designated standard burdens to be used in the evaluation process (refer to Table 13).

4.1.6 *Secondary Excitation Characteristics and Overcurrent Ratio Curves*. Secondary excitation characteristics, as published by the manufacturers, are in the form of current versus voltage (Fig 38). The values are obtained either by calculation from the transformer design data and core-loss

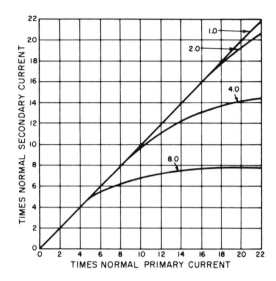

Fig 39
Typical Overcurrent Ratio Curves for Class T
Transformers for Burdens of 1—8 VA

curves or by average test values from a sample of current transformers. The test is an open-circuit excitation current test on the secondary terminals, using a variable voltage rated frequency sine wave and recording rms current versus rms voltage.

For class T transformers, typical overcurrent ratio curves are plotted between primary and secondary current over the range from 1 to 22 times normal primary current for all the standard burdens (Fig 39).

4.1.7 *Polarity*. Polarity marks designate the relative instantaneous directions of currents. At the same instant of time that the primary current is entering the marked primary terminal, the corresponding secondary current is leaving the marked secondary terminal, having undergone a magnitude change within the transformer (Fig 40). The H_1 and X_1 terminals are usually marked either with white dots or labels. As can be seen in Fig 40, one can consider the marked secondary conductor a continuation of the marked primary line as far as current direction is concerned.

4.1.8 *Connections*. There are three ways that current transformers are usually connected on three-phase circuits, (1) wye, (2) vee or open delta, and (3) delta.

(1) *Wye Connection.* In the wye connection a current transformer is placed in each phase with phase relays in two or three secondaries to detect phase faults. On grounded systems, a relay in the current transformer common wire detects any ground or neutral currents. On ungrounded systems, a relay in the current transformer common wire detects multiple ground faults on different feeders. Secondary currents are in phase with primary currents (Fig 41).

(2) *Vee Connection.* This current transformer connection is basically a wye with one leg omitted, using only two current transformers. Applied as shown in Fig 42, this connection gives protection against phase-to-phase faults on all phases of a three-phase system, but will only protect against those phase-to-ground faults that are above the relay settings on the phases in which the current transformers are installed. A third-phase overcurrent relay may be applied in the common connection of the current transformers for backup phase relaying if desired. In this connection the secondary currents are in phase with the primary currents.

Since there is no way of sensing zero-sequence currents with this connection, it is rarely used as the only means of protection in a circuit. It is, however, frequently applied with the addition of a zero-sequence current transformer (doughnut or ring type). The zero-sequence current transformer can be applied to either grounded or ungrounded power circuits, and since the transformer and its associated protective relay are not sensitive to load currents of any magnitude, they may be of relatively low current ratings, thereby providing sensitive ground-fault detection.

(3) *Delta Connection.* This connection uses three current transformers, but unlike the wye connection, the secondaries are interconnected before the connections are made to the relays. The delta connection is used for power transformer differential-relay protection schemes where the power transformer has delta—wye connected windings. The current transformers on the delta side are connected in wye, and the current transformers on the wye side are connected in delta. Any zero-sequence currents associated with an external ground fault on the wye side will circulate in the delta current trans-

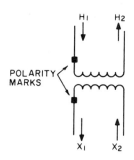

Fig 40
Current Transformer Polarity Diagram

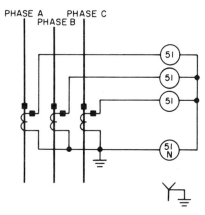

Fig 41
Wye-Connected Current Transformers

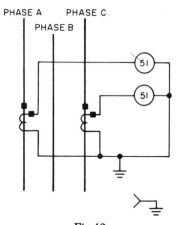

Fig 42
Vee- or Open-Delta-Connected
Current Transformers

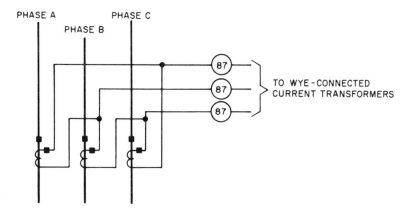

Fig 43
Delta-Connected Current Transformers

former connection and be kept from causing false differential relay operation. Figs 41 and 43 would be combined to complete the differential-relay connections with the ground wire connected to each relay-operating coil.

In addition, delta-connected current transformers connected to overcurrent relays are used to provide complete phase protection for zig-zag grounding transformers.

4.1.9 *Examples of Accuracy Calculations*

Example 1. Calculation of a 600/5 Multiratio Bushing-Type Current Transformer. Consider a 600/5 multiratio bushing-type current transformer with excitation characteristics as shown in Fig 38. It is connected for a 600/5 ratio and to a secondary circuit containing an overcurrent relay with instantaneous attachment, a watthour meter, and an ammeter. The circuit contains 50 ft of No 12 wire, and the primary circuit has a capability of 24 000 A of fault current.

From instruction books for the devices and wire resistance tables, the following data are obtained.

(1) Relay, time unit, 4—12 A, has a burden of 2.38 VA at 4 A at 0.375 power factor (146 VA at 40 A at 0.61 power factor)

(2) Relay, instantaneous unit, 10—40 A, has a burden of 4.5 VA at 10 A setting (40 VA at 40 A setting at 0.20 power factor)

(3) Watthour meter has a burden of 0.77 W at 0.54 power factor at 5 A

(4) Ammeter has a burden of 1.04 VA at 5 A at 0.95 power factor

(5) Wire burden equals 0.08 Ω at 1.0 power factor

(6) Current transformer secondary resistance = 0.298 Ω at 25°C

The steps for determining the performance of the transformer for this application are as follows:

(1) Determine the current transformer secondary burden.

(2) Determine the voltage necessary to operate the relay at its maximum applicable current.

(3) Determine the current transformer exciting current from Fig 38 and calculate the error.

Step (1). As stated previously, the burden is expressed in volt-amperes at a given power factor (PF), or as impedance with a power factor. Since most of the apparatus connected to current transformers contain magnetic paths which become saturated, the burden should be calculated for the maximum specific current involved. On circuits where instantaneous elements are involved, the setting of the instantaneous element is the determining factor for establishing the maximum significant current. When instantaneous elements are not involved, the maximum current available is the determining factor.

Since the relay is equipped with an instantaneous element, it is assumed that it can be set at its maximum tap setting of 40 A. In this case, it will amount to 8 × 600 or 4800 A. Thus the secondary burden should be determined for this value of current.

Device 1, Relay, Time Unit. 146 VA at 40 A at 53°.

$$Z = 146/(40)^2 = 0.091\Omega$$

$$0.091\angle\underline{53}° = 0.0546 + j\,0.0728$$

Device 2, Relay, Instantaneous Unit. 40 VA at 40 A at 20°.

$$Z = 40 / (40)^2 = 0.025 \ \Omega$$

$$0.025\angle\underline{20}° = 0.023 + j\,0.008$$

Device 3, Watthour Meter. 0.77 W at 5 A at 57.3°.

$$VA = W/PF = 0.77/0.54 = 1.43 \ VA$$

$$Z = 1.43 / (5)^2 = 0.057 \ \Omega$$

$$0.057\angle\underline{57.3}° = 0.031 + j\,0.048$$

Since a watthour meter also has an iron-core magnetic circuit, the power factor at 8 times current is approximately 0.94. Thus, at 40 A,

$$Z = \text{resistance/power factor}$$

$$= 0.031/0.94 = 0.033 \ \Omega$$

$$R + jX = 0.031 + j0.011$$

$$VA = I^2R = (40)^2 \times 0.033 = 52.8 \ VA$$

Device 4, Ammeter. 1.04 VA at 5 A at 18°.

$$Z = 1.04/5^2 = 0.041 \ \Omega$$

$$0.041\angle\underline{18}\ ° = 0.039 + j\,0.012$$

Since an ammeter applies basically an air-core magnetic circuit, no saturation is present at 8 times current. Thus, at 40 A,

$$VA = I^2Z = (40)^2 \times 0.041 = 65.5 \ VA$$

Device 5, Wire. 0.08 Ω at 1.00 PF. Thus, at 40 A,

$$VA = I^2R = (40)^2 \times 0.08 = 128 \ VA$$

Device 6, Current Transformer Secondary Resistance. 0.298 Ω at 1.00 PF. Thus, at 40 A,

$$VA = I^2R = (40)^2 \times 0.298 = 476 \ VA$$

Totalizing for devices 1—6 at 40 A,

Device	VA	Impedance
1	146	$0.0546 + j\,0.0728$
2	40	$0.023 \ + j\,0.008$
3	52.8	$0.031 \ + j\,0.011$
4	65.5	$0.039 \ + j\,0.012$
5	128	0.08
6	476	0.298
	908.3	$0.525 \ + j\,0.103$

$$Z_1 = 908.3/(40)^2 = 0.566 \ \Omega$$

$$Z_2 = 0.525 + j\,0.103 = 0.542 \ \Omega$$

Note that Z_1 compares favorably with the more accurate Z_2.

Step (2). The current transformer voltage necessary to produce a secondary current of 40 A through the above burden is IZ.

$$IZ_1 = 40 \times 0.566 = 22.6 \ V$$

$$IZ_2 = 40 \times 0.542 = 21.6 \ V$$

Step (3). From Fig 38 find the secondary exciting current I_e. At 22.6 V $I_{e1} = 0.032$ A, and at 21.6 V $I_{e2} = 0.032$ A. The percent ratio error is given by

$$(I_e /I_s) \times 100 = (0.032/40) \times 100$$

$$= 0.08 \text{ percent}$$

Thus, for this application the current transformer is much more than adequate.

Example 2. Using the 100/5 A Tap on the 600/5 Multiratio Current Transformer. The total requirement of apparent power in volt-amperes would change as the current transformer secondary burden is reduced. The current transformer secondary resistance is 0.066 Ω at 25°C. The apparent power of the current transformer is

$$VA = I^2R = (40)^2 \times 0.066 = 105 \ VA$$

Thus the total value is

$$VA = 908.3 - (476 - 105) = 537.3 \text{ VA}$$

The voltage required for the 100/5 ratio is

$$537.3/40 = 13.4 \text{ V}$$

and

$$Z = 537.3/(40)^2 = 0.335 \ \Omega$$

From Fig 38, for 13.4 V $I_e = 0.5$ A and the percent error is

$$(I_e/I_s) \times 100 = (0.5/40) \times 100 = 1.25 \text{ percent}$$

Thus, again, for the assumed application, the 100/5 ratio is more than adequate for the relaying involved.

Example 3. Using the 100/5 Tap with 100 A Instantaneous Setting. If the 100/5 ratio were applied with a relay that would require operation at 100 A instead of 40 A, the current transformer burden would be approximately as much as that calculated for the 40 A current (0.335 Ω). The current transformer voltage necessary to produce 100 A would be approximately 33 V. As evident from Fig 38, the 100/5 ratio would not develop 33 V, except at much higher exciting currents than shown. Thus the 100/5 current transformer would not be applicable on a circuit where operation at 100 A secondary current is necessary for correct system operation.

4.1.10 *Saturation.* Abnormally high primary currents, high secondary burden, or a combination of these factors will result in the creation of a high flux density in the current transformer iron core. When this density reaches or exceeds the design limits of the core, saturation results. At this point, the accuracy of the current transformer becomes very poor, and the output waveform may be distorted by harmonics. The total result is the production of a secondary current lower in magnitude than would be indicated by the current transformer rate. Saturation effects in themselves are not usually dangerous to properly designed equipment. The greatest danger is loss of protective device coordination. Since low-ratio current transformers will saturate before high-ratio current transformers, the result may be operation of a main circuit breaker and outage of an entire plant system on a fault which should have been cleared by a branch circuit breaker. To avoid or minimize saturation effects, secondary burden should be kept as low as possible. Where fault currents of more than 20 times the current transformer nameplate rating are anticipated, the manufacturer should be consulted for data on high overcurrent performance.

4.1.11 *Safety Precautions.* The most important precaution with respect to current transformers is the admonition to never open the secondary circuit while in service. If the circuit is opened, the entire primary current becomes magnetizing current, and an excessive voltage will be induced in the secondary. This voltage, which can rise above 1000 V, constitutes a real hazard to personnel and can cause a failure of the transformer insulation.

4.2 Voltage (Potential) Transformers. A voltage (potential) transformer is basically a conventional transformer with primary and secondary windings on a common core. Standard voltage transformers are single-phase units designed and constructed so that the secondary voltage maintains a fixed relationship with the primary voltage. The required ratio of a voltage transformer is determined by the voltage of the system to which it is to be connected and by the way in which it is to be connected. Most voltage transformers are designed to provide 120 V at the secondary terminals when nameplate-rated voltage is applied to the primary. Standard ratings are shown in Table 14. Special ratings are available for applications involving unusual connections. These ratings are not included in standard listings.

Voltage transformers are normally applied to systems having nominal voltages within 10 percent (plus or minus) of the transformer nameplate voltage. This may mean system voltage either line to line or line to neutral, depending on the connection to be used.

Standard accuracy classifications of voltage transformers range from 0.3 to 1.2, representing percent ratio corrections to obtain a true ratio. These accuracies are high enough so that any standard transformer will be adequate for protective relaying purposes as long as it is applied within its open-air thermal and voltage limits.

Table 14
Standard Voltage Transformer Ratings

Primary (amperes)	Secondary (amperes)	Ratio
120	120	1/1
240	120	2/1
480	120	4/1
600	120	5/1
2400	120	20/1
4200	120	35/1
4800	120	40/1
7200	120	60/1
8400	120	70/1
12 000	120	100/1
14 400	120	120/1
24 000	120	200/1
36 000	120	300/1
48 000	120	400/1
72 000	120	600/1
96 000	120	800/1
120 000	120	1000/1
144 000	120	1200/1
168 000	120	1400/1
204 000	120	1700/1
240 000	120	2000/1
300 000	120	2500/1
360 000	120	3000/1

Fig 44
Voltage Transformer

Thermal burden limits as given by transformer manufacturers should not be exceeded in normal practice since transformer accuracy and life will be adversely affected. Thermal burdens are given in volt-amperes and may be calculated by simple arithmetic addition of the volt-ampere burdens of the devices connected to the transformer secondary. If the sum is within the rated burden, the transformer should perform satisfactorily over the range of voltages from zero to 110 percent of nameplate voltage.

Voltage transformers are normally identified for polarity by marking a primary bushing H and a secondary terminal X. Alternatively, these points may be identified by distinctive color markings. The standard voltage relationship provides that the instantaneous polarities of H and X are the same.

Where balanced system load and therefore balanced voltages are anticipated, voltage trans-

formers are usually connected in open delta. Where line-to-neutral loading is expected, voltage transformers are more often connected wye—wye, particularly where metering is required. Many protective devices require specific delta or wye voltages; therefore, it is desirable to make a study of requirements before choosing the connection scheme. Wye—delta or delta—wye connections are occasionally used with certain special relays, but these connections are infrequent in industrial use. Where ungrounded power systems are in use, voltage transformers connected wye—broken delta are sometimes used for ground detection. When so connected, the transformers can seldom be used for any other purpose.

The application of fuses to voltage-transformer circuits has been a subject of discussion for many years. General practice now calls for a current-limiting fuse or equivalent in the primary connection where this connection is made to an ungrounded conductor of the system. Fig 44 shows a typical voltage transformer with fuses.

Voltage-transformer secondary fusing practices cannot be so clearly defined. Since manufacturer recommended primary fuses are normally sized for transformer full-load rating, it is difficult or impossible to secure coordination with full-capacity secondary fuses. Where branch circuits are tapped from voltage-transformer secondaries to supply devices located at a distance from the voltage transfomer, it may be desirable to fuse the branch at a reduced rating.

4.3 Standards References. The following standards publication was used as reference in preparing this chapter.

ANSI C57.13-1968, Requirements for Instrument Transformers

4.4 Bibliography

[1]   FINK, D.G., and CARROLL, J.M. *Standard Handbook for Electrical Engineers*. New York: McGraw-Hill, 1968.

[2]   *Applied Protective Relaying*. Newark, NJ: Westinghouse Electric Corporation, 1957.

[3]   *Electric Utility Engineering Reference Book*; vol 3, *Distribution Systems*. Trafford, PA: Westinghouse Electric Corporation, 1965.

[4]   *Manual of Instrument Transformers*. Schenectady, NY: General Electric Company, GET-97, 1969.

[5]   *Instrument Transformer Burden Data*. Schenectady, NY: General Electric Company, GET-1725, 1961.

[6]   MASON, C.R. *The Art and Science of Protective Relaying*. New York: Willey, 1956.

5. Fuse Characteristics

5.1 General Discussion. A fuse may be defined as "a device which protects a circuit by fusing open its current-responsive element when an overcurrent or short-circuit current passes through it." A fuse has these functional characteristics.

(1) It combines both the sensing and interrupting element in one self-contained device.

(2) It is direct acting in that it responds only to a combination of magnitude and duration of circuit current flowing through it.

(3) It normally does not include any provision for making or breaking the connection to an energized circuit but requires separate devices such as an interrupter switch to perform this function.

(4) It is a single-phase device. Only the fuse in the phase or phases subjected to overcurrent will respond to de-energize the affected phase or phases of the circuit or equipment that is faulty.

(5) After having interrupted an overcurrent it is renewed by replacing the fuse before restoring service.

A. Low-Voltage Fuses

5.2 Low-Voltage Fuse Ratings. Low-voltage fuses have current, voltage, and interrupting ratings which should not be exceeded in application. In addition, some fuses are rated according to their current-limiting capability as established by standards from the National Electrical Manufacturers Association (NEMA) or Underwriters Laboratories Inc (UL). They are so designated by a class marking on the label such as class G, J,

K1, K5, K9, L, and R. Current-limiting capabilities are established by NEMA or UL, or both, according to the maximum peak let-through current and the maximum let-through energy I^2t by the fuse when clearing a fault.

(1) The current rating of a fuse is the maximum direct or rms alternating current, in amperes, at rated frequency which it will carry without exceeding specified limits of temperature rise. Current ratings which are available on the low-voltage fuse market range from milliamperes to 6000 A.

(2) The voltage rating is the maximum alternating- or direct-current voltage at which the fuse is designed to operate. Low-voltage fuses are usually given voltage ratings of 600, 300, 250, or 125 V alternating or direct current or both.

(3) The interrupting rating is the assigned maximum short-circuit current (usually alternating current) at rated voltage which the fuse will safely interrupt. Low-voltage fuses may have interrupting ratings of 10 000, 50 000, 100 000, or 200 000 A, symmetrical rms.

(4) The current-limiting ratings according to class will be discussed in the next section covering fuse standards.

5.3 Available Standards Pertaining to Fuses

5.3.1 *National Electrical Code.* NFPA No 70, National Electrical Code (1975) (NEC) (ANSI C1-1975) recognizes two principal categories of fuses, (1) plug fuses and (2) cartridge fuses. Figs 45 and 46 illustrate the various plug and cartridge type fuses mentioned in the NEC.

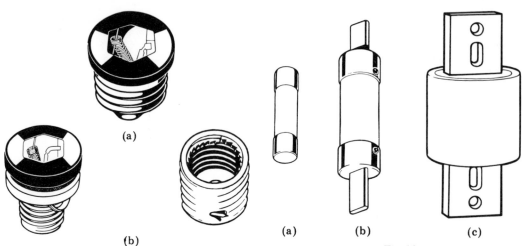

Fig 45
Plug Fuses. (a) Edison Base Fuse
(b) Type S Fuse and Adapter

Fig 46
Cartridge Fuses. (a) Ferrule Type, 0—60 A
(b) Knife-Blade Type, 70—600 A
(c) Bolt Type, 601—6000 A

Table 15
Fuse and Fuseholder Case Sizes According to Current and Voltage

0—600 A			601—6000 A
250V or Less (amperes)	300V or Less (amperes)	600V or Less (amperes)	600V or Less (amperes)
0—30	0—15	0—30	601—800
31—60	16—20	31—60	801—1200
61—100	21—30	61—100	1201—1600
101—200	31—60	101—200	1601—2000
201—400		201—400	2001—2500
401—600		401—600	2501—3000
			3001—4000
			4001—5000
			5001—6000

In addition the NEC mentions the following types: non-time-delay fuses, time-delay fuses, dual-element fuses, current-limiting fuses, non-current-limiting fuses, miscellaneous fuses, fuses over 600 V, and primary fuses.

(1) *Plug Fuses.* Plug fuses are rated 125 V and are available with current ratings up to 30 A. Their use is limited to circuits rated 125 V or less. However, they may also be used in circuits supplied from a system having a grounded neutral, and in which no conductor operates at more than 150 V to ground. Fig 45 shows both types of plug fuses mentioned in the NEC. The NEC requires type S fuses in all new installations of plug fuses because they are tamper resisting and size limiting, thus making it difficult to overfuse.

NOTE: To be listed as time-delay fuses, plug fuses are required by UL to have a minimum opening time of 12 s at 200 percent of rated current.

(2) *Cartridge Fuses.* Cartridge fuses and fuseholders are classified by the NEC in ratings of 0 to 600 A and 601 to 6000 A. Fuses rated 600 A and below have three voltage classifications,

0—600 A   Not over 250 V
0—60 A   Not over 300 V
0—600 A   Not over 600 V

There are no 250 V ratings over 600 A. However, 600 V fuses may be used on lower voltages. Table 15 shows cartridge fuses and fuseholder case sizes according to current and voltage.

All fuses recognized by the NEC which have interrupting ratings exceeding 10 000 A must be marked on the fuse label with the interrupting rating.

5.3.2 *Underwriters Laboratories Inc.* Underwriters Laboratories Inc (UL) standards on fuses UL 198.1-1973, UL 198.2-1970, UL 198.3-1970, UL 198.4-1973, UL 198.5-1973, and UL 198.6-1974 require the following for listing.

(1) Fuses must carry 110 percent of their rating continuously when installed in the test circuit specified in the standard.

(2) Fuses rated from 0 through 60 A must open within 1 h and fuses rated from 61 through 600 A within 2 h when carrying 135 percent of rating in the specified test circuit. Fuses rated above 600 A must open within 4 h when carrying 150 percent of rated current in the specified test circuit.

(3) Different current and voltage rated fuses must have specified dimensional tolerances to prevent unsafe interchangeability.

(4) All fuses listed by UL must be capable of interrupting fault currents up to 10 000 A alternating current at rated voltage, and may carry much higher interrupting ratings such as 100 000 or 200 000 A, symmetrical rms.

(1) *Fuse Class Designations.* UL, in conjunction with NEMA, has established standards for the classification of fuses by letter rather than by type. The class letter may designate interrupting rating, physical dimensions, degree of current limitation (maximum peak let-through current), and maximum clearing energy (in amperes squared times seconds) under specific test conditions, or they may indicate combinations of these characteristics. The descriptions of these classes are as follows.

(a) *Class G Fuses* (0—60 A). Class G fuses are miniature fuses rated 300 V, used mainly on 480 Y/277 V systems connected phase to ground. They are available in ratings up to 60 A and have an interrupting rating of 100 000 A, symmetrical rms. Case sizes for 15, 20, 30, and 60 A are each of different length. Fuseholders designed for a specific case size will reject larger fuses.

Class G fuses are considered by UL if they have a minimum time delay fuses s at 200 percent of their current rating. of 12

(b) *Class H Fuses* (0—600 A). Class H fuses have dimensions which were listed in the NEC prior to 1959. These fuses are often referred to as "Code fuses" or "NEC fuses." Although these fuses are not marked with an interrupting rating, they are tested by UL on circuits which can deliver 10 000 A alternating current. They are rated 600 V or 250 V. The most common class H fuses are (a) one-time fuses (nonrenewable) and (b) renewable fuses.

The ordinary one-time cartridge fuse is the oldest type of cartridge fuse in common use today. It utilizes a zinc or copper link and has limited interrupting capabilities. The usage of one-time fuses is decreasing due to their limited interrrupting rating and lack of intentional time delay.

Renewable fuses are similar to one-time fuses except that they can be taken apart after interrupting a fault and the fusible element replaced. Renewal links are usually made of zinc. Their ends are clamped or bolted to the fuse terminals.

To be listed as time-delay fuses, Class H nonrenewable fuses are required by UL to have a minimum opening time of 10 s at 500 percent of rated current.

(c) *Class J Fuses* (0—600 A). Class J fuses have specific physical dimensions which are smaller than 600 V class H fuses. Class H fuses cannot be installed in fuseholders designed for class J fuses. Class J fuses are current limiting and carry an interrupting rating of 200 000 A, symmetrical rms. UL has established maximum allowable limits for peak let-through current I_p and let-through energy I^2t which are slightly less than those for class K1 fuses of the same current rating. (See Table 16 for a comparison.)

Time-delay standards have not been established for class J fuses; therefore none are listed by UL as time-delay fuses. Fuses having class J dimensions are available with varying degrees of time delay in the overload range, and at least one make is available that has a minimum opening time of 10 s at 500 percent of rating.

(d) *Class K Fuses* (0—600 A). Class K designates a specific degree of peak let-through current I_p and maximum clearing energy I^2t. Present class K fuses have the same dimensions as

Table 16
Class J and Class K Fuse Maximum Allowable Peak Let-Through Current I_p and I^2t as Established by UL at an Available Symmetrical Current of 100 000 A

Class	Fuse Rating (amperes)	I_p (amperes)	I^2t (amperes squared-seconds)
J	30	7 500	7×10^3
	60	10 000	30×10^3
	100	14 000	80×10^3
	200	20 000	300×10^3
	400	30 000	1100×10^3
	600	45 000	2500×10^3
K1	30	10 000	10×10^3
	60	12 000	40×10^3
	100	16 000	100×10^3
	200	22 000	400×10^3
	400	35 000	1200×10^3
	600	50 000	3000×10^3
K5	30	11 000	50×10^3
	60	21 000	200×10^3
	100	25 000	500×10^3
	200	40 000	1600×10^3
	400	60 000	5000×10^3
	600	80 000	$10 000 \times 10^3$
K9	30	14 000	50×10^3
	60	28 000	250×10^3
	100	35 000	650×10^3
	200	60 000	3500×10^3
	400	80 000	$15 000 \times 10^3$
	600	130 000	$40 000 \times 10^3$

class H fuses, but have interrupting ratings higher than 10 000 A, that is, 50 000, 100 000, or 200 000 A, symmetrical rms. UL has established three levels designated K1, K5, and K9, with class K1 having the greatest current-limiting ability and K9 the least (Table 16). Class K9 fuses are no longer manufactured.

Some manufacturers of electrical equipment are furnishing equipment having a withstand rating based on a particular class of fuse.

To be listed as time-delay fuses, class K fuses are required by UL to have a minimum opening time of 10 s at 500 percent of rated current.

(e) *Class L Fuses* (601—6000 A). Class L fuses have specific physical dimensions and bolt-type terminals. They are rated 600 V and carry an interrupting rating of 200 000 A, symmetrical rms. Class L fuses are current limiting and UL has specified maximum values of I_p and I^2t for each rating according to Table 17.

Standards for time-delay characteristics, in the overload range, have not been established for class L fuses. Some available class L fuses have a minimum time delay of approximately 4 s at 500 percent of rated current. Class L fuses are not listed by UL as time-delay fuses.

(2) *Miscellaneous Fuses.* There are other fuses with special characteristics and dimensions designed for supplementary overcurrent protection, some of which conform to UL standards.

Table 17
Class L Fuse Maximum Allowable Peak
Let-Through Current I_p and $I^2 t$ as Established by
UL at an Available Symmetrical Current of 100 000 A

Fuse Rating (amperes)	I_p (amperes)	$I^2 t$ (amperes squared-seconds)
601—800	80 000	$10\ 000 \times 10^3$
801—1200	80 000	$12\ 000 \times 10^3$
1201—1600	100 000	$22\ 000 \times 10^3$
1601—2000	120 000	$35\ 000 \times 10^3$
2001—2500	165 000	$75\ 000 \times 10^3$
2501—3000	175 000	$100\ 000 \times 10^3$
3001—4000	220 000	$150\ 000 \times 10^3$
4001—5000		$350\ 000 \times 10^3$
5001—6000		$350\ 000 \times 10^3$

Supplementary fuses cannot be used for protecting branch circuits. (Refer to Section 5.4.5.)

(3) *Non-Time-Delay and Time-Delay Fuses.* Non-time-delay fuses are fuses that have no intentional built-in time delay. They are generally employed in other than motor circuits or in combination with circuit breakers where the circuit breakers provide protection in the overload current range and the fuses provide protection in the short-circuit current range. They are not generally used in motor circuits except in combination starters where an overload relay provides protection in the overload range and the fuse provides short-circuit protection only.

Time-delay fuses have intentional built-in time delay in the overload range. This time-delay characteristic often permits the selection of fuse ratings closer to full-load motor currents.

The dual-element-type time-delay fuse is widely used as it has adequate time delay to permit its use as motor overcurrent running protection without the possibility of needless opening due to starting surge currents.

5.3.3 *National Electrical Manufacturers Association.* NEMA classification and performance requirements are similiar to UL requirements.

5.4 Fuses Not Covered by Existing Standards. There are many fuses which are called "nonstandard" fuses due to the fact that they have special applications or special dimensions and are not general-purpose fuses.

5.4.1 *Cable Limiters.* Cable limiters or protectors are available for use in multiple-cable circuits to provide short-circuit protection for cables. Cable limiters are rated 600 V with interrupting ratings as high as 200 000 A, symmetrical rms.

They are rated according to cable size, that is, No 4/0, 500 kcmil, etc, and have numerous types of terminations. Fig 47 shows typical cable limiters with two different terminations, (a) tubular-type terminals for crimping onto cables and (b) tubular-to-bolt type terminals for connecting cables to bus bars. Many other types and combinations of terminations are available to match individual specifications.

These limiters are designed to provide short-circuit-current protection for cables. They are used primarily in low-voltage networks or in service-entrance circuits where more than two cables per phase are brought into a switchboard. A typical one-line diagram representing a cable limiter installation is shown in Fig 48. (Note that for isolation of a faulted cable the limiters must be located at each end of each cable.)

5.4.2 *Fuses for Semiconductors.* Germanium and silicon rectifiers have very high current ratings considering their relatively small size. However, their low mass causes them to have poor overload surge capacity and therefore little ability to sustain high overload currents for any peri-

(a)   (b)

Fig 47
Typical Cable Limiters
(a) Tubular Terminations. (b) Tubular-Bolt Terminations

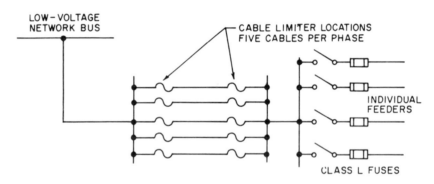

Fig 48
Typical Circuit for Cable Limiter Application

od of time. The problem of protecting semiconductors is, therefore, quite considerable. This
protection can only be achieved by using devices
that are extremely fast in clearing short-circuit
conditions. Fuses for semiconductors are designed to respond quickly to short-circuit conditions.

Most diode and thyristor manufacturers provide cell data sheets which spell out current and
voltage ratings as well as short-time withstand
data. All of these data are pertinent from a fuse
selection standpoint; they must be compared
with available fuse data.

Current and voltage ratings of fuses are usually
given on an rms basis and must be related to data
for specific rectifier circuits. One of the most

popular rectifier circuits is the three-phase bridge
rectifier shown in Fig 49. Rectifier fuses are
selected on the basis of (1) the no-load rms
voltage of the transformer, (2) the rms current of
the rectified wave through the specific cell, (3)
the maximum short-circuit current the transformer can deliver, and (4) the short-time characteristics of the cell, including peak current, I^2t
withstandability, and peak inverse voltage.

The short-time let-through data, which are
available from fuse manufacturers, usually come
in the form of peak let-through current I_p versus
available short-circuit current and let-through energy I^2t versus available short-circuit current.
Figs 50 and 51 show typical let-through data for
one family of rectifier fuses.

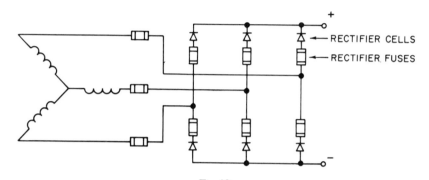

Fig 49
Three-Phase Bridge Rectifier Showing Location of Rectifier Fuses

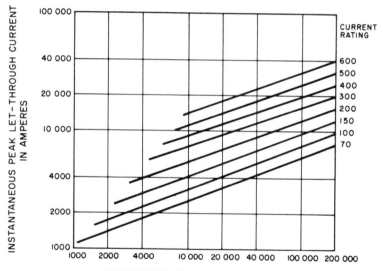

SYMMETRICAL RMS CURRENT IN AMPERES

Fuse Voltage Rating	600 V rms
Test Voltage	600 V rms
Frequency	60 Hz
Power Factor	15 Percent or Less
Arcing Angle	Within 30° Prior to Peak of Voltage Wave

Fig 50
Typical Peak Let-Through Current versus Available
Short-Circuit Current for 600 V or Less

NOTE: Check specific manufacturer for exact data.

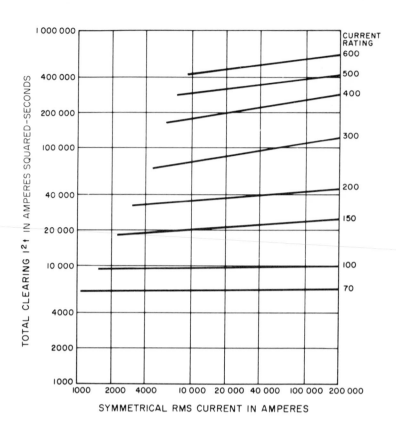

For I^2t Values at Voltages Less than 600 V
Use the Following Multipliers:

Voltage (V)	Multiplier
600	1.00
500	0.85
300	0.65
250	0.50
125	0.30

Fuse Voltage Rating	600 V rms
Test Voltage	600 V rms
Frequency	60 Hz
Power Factor	15 Percent or Less
Arcing Angle	Within 30° Prior to Peak of Voltage Wave

Fig 51
Typical Total Clearing Energy I^2t versus Available Fault Current

NOTE: Check specific manufacturer for exact data.

Fig 52
Capacitor Fuse with Indicator for Protection
of Power-Factor Correction Capacitors

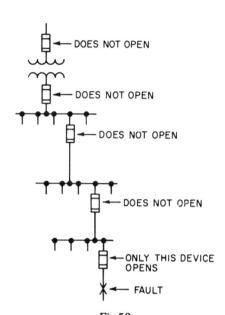

Fig 53
Selective Operation of Overcurrent Protective
Devices. All Three Phases Must Be Opened
Whenever Three-Phase Loads Are Involved

5.4.3 *Welder Limiters.* Welder limiters are special current-limiting devices for use on welder circuits only. Their time—current characteristics are such that they will carry on the intermittent overloading encountered in welder operation, while providing short-circuit protection to the circuit and equipment. They have excess current capacity in the operating range as needed for this type of service.

Welder limiters are available in nominal current ratings from 150 to 600 A and can be used on alternating-current circuits of 600 V or less.

5.4.4 *Capacitor Fuses.* Capacitor fuses are available in 600 and 250 V alternating-current ratings. These fuses are usually applied in circuits with power-factor correction capacitors to isolate short-circuited capacitors from the remainder of the bank. Capacitor fuses have current ratings ranging from 25 to 250 A and are usually installed at about 150—200 percent of a capacitor's current ratings.

Indicators are usually provided as an integral part of these fuses for ease in finding a blown fuse upon visual inspection. These fuses are available with a variety of mounting terminals. Fig 52 shows a typical capacitor fuse with indicator.

5.4.5 *Miscellaneous Fuses.* Miscellaneous fuses are most commonly used to protect equipment which is plugged into branch circuits. Some examples of this type of equipment are individual lighting fixtures (that is, fluorescent), outdoor lighting, electronic circuits, control panels, and communication equipment.

The control power circuit of motor-control equipment also uses miscellaneous fuses. Improper protection of control transformers has been a serious problem in industry and should be given serious thought. Midget-type time-delay fuses are available for motor controllers rated 480 V or lower and can be sized very close to the control transformer's full-load current rating.

Fluorescent fixtures should incorporate individual protection within the fixture but external to the ballast so that a faulted ballast can be isolated from the remaining fixtures on a branch circuit to prevent an area blackout. This can be accomplished by having miscellaneous fuses installed by the fixture manufacturer in each fixture.

5.5 Selectivity of Fuses

5.5.1 *General Considerations.* Since the electrical distribution system is the heart of most industrial, commercial, and institutional type installations, it is imperative that any unnecessary shutdowns of electrical power be prevented. Unnecessary blackouts can be avoided by the proper selection of overcurrent protective devices. Se-

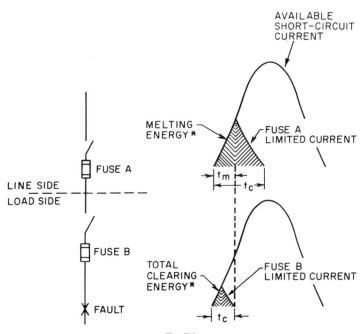

**Fig 54
Selectivity of Fuses. Total Clearing Energy of Fuse B Must Be
Less than Melting Energy of Fuse A**

*Indicates, but does not equal, energy.

lectivity (often referred to as selective coordination) may be defined as the complete isolation of a faulted circuit to the point of fault without disturbing any of the other protective devices in the system. To obtain complete isolation of a fault (except in a single-phase circuit) all phase conductors should be opened. If only one or two phases are opened in a three-phase circuit, the fault remains connected to the system through the multiphase load circuits. Under this condition, the fault current is often reduced but not cleared. Anti-single-phasing devices may be required. Fig 53 shows how a selective system should operate.

Fig 54 illustrates the general principle by which fuses coordinate with one another for any value of short-circuit current. Note that for selectivity, the total clearing energy of fuse B must be less than the melting energy of fuse A.

In fuse applications, coordination is usually achieved through the use of selectivity ratio tables (Section 5.5.2).

5.5.2 *Selectivity Ratio Tables.* Table 18 shows a typical selectivity schedule for various combinations of fuses. This schedule is very general and may not apply equally from the lowest ratings to the highest ratings in all lines of fuses at all available currents. More specific data are usually available from the fuse manufacturers.

An example of using Table 18 is found in Fig 55 where a 1200 A class L fuse is to be selectively coordinated with a 400 A class K5 time-delay fuse.

Selectivity schedules or tables are used as a simple check for selectivity assuming that identical or reduced fault currents flow through the circuits in descending order, that is, main—feeder—branch. Where closer fuse sizing is desired than indicated, check with the fuse manufacturer as the ratios may be reduced for lower values of short-circuit current. A coordination study may be desired when the simple check as outlined is not sufficient, and can be accomplished by plotting fuse time—current characteristic curves on

Table 18
Typical Selectivity Schedule*

Line Side	Load Side							
	Class L Time-Delay Fuse 601–6000 A	Class L Fuse 601–6000 A	Class K1 Fuse 0–600 A	Class J Fuse 0–600 A	Class K5 Time-Delay Fuse 0–600 A	Class K5 Time-Delay Current-Limiting Fuse 0–600 A	Class J Time-Delay Fuse 15–600 A	Class G Fuse 0–60 A
Class L Time-Delay Fuse 601–6000 A	2:1	2:1	2:1	2:1	4:1	3:1	3:1	—
Class L Fuse 601–6000 A	2:1	2:1	2:1	2:1	6:1	5:1	5:1	—
Class K1 Fuse 0–600 A			3:1	3:1	8:1	4:1	4:1	4:1
Class J Fuse 0–600 A			3:1	3:1	8:1	4:1	4:1	4:1
Class K5 Time-Delay Fuse 0–600 A			1.5:1	1.5:1	2:1	1.5:1	1.5:1	1.5:1
Class K5 Time-Delay Current-Limiting Fuse 0–600 A			1.5:1	1.5:1	4:1	2:1	2:1	2:1
Class J Time-Delay Fuse 15–600 A			1.5:1	1.5:1	4:1	2:1	2:1	2:1

*Check specific manufacturer for exact data.

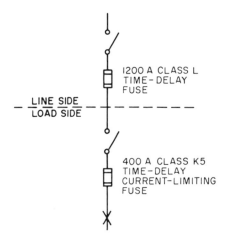

1200 A CLASS L
TIME-DELAY
FUSE

LINE SIDE
LOAD SIDE

400 A CLASS K5
TIME-DELAY
CURRENT-LIMITING
FUSE

Current Rating Ratio = 1200/400 = 3/1
Maximum Ratio for Coordination = 3/1
(Table 18)

Fig 55
Typical Application Example of Table 18. Selective Coordination Is Apparent
as Fuses Meet Coordination-Ratio Requirements

NOTE: Check specific manufacturer for exact data.

standard NEMA log—log graph paper. Since fuse ratios for high- or medium-voltage fuses to low-voltage fuses are not available, it is recommended that the fuse values in question be plotted on log—log paper.

Time—current characteristic fuse curves are available in the form of melting and total clearing time curves on transparent paper which is easily adapted to tracing. A typical example of coordinating high-voltage and low-voltage fuses using graphic analysis is shown in Fig 56. Note that the total clearing time curve of the 1200 A fuse is plotted against the minimum melting time curve of the 125 E rated 5 kV fuse. The curves are referred to 240 V for a study of secondary faults.

5.6 Current-Limiting Characteristics. Due to the speed of response to short-circuit currents, fuses have the ability to cut off the current before it reaches dangerous proportions. Fig 57 shows the current-limiting action of fuses. The available short-circuit current would flow if there were no fuse in the circuit or if a non-current-limiting protective device were in the circuit. A fuse will limit the available peak current to a value less than that available and will open in one half-cycle or less in its current-limiting range, therefore

letting through only a portion of the dangerous available short-circuit energy. The degree of current limitation is usually represented in the form of peak let-through-current charts.

5.6.1 Peak Let-Through-Current Charts. Peak let-through-current charts (sometimes referred to as current-limiting effect curves) are useful from the standpoint of determining the degree of short-circuit protection that a fuse provides to the equipment beyond it. These charts show fuse instantaneous peak let-through current as a function of available symmetrical rms current, as shown in Fig 58. The straight line running from the lower left to the upper right shows a 2.3 relationship for 15 percent power factor between the instantaneous peak current which would occur without a current-limiting device in the circuit and the available symmetrical rms current. The following data can be determined from the let-through-current charts and are useful in relating to equipment withstandability:

(1) Peak let-through current (magnetic effect)

(2) Apparent equivalent symmetrical rms let-through current (heating effect)

(3) Less than one half-cycle clearing time (fuse is operating in current-limiting range)

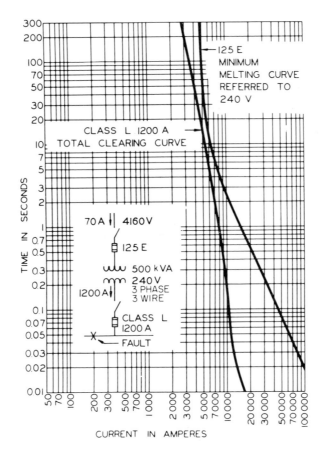

Fig 56
Typical Coordination Study of Primary and Secondary Fuses
Showing Selective System

NOTE: Check specific manufacturer for exact data.

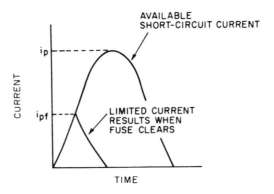

Fig 57
Current-Limiting Action of Fuses

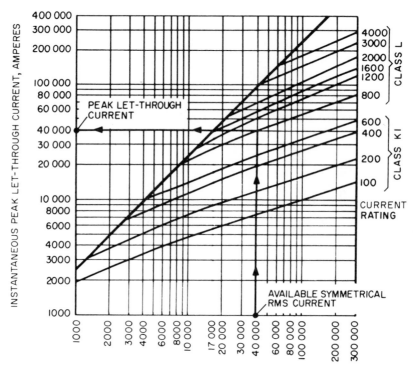

Fig 58
Typical 60 Hz Peak Let-Through Current as a Function of Available Symmetrical
RMS Current. Power Factor Below 20 Percent

NOTE: Check specific manufacturer for exact data.

NOTE: This procedure will yield a conservative value of symmetrical rms let-through current if the component has been given a withstand time rating of one half-cycle or longer under a test power factor of 15 percent or larger.

These data may be compared to short-circuit withstand data of those components which are static in nature such as wire, cable, bus, etc, so that engineering decisions can be made concerning short-circuit protection.

An example showing the application of the let-through-current charts is represented in Fig 59 where the component is protected by an 800 A current-limiting fuse. It is desired to determine the fuse let-through-current values with 40 000 A, symmetrical rms, available at the line side of the fuse. Using the let-through-current chart of Fig 58 we enter at an available current of 40 000

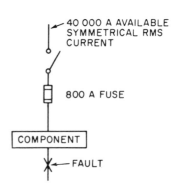

Fig 59
Example for Applying Fuse
Let-Through Charts

A, symmetrical rms, and find a fuse peak let-through current of 40 000 A. The clearing time will be less than one half-cycle. The downstream circuit components will not be subjected to an I^2t duty greater than the total clearing let-through current of the fuse (see Tables 16 and 17).

5.6.2 *Maximum Clearing Energy I^2t and Peak Let-Through Current I_p.* I^2t is a measure of the energy which a fuse lets through while clearing a fault. Every piece of electrical equipment is limited in its capability to withstand electrical destruction. When equipment has an I^2t withstand rating, it can be compared with maximum clearing I^2t values for fuses. These values are available from fuse manufacturers.

Magnetic forces can be substantial under short-circuit conditions and should also be examined. These forces vary with the square of the peak current $I_p{}^2$ and can be reduced considerably when current-limiting fuses are used, recognizing that the duration of the short-circuit current is an important factor. Some types of electrical equipment should be examined from the standpoint of peak current withstand as well as I^2t withstand.

5.7 Application of Fuses

5.7.1 *Bus Bracing Requirements.* Reduced bus-bracing requirements may be attained when current-limiting fuses are used. Fig 60 shows an 800 A motor-control center being protected by 800 A class L fuses. The maximum available fault current to the motor-control center (taking into consideration future growth) is 40 000 A, symmetrical rms. If a non-current-limiting device were used ahead of the motor-control center, the bracing requirement would be a minimum of 40 000 A, symmetrical rms. Since current-limiting fuses are used, however, a substantial reduction in bracing is possible.

5.7.2 *Circuit Breaker Protection.* Circuit breakers equipped with current-limiting fuses may be applied in circuits where the available short-circuit current exceeds the interrupting rating of the circuit breakers. The short-circuit withstand data for circuit breakers can be compared to fuse let-through current values to determine the degree of protection provided. Circuit breaker installations which are several years old may not meet present-day short-circuit current requirements because of increased power demand.

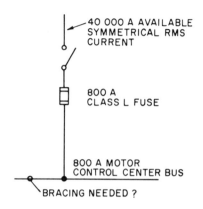

Fig 60
Example for Determining Bracing Requirements of 800 A Motor-Control Center

These types of installations may also be protected from excessive short-circuit currents by applying current-limiting fuses. The circuit breaker and fuse manufacturers should be contacted for proper applications.

An example of applying fuses to protect molded-case circuit breakers is shown in Fig 61 where a 225 A lighting panel has circuit breakers with an interrupting capacity of 10 000 A, symmetrical rms. The available fault current at the line side of the lighting panel is 40 000 A, symmetrical rms.

5.7.3 *Wire and Cable Protection.* Fuses must be sized for conductor protection according to the NEC. Where non-current-limiting devices are used, short-circuit protection for small conductors may not be available and reference should be made to the wire damage charts for short-circuit withstands of copper and aluminum cable in IPCEA P 32-382 (Rev Mar 1969), Short-Circuit Characteristics of Insulated Cable.

Due to the current-limiting ability of fuses, small conductors are protected from high-magnitude short-circuit currents even though the fuse may be 300–400 percent of the conductor rating such as allowed by the NEC for non-time-delay fuses for motor branch circuit protection.

5.7.4 *Motor Starter Short-Circuit Protection.* UL tests motor starters under short-circuit conditions. The short-circuit test performed may be used to establish a withstand rating for starters. UL tests starters of 50 hp and under with 5000 A of available short-circuit current and uses one-

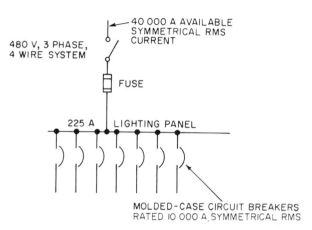

Fig 61
Application of Fuses to Protect Molded-Case Circuit Breakers

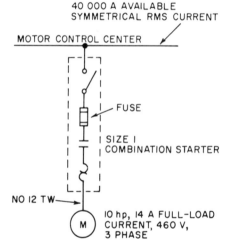

Fig 62
Application of Fuses to Provide
Short-Circuit Protection as well as
Backup Protection for Motor Starters

time fuses sized at 400 percent of the maximum continuous-current rating of the starter. Starters over 50 hp in size are tested in similar fashion, except that 10 000 A is the available fault current. (UL 508-1971, Electric Industrial Control Equipment, paragraphs 121, 131, and 144 A.)

When applying starters in systems with high available fault currents, current-limiting fuses must be used to reduce the let-through energy to a value less than that established by the UL test

procedures described. The starter manufacturer should be contacted for proper applications.

Fig 62 is a typical one-line diagram of a motor circuit where the available short-circuit current has been calculated to be 40 000 A, symmetrical rms, at the motor-control center and the fuse is to be selected such that short-circuit protection is provided.

The 1971 NEC originally recognized motor short-circuit protectors which are fuse-like devices having unique dimensions for use in combination starters. Proper selection is made by referring to motor starter manufacturers' application literature.

5.7.5 *Transformer Protection.* Low-voltage distribution-type transformers are usually equipped on the high-voltage side (above 600 V) with high-voltage fuses sized for short-circuit protection. Transformer overload protection should be provided by fusing the low-voltage secondary with current-limiting fuses sized at 100—125 percent of the transformer secondary full-load current. The transformer in Fig 63 shows proper size low-voltage fuses for a 1000 kVA transformer to provide overload protection. For complete protection of transformers, refer to Chapter 10.

Lighting transformers are quite frequently used in low-voltage electrical distribution systems to transform 480 V to 208Y/120 V. For these types of transformers, time-delay fuses are sized at 100—125 percent of the primary full-load current. Some considerations should be given to the magnetizing inrush current since for

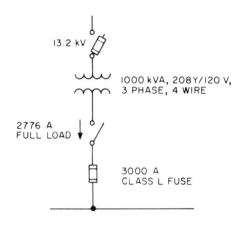

Fig 63
Sizing of Low-Voltage Fuses for Transformer
Secondary Protection

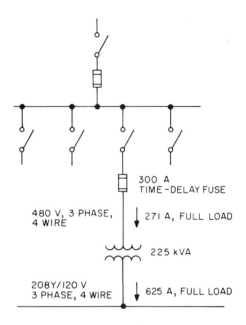

Fig 64
Application of Time-Delay Fuse for
Transformer Protection

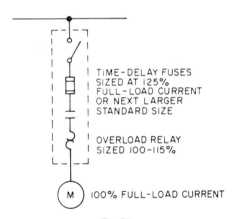

Fig 65
Application of Time-Delay Fuses for
Typical Motor Circuit

dry-type transformers this current may be as high as 20—25 times their rating. These inrush currents can easily be checked against the melting time curve at 0.1 s (usually taken as the maximum duration of inrush current) for needless opening, and a larger size time-delay fuse may need to be selected. Where dry-type and oil-filled transformers have inrush currents of about 12 times rating lasting for 0.1 s, time-delay fuses may be sized at 100—125 percent. Fig 64 shows a 225 kVA lighting transformer with properly applied time-delay fuses.

5.7.6 *Motor Overcurrent Protection.* Single- and three-phase motors can be protected by specifying time-delay fuses for motor running protection according to the NEC. These ratings will depend on service factor or temperature rise, or both. Where overload relays are used in motor starters, a larger size time-delay fuse may be used to coordinate with the overload relays.

Combination motor starters which employ overload relays sized for motor running protection (100—115 percent) can incorporate time-delay fuses sized at 125 percent or the next larger standard size to serve as backup protection. (Time-delay fuses sized up to 175 percent may be used for branch circuit protection only.) A combination motor starter with backup fuses will provide the best all around protection, motor control, and flexibilty. Fig 65 illustrates the use of fuses for protection of a motor circuit.

During the period that motors are operated near full load, single-phasing protection may be provided by time-delay fuses sized at approximately 125 percent of motor full-load current. Loss of one phase, either primary or secondary, will result in an increase in the line current to the motor. This will be sensed by the motor fuses as they are sized at 125 percent and the single-phasing current will open the fuses before damage to the windings results. If the motors are operated at less than full load, anti-single-phasing protection may be required.

B. High-Voltage Fuses

5.8 Types of High-Voltage Fuses. High-voltage fuses (which classification includes the medium-voltage range) suitable for the range of voltages considered (2.3 to 161 kV) fall into two general categories, distribution fuse cutouts and power fuses.

5.8.1 *Distribution Fuse Cutouts.* According to ANSI C37.100-1972, Definitions for Power Switchgear, the distribution cutout is identified by the following characteristics:

(1) Dielectric withstand (basic impulse insulation level) strengths at distribution levels

(2) Application primarily on distribution feeders and circuits

(3) Mechanical construction basically adapted to pole or crossarm mounting except for the distribution oil cutout

(4) Operating voltage limits corresponding to distribution system voltage

Characteristically, a distribution fuse cutout consists of a special insulating support and fuseholder. The fuseholder, normally a disconnecting type, engages contacts supported on the insulating support and is fitted with a simple inexpensive fuse link (Fig 66). The fuseholder is lined with an organic material, usually horn fiber. Interruption of an overcurrent takes place within the fuseholder by the action of deionizing gases liberated when the liner is exposed to the heat of the arc established when the fuselink melts in response to the overcurrent.

Distribution fuse cutouts were developed for use in overhead distribution circuits. They are commonly applied on such circuits, where their principal application is in connection with distribution transformers supplying a residential area or a small commercial or industrial plant. Cutouts provide protection to the distribution circuit by de-energizing and isolating a faulted transformer. Another application is the fault protection of small pole-mounted capacitor banks used for power-factor correction or voltage regulation.

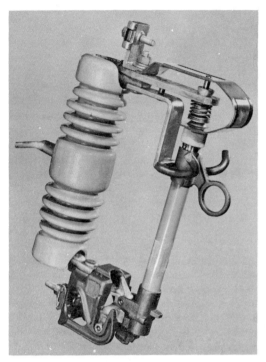

Fig 66
Open Distribution Fuse Cutout, Rated 100 A

Table 19
Maximum Short-Circuit-Current Interrupting
Ratings for Distribution Fuse Cutouts

Nominal Rating (kV)	Short-Circuit-Current Interrupting Rating (amperes)
4.8	12 500
7.2	8600
14.4	7100
25.0	4000

Distribution fuse cutouts are available for use on outdoor overhead distribution circuits with operating voltages up to 14.4 kV in maximum continuous-current ratings of 100 and 200 A, and up to 25 kV in a maximum rating of 100 A. The maximum interrupting ratings, expressed in symmetrical rms amperes, are given in Table 19.

5.8.2 *Power Fuses.* According to ANSI C37.100-1972 the power fuse is identified by the following characteristics:

(1)   Dielectric withstand (basic impulse insulation level) strengths at power levels

(2)   Application primarily in stations and substations

(3)   Mechanical construction basically adapted to station and substation mountings

Power fuses have other characteristics that differentiate them from distribution fuse cutouts in that they are available in higher voltage ratings, higher load-current ratings, higher interrupting-current ratings, and in forms suitable for indoor and enclosure application as well as all types of outdoor applications.

A power fuse consists of a fuse support plus a fuse unit, or alternately a fuseholder which accepts a refill unit or fuse link.

The two basic types of power fuses, expulsion type and current-limiting type, effect interruption of overcurrents in a radically different manner. The expulsion type, like the distribution cutout, interrupts overcurrents through the deionizing action of the gases liberated from the lining of the interrupting chamber of the fuse by the heat of the arc established when the fusible element melts. The current-limiting type interrupts overcurrents when the arc established by the melting of the fusible element is subjected to the mechanical restriction and cooling action of a powder or sand filler surrounding the fusible element.

The earliest form of high-voltage power fuse was the expulsion type, an outgrowth of the distribution fuse cutout, employing longer and heavier fuseholder tubes to cope with higher circuit voltages and short-circuit interrupting requirements. The expulsion-type power fuse possesses operating characteristics similar to those of a distribution cutout, except that the noise and emission of gases and flame are greatly magnified as the holder tubes are lengthened and strengthened to handle higher voltages and fault currents. Therefore, this type of fuse has been restricted to outdoor usage only and generally in substations that are remotely located from human habitation. Expulsion-type power fuses are used for the fault protection of small- and medium-size power transformers or substation capacitor banks.

The limited interrupting capacity of these early expulsion-type power fuses, coupled with their inability to be used within buildings or enclosures, led to the development in the United States during the 1930s of a new form of expulsion fuse known as the boric-acid or solid-material fuse. In this fuse the material used to obtain the deionizing action necessary to interrupt overcurrents was not organic, but solid boric-acid powder molded into a dense lining for the interrupting chamber of the fuse. The benefits gained by the use of boric acid are listed below.

(1)   For identical physical dimensions of interrupting chambers a fuse lined with boric acid can interrupt (a) a higher voltage circuit, (b) a higher value of overcurrent in a structure of the same strength, (c) the full range of currents from minimum melting to the interrupting rating, with (d) lower arc energy and reduced emission of gases and flame than is possible in a fuse where organic fiber is used as the liner.

(2)   The gas liberated from the boric acid is noncombustible and highly deionized, and thus reduces the amount of flame discharged. As a result, the clearances in the path of the exhaust gases necessary to prevent an electrical flashover of circuits or devices are often reduced to values approaching the clearance distances required in air.

(3)   The most significant feature obtained by the use of boric acid is the ability to control the gas discharge from the fuse on operation. When exposed to the heat of the arc, boric acid liberates steam, which can be condensed to the liquid stage by venting the gas into a suitably designed cooling device.

The solid-material-type power fuse permitted wider use of power fuses. It can operate with negligible or harmless noise levels and emission of flame and gases in a greatly expanded range of current and interrupting ratings.

Virtually simultaneously with the development of the solid-material boric-acid power fuse, an American version of the European current-limiting or "silver-sand" fuse was introduced. Two forms evolved. One was a line of current-limiting fuses to be used with, and coordinated with, high-voltage motor starters for high-capacity 2400 and 4160 V distribution circuits. The second was a line of current-limiting fuses suitable for use with 2400 to 34 500 V potential, distribution, and small power transformers.

The high-voltage current-limiting power fuse has three features that have led to its extensive usage on high-capacity medium- and high-voltage power-distribution circuits.

(1) Interruption of overcurrents is accomplished quickly without the expulsion of arc products or gases as all the arc energy of operation is absorbed by the sand filler of the fuse and subsequently released as heat at relatively low temperatures. This enables the current-limiting fuse to be used indoors or in enclosures of small size as there is no noise accompanying operation. Furthermore, since there is no discharge of hot gases or flame, only normal electrical clearances need be provided.

(2) Current-limiting action or reduction of current through the fuse to a value less than that available from the power-distribution system at the location of the fuse occurs if the value of overcurrent greatly exceeds the continuous-current rating of the fuse. Such a current reduction reduces the stresses and possible damage to the circuit up to the fault or to the faulted equipment itself, and also reduces the shock to the power system. In the case of current-limiting fuses used with a high-voltage motor starter the contactor is only required to have momentary-current and making-current capabilities equal to the maximum let-through current of the largest current rating of the fuse that will be used in the starter.

(3) Very high interrupting ratings are achieved by virtue of its current-limiting action so that current-limiting power fuses can be applied on medium- or high-voltage distribution circuits of very high short-circuit capacity.

5.9 Fuse Ratings

5.9.1 *High-Voltage Expulsion-Type Power Fuses.* This category has its principal usage in outdoor applications at the subtransmission voltage level. The available ratings of this fuse are given in Table 20.

5.9.2 *High-Voltage Solid-Material Boric-Acid Fuses.* High-voltage solid-material boric-acid fuses are available in two styles, (1) the fuse-unit style in which the fusible element, interrupting element, and operating element are all combined in an insulating tube structure, with the entire unit being replaceable, and (2) the fuseholder and refill-unit style of which only the refill unit is replaced after operation.

Table 20
Ratings of Expulsion-Type Power Fuses

Nominal Rating (kV)	Maximum Continuous Current (amperes)	Maximum Three-Phase Symmetrical Interrupting Rating (MVA)
7.2	100, 200, 300, 400	162
14.4	100, 200, 300, 400	406
23	100, 200, 300, 400	785
34.5	100, 200, 300, 400	1174
46	100, 200, 300, 400	1988
69	100, 200, 300, 400	2350
115	100, 200,	3110
138	100, 200,	2980
161	100, 200,	3480

Table 21
Ratings of Solid-Material Boric-Acid Power Fuses (1)

Nominal Rating (kV)	Maximum Continuous Current (amperes)	Maximum Three-Phase Symmetrical Interrupting Rating (MVA)
34.5	100, 200, 300	2000
46	100, 200, 300	2500
69	100, 200, 300	2000
115	100, 250	2000
138	100, 250	2000

Table 22
Ratings of Solid-Material Boric-Acid Power Fuses (2)

Nominal Rating (kV)	Maximum Continuous Current (amperes)	Maximum Three-Phase Symmetrical Interrupting Rating (MVA)
2.4	200, 400, 720	155
4.16	200, 400, 720	270
7.2	200, 400, 720	325
14.4	200, 400, 720	620
23	200, 300	750
34.5	200, 300	1000

Fig 67
Substation Serving Large Industrial Plant at
69 kV. Combination of Interrupter Switch
with Shunt Device Trip and Solid-Material
Boric-Acid Power Fuses Provides Protection
for Power Transformer

Table 23
Ratings of Current-Limiting Power Fuses

Nominal Rating (kV)	Maximum Continuous Current (amperes)	Maximum Three-Phase Symmetrical Interrupting Rating (MVA)
2.4	100, 200, 450	155—210
2.4/4.16Y	450	360
4.8	100, 200, 300, 400	310
7.2	100, 200	620
14.4	50, 100, 175, 200	780—2950
23	50, 100	750—1740
34.5	40, 80	750—2600

The ratings of fuses of style (1), which are
principally used outdoors at subtransmission
voltages (Fig 67), are given in Table 21.

The fuses of style (2) are used either indoors or
outdoors at medium and high distribution voltages (Fig 68), and their ratings are given in Table
22.

5.9.3 *Current-Limiting Fuses.* Current-limiting
power fuses suitable for the protection of potential transformers, auxiliary power transformers,
power transformers, and capacitor banks are
available in the ratings given in Table 23.

The interrupting rating of fuses for potential
transformers is about 2000 MVA, three-phase
symmetrical, for voltages of 14.3 through 34.5
kV.

Some current-limiting power fuses are E rated
instead of current rated. Fuses rated 100 E and
less open in 300 s at currents between 200 and
240 percent of their E ratings. Fuses rated above
100 E open in 600 s at currents between 220 and
264 percent of their E ratings. E-rated fuses are
classified as general-purpose or backup fuses.

Some current-limiting distribution fuses are C
rated. These fuses open in 1000 s at currents between 170 and 240 percent of their C ratings.

Current-limiting fuses suitable for use with
high-voltage motor starters are designated by an
R rating. These R-rated fuses are available in
ratings from 2R to 36R up to 5500 V, and have
an asymmetrical interrupting rating up to 80 000
A rms at 5000 V. Some manufacturers of these
fuses have assigned maximum continuous-

Fig 68
Solid-Material Boric-Acid Power Fuse
Rated 14.4 kV with Controlled Venting
Device; 200 A Indoor Disconnecting Type

Fig 69
High-Voltage R-Rated Fuses in
Drawout-Type Motor Starter,
Rated 180 A

Fig 70
Metal-Enclosed Switchgear
with Fused Interrupters; 200 A
125 MVA 4.16 kV Outdoor Manual Type

current ratings of 70 A for the 2R fuse and up to 650 A for the 36R fuse. These ratings are at an ambient temperature of 55°C. Other manufacturers have used the R rating to indicate a 20 s blow point at 100 times the R rating of the fuse. Thus a 12 R fuse opens in 20 s at 1200 A. These fuses are selected by their characteristic to coordinate with the motor and motor controller to provide short-circuit protection. Fig 69 shows how R-rated fuses are mounted in a typical fused high-voltage drawout-type motor starter.

5.10 Areas of Application of Fuses. Power fuses can be utilized on systems rated up to 161 kV to protect the transmission circuits, power and potential transformers, and capacitor banks. Power fuses can also provide a degree of backup protection for transformer secondary faults.

5.10.1 *Power Distribution.* From the basic concepts of overcurrent protection applicable to the medium- and high-voltage distribution systems

used in industrial plants and commercial buildings it becomes apparent that the principal functions of overcurrent protective devices at these voltages are to detect fault conditions in elements of the high-voltage circuits and to interrupt these high values of overcurrent. Fuses provide fast clearing of high-magnitude fault currents. Their speed of operation for low and medium magnitudes of fault currents should be checked to determine if supplementary protection (such as ground-fault protection) is needed to clear these arcing-type faults. Their secondary function is to act as backup overcurrent protection in the event of a malfunction of the next overcurrent protective device closer to the fault location on the system.

Modern high-voltage power fuses have the capabilities and characteristics to provide this vital overcurrent protection for virtually all types and sizes of distribution systems ranging from a simple radial circuit where power is purchased at a medium or high voltage with a transformer to provide utilization voltage, up to extensive primary-selective circuits supplying several transformer substations. Such fuses used with load-interrupter switches may be applied outdoors, in vaults, or in metal-enclosed interrupter switchgear (Fig 68).

Metal-enclosed interrupter switchgear, in addition to power fuses and interrupter switches, may include bus and connections, instrument transformers, control wiring, and accessories, all completely enclosed with sheet metal (except for ventilating openings and inspection windows). This class of switchgear is available in either indoor or outdoor forms in voltage classifications from 4.16 to 34.5 kV and with buses rated 600, 1200, or 2000 A (Fig 70).

Within the medium- or high-voltage distribution system of "industrial" power users, which includes shopping centers, schools, hospitals, large commercial office and apartment buildings, and transportation terminals, as well as all sizes of strictly industrial plants, there are four basic applications for fused interrupter switchgear.

(1) *Switching Centers.* These are metal-enclosed switchgear assemblies applied as the originating points for a system of feeders.

(2) *Substation Primaries.* These are metal-enclosed switchgear assemblies applied to the primary side of a substation with voltage transformation.

(3) *Substation Secondaries.* These are metal-enclosed switchgear assemblies applied to the secondary side of a substation with voltage transformation.

(4) *Service Entrances.* These are metal-enclosed switchgear assemblies or units applied to a single or dual incoming service and only one outgoing feeder. Usually there is a bay for utility metering equipment. There is no transformation. The essential difference between service entrances and switching centers is that service entrances have only one outgoing feeder.

These defined points of application have nothing to do with whether the gear is manual or automatic nor with the circuit arrangement that has been selected for the particular system. Over the years three basic distribution systems that utilize fused interrupter switches have become recognized as typical or standard. They are de-

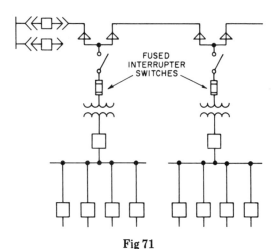

Fig 71
One-Line Diagram of Simple Radial System

scribed below, in the order of ability to provide maximum continuity of service.

When fused interrupter switches are used, care should be taken to open all phase conductors to obtain complete isolation of a fault. If only one or two phases are opened in a three-phase circuit, the fault remains connected to the system through the multiphase load circuits. Under this condition, the fault current is often reduced but not cleared. Anti-single-phasing devices are recommended.

(1) *Radial System.* The simplest radial system might consist of a single incoming fused interrupter switch, a single feeder, and one transformer bank stepping down from the utility's distribution voltage to the utilization voltage (Fig 71). Fuses or circuit breakers would protect the several circuits on the secondary side of the transformer. With this simple system both primary and backup protection is provided. For instance, a fault on a secondary feeder should be cleared by the secondary protective device. However, if this device should fail, the primary fuse will serve as partial backup protection.

(2) *Secondary-Selective System.* The secondary-selective system has been defined as two radial systems tied together on the secondary or low-voltage side of the transformers. The application of power fuses in this type of system is shown in Fig 72.

(3) *Primary-Selective System.* The basic advantage of this system is that it provides continu-

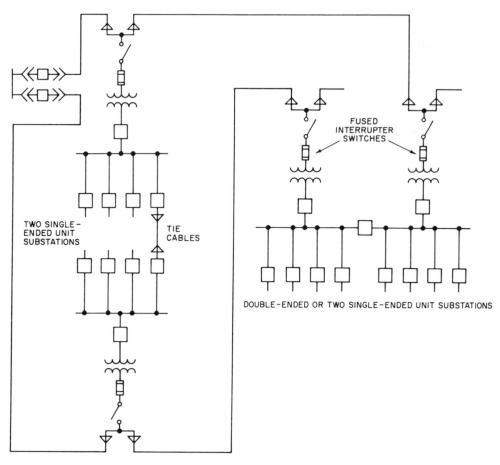

Fig 72
One-Line Diagram of Typical Secondary-Selective Arrangements
of Load-Center Distribution System

ity in the primary source feeding the load. The requirement imposed on the primary switchgear at each load tap is that it must be capable of alternately switching from one source to the other as well as providing short-circuit protection.

This circuit arrangement provides means of reducing both the extent and duration of an outage caused by a primary-feeder fault as compared to the radial circuit arrangement. This operating feature is provided through the use of duplicate primary-feeder circuits and fused interrupter switches that permit connection of each secondary substation transformer to either of the two primary-feeder circuits (Fig 73). Each prima-

ry-feeder circuit must have sufficient capacity to carry the total load.

5.10.2 *Fuses for Medium Voltage Motor Circuits.* The power requirements for large motors driving compressors for air-conditioning systems, pumps, and numerous other loads in industrial plants are frequently too great for low-voltage distribution circuits; so motors operating at 2400 and 4160 V and higher are now commonly used. Since the distribution circuits supplying such a motor usually have many parallel loads, their load-current and short-circuit-current capacity is high. The latter condition imposes a severe duty on the motor starter. A coordinated current-limiting fuse/contactor combination is

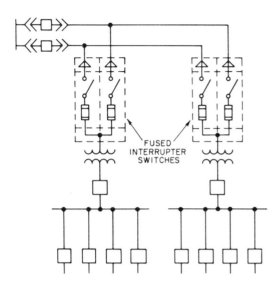

Fig 73
One-Line Diagram of Primary-Selective Load-
Center System with Two Single-Throw Inter-
locked Primary Fused-Interrupter Switches

available for the motor-starting and control devices for motors up to about 2500 hp on systems
rated up to 13 800 V.

The current-limiting fuses for this service fall
into a special performance category and are designated by an R rating. In addition to carrying
normal motor load current plus harmless transient overloads, they must repeatedly withstand
starting inrush currents without deteriorating or
operating. Such inrush currents occasionally exceed six times the motor full-load current and
last for several seconds until the motor and load
are brought up to speed.

When an insulation failure occurs in the motor
(or in the circuit from the motor starter to the
motor) and the current could reach a very high
value, the fuse must detect such a fault in a few
milliseconds. The fast melting of the fusible element and its subsequent current-limiting action
prevent the current from rising to the maximum
available magnitude, and then in a few more
milliseconds this restricted value of overcurrent
is interrupted.

In a properly designed and coordinated current-limiting fuse/contactor type of motor starter, the overload relays and the contactor should

relieve the fuses of all overload and locked-rotor
overcurrent interrupting duty and provide protection from undervoltage or single phasing of
the supply. Combinations of this type are generally restricted to motors having less than a 600
A load current as fuses of any higher current rating do not have appreciable current-limiting action at the fault current levels available on these
medium-voltage circuits.

5.11 Selection of Fuse Type and Rating. The
selection of the most suitable type and rating of
high-voltage power fuses for the various applications that have been enumerated for medium- or
high-voltage industrial distribution circuits de-
serves very careful consideration and study.

The basic rules for the application of a power
fuse are that it must be selected for voltage
rating, current-carrying capacity, and interrupting capacity. More important, however, is the
selection of the type of fuse that will meet the
requirements of the distribution system and its
components in order to provide the proper balance of primary overcurrent protection for the
system with adequate backup of the next load-
side overcurrent protective device, to optimize
system protection and continuity of service. No
less important is the selection of the fuse type
that will also result in a minimum investment of
capital, a minimum space requirement, and a
minimum of maintenance. Security of personnel
is paramount whether it is associated with fuse
operation or replacement of fuses.

5.11.1 *Voltage Rating.* The voltage rating
should be selected as the next higher standard
voltage rating above the maximum operating voltage level of the system.

Solid-material boric-acid power fuses are not
voltage critical in that they can safely be applied
at voltages less than their rated voltage with no
detrimental effects. Fuses have a constant-cur-
rent or at best a slightly increased current interrupting ability when applied to systems operating one or more voltage levels below the fuse
rating.

The current-limiting power fuse is voltage critical. This fuse functions by developing a back
electromotive force, and on older fuse designs
which do not use ribbon-type elements, this is a
function of its maximum rated voltage rather
than system voltage. This means that care must

be used in selecting the fuse voltage rating to match the system voltage to avoid subjecting system components to excessive voltages. If necessary, the fuse manufacturer, should be consulted on the application of specific fuse types.

5.11.2 *Current Rating.* The current rating of a fuse should be selected considering many factors.

(1) Normal continuous load. All modern power fuses are so rated that they will carry their rated current continuously if they are applied in locations where the ambient temperature conditions do not exceed the ANSI standard value of 40°C. This should be checked where enclosures are required as temperatures within an enclosure may be 15°C or more above outside ambient.

(2) Transient inrush currents of transformers. Energizing a transformer from the distribution circuit or subtransmission circuit is accompanied by a short-duration inrush of magnetizing current. The integrated heating effect as seen by a fuse in such a circuit is customarily represented as that of a current having a magnitude of either 8 or 12 times (depending on the size of the transformer) the full-load current of the transformer and a duration of 0.1 s. Expulsion power fuses or solid-material boric-acid power fuses are available with a selection of time—current characteristics, all of the inverse-time form, and having sufficient time delay such that a fuse having a current rating equal to the transformer rating will withstand such transient inrush currents without operating and without damage to the element. Current-limiting power fuses have an inherently steeper time—current characteristic producing a "faster" fuse. This frequently requires a fuse current rating that is substantially greater than the transformer full-load current rating.

(3) Motor starting. The inrush current and its duration associated with the starting of a large motor is a function of the motor characteristics and the load it drives. The current-limiting fuses used with motor-starting contactors should be of the type specifically designed for the service. The ratings assigned to such R-rated fuses indicate the maximum starting inrush current and duration of inrush that these fuses can withstand.

(4) Normal repetitive overloads. Transformers and distribution-circuit components inherently have some short-time overload capability that may be used to advantage in planning maximum economy into an industrial distribution system. These overloads may simply be due to the starting inrush of a large motor on a general-purpose circuit, to a shift startup in an industrial plant, or to peaking air-conditioning or heating loads in abnormal weather. Expulsion fuses and solid-material boric-acid fuses have inherent continuous overload capability in all ratings. It is a maximum in the smaller current ratings that are available in a particular physical size of fuse refill or fuse unit and decreases in the larger current ratings in the same fuse refill or fuse unit size. These capabilities are generally published in the form of a time—current curve in the 1/2 to 8 h range.

(5) Emergency overloads. With the use of the more sophisticated distribution systems, continuity of service is achieved by switching the loads from a faulty circuit or transformer to another circuit or transformer until repairs or replacement can be made. Emergency-overload characteristic curves similar to the continuous overload characteristic curves are available for the solid-material boric-acid power fuses. It is imperative to consider the effect of such overloads on the melting-time—current characteristics of the fuse as it affects selectivity with other overcurrent protective devices.

5.11.3 *Interrupting Rating.* The interrupting rating of a fuse relates to the value of the maximum symmetrical rms current available in the first half-cycle after the occurrence of a fault. The rating may be expressed in any of three ways:

(1) Maximum symmetrical rms current
(2) Maximum asymmetrical rms current
(3) Equivalent three-phase symmetrical interrupting power in kilovolt-amperes

Symmetrical rms current ratings are particularly useful if a careful system short-circuit study has been made, since the results obtained are expressed in these terms.

Asymmetrical rms current ratings merely represent the maximum current that the fuse may have to interrupt because of its fast-acting characteristic. For power fuses it is 1.6 times the symmetrical rms current-interrupting rating.

Equivalent three-phase symmetrical interrupting ratings are given as a reference for comparison with circuit-breaker capabilities. Also, simplified short-circuit duty requirements can be calculated in these terms.

In selecting a fuse for proper interrupting rating it is, of course, important to make sure that it is adequate for the short-circuit duty required. Generally, however, other requirements automatically ensure that the fuses used have inherently more interrupting capability than the system requires. For example, fuses of the higher continuous-current ratings, such as 300 or 400 A, normally have a higher interrupting rating than fuses with maximum ratings of 100 or 200 A. Current-limiting fuses used for the protection of potential, distribution, or small power transformers have interrupting ratings that exceed the requirements of the medium- and high-voltage distribution circuits used in industrial systems.

Distribution circuits supplying a number of large high-voltage motors must be designed so that the short-circuit interrupting duty does not exceed the interrupting rating of the fuse—contactor combinations.

5.11.4 *Selectivity, General Considerations.* A minimum melting time—current curve indicates the time that a fuse will carry a designated current without blowing (assuming no initial load). Fig 74 illustrates a typical family of minimum melting time—current characteristics for current-limiting power fuses. Fig 75 shows the minimum melting time—current characteristics for a representative line of solid-material boric-acid fuses. The curve should indicate the tolerance applicable to these times. This curve should further indicate whether the fuse is nondamageable, that is, whether it can carry without damage the designated current for a time that immediately approaches the time indicated by the curve. Since this minimum melting time—current curve is based on no initial load current through the fuse, it is necessary to modify this curve to recognize the temporary change in melting time due to load current. This modification will permit precise coordination with other overcurrent protective devices nearer the load. Further aids are available in the form of curves that evaluate the temporary change in melting characteristics in the event that the fuse has been carrying a heavy emergency overload or in the case of fuses being applied in a location having an exceptionally high ambient temperature.

If the fuse is susceptible to a change in melting time—current characteristics when exposed to currents for times less than the minimum melting time, the manufacturer's published "safety-zone" allowance or set-back curve should be used in any coordination or scheme of overcurrent protective devices.

Maximum system protection from primary faults that should cause the fuse to operate, and maximum backup protection to the system and equipment on the load side of the fuse where overcurrent protection malfunctions, require the smallest current rating of fuse that meets the requirements stated earlier for the selection of current rating. Fuses having a small tolerance or narrow operating band and utilizing nondamageable elements are best suited to accurate selective application. To assist in this evaluation, total clearing time—current characteristic curves are used. They represent the maximum total operating time of the fuses taking into account maximum manufacturing tolerances of the fusible element, the operating voltage, and recognizing that the fuse may not be carrying any initial load to reduce the melting time of the fuse element. Fig 76 indicates the corresponding total clearing time characteristic for the current-limiting fuses of Fig 74. Fig 77 shows the total clearing time characteristics for the solid-material boric-acid fuse of Fig 75.

For the very simple radial system with a single transformer, manufacturers frequently supply a table of recommended fuse current ratings for transformers of various voltage and power classes. The bases for such tables should be carefully scrutinized to see that they are adequate for the necessary requirements previously outlined, but more particularly, that the recommended rating is not too large, as the system protection may be jeopardized and the full benefits of backup overcurrent protection possible with fuses will not be achieved.

A good axiom in selecting a fuse to fit into an overcurrent protective scheme is to use the very smallest current rating that will carry, and not be damaged by, any load that should be maintained without interruption. The effort required to make a precise selection of fuse current rating and speed characteristics is small compared to the benefits obtained in overall system overcurrent protection.

5.11.5 *Selectivity for Motor Controllers.* The application of R-rated fuses in a high-voltage motor controller is basically one of comparing

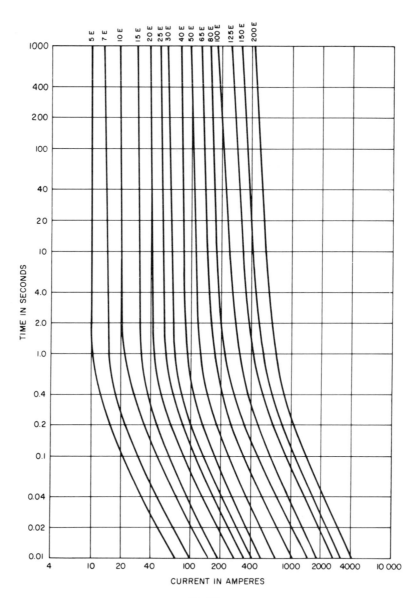

Fig 74
Typical Minimum Melting Time—Current Characteristics for High-Voltage
Current-Limiting Power Fuses

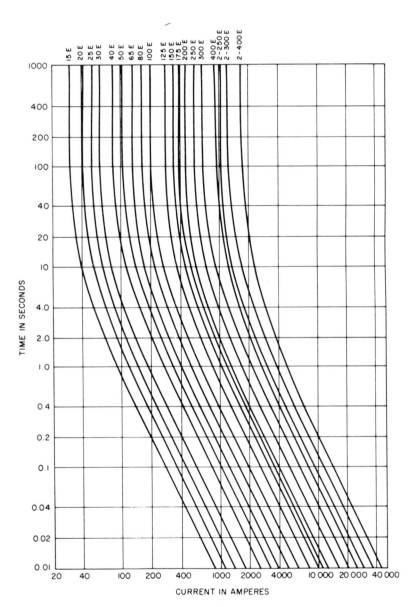

CURRENT IN AMPERES

Fig 75
Typical Minimum Melting Time—Current Characteristics for High-Voltage
Solid-Material Boric-Acid Power Fuses

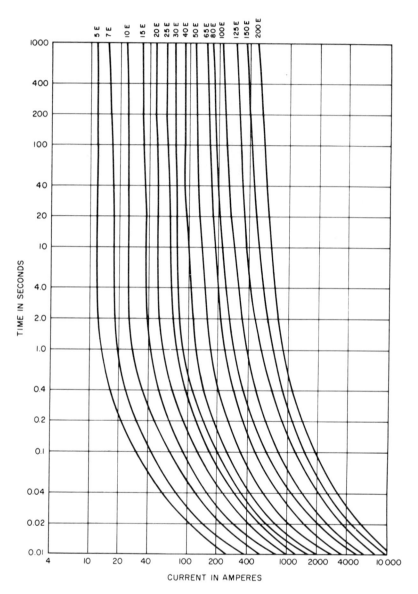

Fig 76
Typical Total Clearing Time—Current Characteristics for High-Voltage
Current-Limiting Power Fuses (See Fig 74)

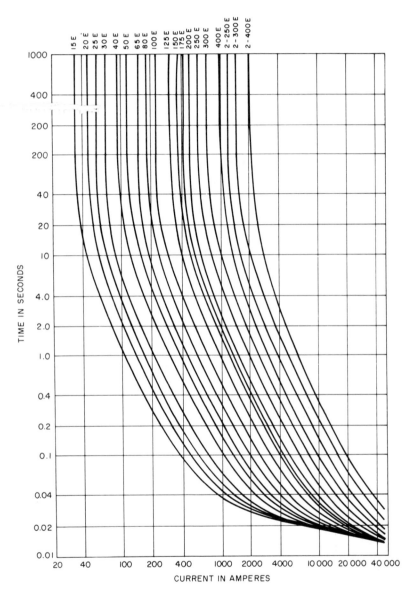

Fig 77
Typical Total Clearing Time—Current Characteristics for High-Voltage
Solid-Material Boric-Acid Power Fuses (See Fig 75)

the minimum melting time—current characteristics of the R-rated fuses with the time—current characteristics of the overload relay curve. The size fuse which is selected should be such that short-circuit protection is provided by the fuse and overload protection is provided by the controller overload relays. The following data are required for proper application:

(1) Motor full-load current rating

(2) Motor locked-rotor current

(3) R-rated fuse minimum melting time—current curves

(4) Overload relay time—current characteristic

For example, a 2300 V motor has a 100 A full-load current rating and locked-rotor current of 600 A. The overload relay is to be sized at 125 percent of motor full-load current and system voltages could vary enough to cause a 10 percent increase in locked-rotor current. Taking these percentages into consideration, the overload relay is sized at 125 A and the locked-rotor current would be taken as 660 A. Fig 78 shows these adjusted values (dashed lines). Also shown is the thermal overload relay curve with the multiple 1 of the relay current setting lining up with the 125 A adjusted full-load current.

The fuse selected for the application is the smallest fuse whose minimum melting time characteristic does not cross the relay curve for currents less than the adjusted locked-rotor value. In this example, the 9R would be the proper choice. Since the application is at 2300 V, 2300 V 9R fuses would be recommended.

The largest size fuse that could be applied is that rating which still protects the controller when the fuse clears the maximum available fault current. Maximum size fuses can best be chosen by the motor controller manufacturer.

5.11.6 *Selection of Fuse Type.* Since there is absolutely no noise, flame, or any expulsion of gases, the current-limiting fuse is especially suited to indoor industrial applications in close-fitting enclosures. The only electrical clearances required are those determined by the insulation class corresponding to the voltage rating of the fuse. These fuses are often used for the protection of potential transformers, capacitors, distribution-class transformers, and small to medium power transformers. The fault-current limiting action inherent in the operation of these fuses

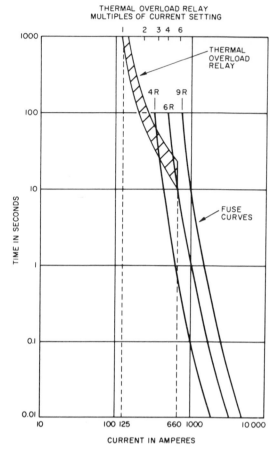

Fig 78
Time—Current Coordination of R-Rated Fuses and Thermal Overload Relay in High-Voltage Fused Motor Starter

upon the occurrence of high fault currents reduces the shock to the distribution circuit and simplifies selectivity with overcurrent protective devices that are closer to the source of power. However, their fast fusing action may require selecting a current rating which is several times the transformer full-load current rating. This is necessary in order to accommodate transient inrush current conditions and for coordination with secondary protective devices. This fuse application is in line with the most modern fusing practice which is based on protecting the continuity of the electrical service with the possible loss of some transformer service life. Current-limiting fuses are available in a range of voltage, current, and interrupting ratings (as shown in Table 23) to cover most of the required applications. High-voltage current-limiting fuseholders are not of the rejection type and precautions should be taken to assure that replacement fuses are of the correct size and type.

The solid-material boric-acid fuse was developed initially for high load and interrupting capacity in indoor or confined locations. Thus it is suitable for use in medium- and high-voltage industrial distribution systems. It is available in a wide range of voltage and current ratings with corresponding high short-circuit interrupting ratings (Table 21 and 22) in order to cover the required applications. An important factor to be considered in making the selection of fuse type is that the solid-material boric-acid fuse makes use of fuse refills which include the complete fuse element and interrupting element necessary to restore the fuse to original operation condition after it has been called on to function.

Working tools for each of the fuse types have been provided for the system engineer who is developing these distribution systems. Time—

current characteristic curves, preloading adjustment factors, and ambient temperature adjustment factors are provided for each fuse type. In addition, overload capability curves are provided for solid-material boric-acid fuses, and peak let-through current curves, minimum melting and maximum total I^2t charts and peak arc voltage data are provided for current-limiting fuses.

5.12 Standards References. The following standards publications were used as references in preparing this document.

ANSI C37.100-1972, Definitions for Power Switchgear

IEC No 291 (1969) Fuse Definitions

IPCEA P 32-382 (Rev Mar 1969), Short-Circuit Characteristics of Insulated Cable

NEMA BU 1-1969, Busways

NFPA No 70, National Electrical Code (1975) (ANSI C1-1975)

UL 198-1970—1974, Fuses

UL 508-1971, Electric Industrial Control Equipment

6. Low-Voltage Circuit Breakers

6.1 General Discussion. Low-voltage circuit breakers fall into two basic classifications, (1) molded-case circuit breakers and (2) low-voltage power circuit breakers. ANSI C37.100-1972, Definitions for Power Switchgear, defines them in these ways.

(1) A molded-case circuit breaker is one which is assembled as an integral unit in a supporting and enclosing housing of insulating material.

(2) A low-voltage power circuit breaker is one for use on circuits rated 1000 V alternating current and below, or 3000 V direct current and below, but not including molded-case circuit breakers.

The term "air circuit breaker" is often used in speaking of low-voltage power circuit breakers. Since the arc interruption takes place in air in both molded-case circuit breakers and low-voltage power circuit breakers, this term really applies to both types.

The ratings which apply to circuit breakers and the actual assigned numerical values reflect the mechanical, electrical, and thermal capabilities of their major components. Basic ratings are

(1) Rated voltage
(2) Rated frequency
(3) Rated continuous current
(4) Rated interrupting current
(5) Rated short-time current (low-voltage power circuit breakers only)

Table 24 lists typical interrupting-current ratings for molded-case circuit breakers. Tables 25 and 26 (from ANSI C37.16-1973, Preferred Rat-ings, Related Requirements, and Application Recommendations for Low-Voltage Power Circuit Breakers and AC Power Circuit Protectors) show standard ratings for low-voltage power circuit breakers with and without instantaneous overcurrent trip devices, respectively. An examination of Tables 24—26 shows the differences between low-voltage circuit breakers. Interrupting ratings or short-circuit-current ratings, as they are referred to for low-voltage power circuit breakers, are a function of the voltage ratings, which range from 120 through 635 V alternating current. Short-circuit-current ratings are also a function of the trip device characteristics. Note that the short-circuit-current ratings for circuit breakers with instantaneous trip devices are in most cases higher than for those without instantaneous trip, but which are equipped with short-time tripping devices. Standards for molded-case circuit breakers do not list ratings for units without instantaneous trip elements. Low-voltage power circuit breakers are rated to carry their continuous current within their enclosures.

Individual manufacturer's literature will indicate forms of molded-case circuit breakers which go beyond the standard ratings shown in these tables, in terms of both continuous- and short-circuit-current ratings. Molded-case circuit breakers and low-voltage power circuit breakers are both available with integrally mounted fuses for applications on systems with high available short-circuit currents. Ref [1] provides detailed information on physical and electrical characteristics of low-voltage molded-case and power circuit breakers.

Table 24
Typical Interrupting Current Ratings for Molded-Case Circuit Breakers, in Amperes

Frame Sizes	Rated Continuous Current	Single-Pole Alternating Current				Two- and Three-Pole Alternating Current 120/240 V		Two- and Three-Pole Alternating Current — 600 V AC Rated Circuit Breakers					
(amperes)	(amperes)	120 V (120/240 V)		277 V		240 V		240 V		480 V		600 V	
		Sym	Asym	Sym	Asym	Sym	Asym	Sym	Asym	Sym	Asym	Sym	Asym
100	0–100	5000	5000	—	—	5000	5000	—	—	—	—	—	—
100	0–100	7500	7500	10 000	10 000	7500	7500	—	—	—	—	—	—
100	0–100	—	—	—	—	—	—	18 000	20 000	14 000	15 000	14 000	15 000
100	0–100	—	—	—	—	—	—	65 000	75 000	25 000	30 000	18 000	20 000
100	0–100	—	—	—	—	—	—	100 000	—	100 000	—	100 000	—
200	125–200	—	—	—	—	10 000	10 000	—	—	—	—	—	—
225	125–225	—	—	—	—	10 000	10 000	—	—	—	—	—	—
225	70–225	—	—	—	—	—	—	22 000	25 000	18 000	20 000	14 000	15 000
225	70–225	—	—	—	—	—	—	25 000	30 000	22 000	25 000	22 000	25 000
225	70–225	—	—	—	—	—	—	65 000	75 000	35 000	40 000	25 000	30 000
225	70–225	—	—	—	—	—	—	100 000	—	100 000	—	100 000	—
400	200–400	—	—	—	—	—	—	35 000	40 000	25 000	30 000	22 000	25 000
400	200–400	—	—	—	—	—	—	65 000	75 000	35 000	40 000	25 000	30 000
600	300–600	—	—	—	—	—	—	42 000	50 000	30 000	35 000	22 000	25 000
600	300–600	—	—	—	—	—	—	100 000	—	100 000	—	100 000	—
800	300–800	—	—	—	—	—	—	42 000	50 000	30 000	35 000	22 000	25 000
800	600–800	—	—	—	—	—	—	65 000	75 000	35 000	40 000	25 000	30 000
800	600–800	—	—	—	—	—	—	100 000	—	100 000	—	100 000	—
1000	600–1000	—	—	—	—	—	—	42 000	50 000	30 000	35 000	22 000	25 000
1200	700–1200	—	—	—	—	—	—	42 000	50 000	30 000	35 000	22 000	25 000

Table 25
Ratings for Low-Voltage Alternating-Current Power Circuit Breakers with Direct-Acting Instantaneous Trip Devices

Line Number	Voltage Ratings		Insulation Level, Dielectric Withstand (volts)	Three-Phase Short-Circuit Rating* (symmetrical amperes)	Frame Size (amperes)	Continuous Current Ratings (amperes) Range of Trip Ratings † Dual Overcurrent Trip or Instantaneous Overcurrent Trip
	Rated Voltage (volts)	Rated Maximum Voltage (volts)				
1	600	635	2200	14 000	225	40–225
2	600	635	2200	22 000	600	40–600
3	600	635	2200	42 000	1600	200–1600
4	600	635	2200	42 000	2000	200–2000
5	600	635	2200	65 000	3000	2000–3000
6	600	635	2200	85 000	4000	4000
7	480	508	2200	22 000	225	40–225
8	480	508	2200	30 000	600	100–600
9	480	508	2200	50 000	1600	400–1600
10	480	508	2200	50 000	2000	400–2000
11	480	508	2200	65 000	3000	2000–3000
12	480	508	2200	85 000	4000	4000
13	240	254	2200	25 000	225	40–225
14	240	254	2200	42 000	600	150–600
15	240	254	2200	65 000	1600	600–1600
16	240	254	2200	65 000	2000	600–2000
17	240	254	2200	85 000	3000	2000–3000
18	240	254	2200	130 000	4000	4000

From ANSI C37.16-1973. This table may also be found in IEEE JH 2112-1[1].

*Single-phase short-circuit-current ratings are 87 percent of these values.

†The continuous-current-carrying capability of some circuit-breaker–trip-device combinations may be higher than the trip device current rating.

Table 26

Ratings for Low-Voltage Alternating-Current Power Circuit Breakers without Direct-Acting Instantaneous Trip Devices

Line Number	Voltage Ratings		Insulation Level, Dielectric Withstand (volts)	Three-Phase Short-Circuit Rating‡ or Short-Time Rating (symmetrical amperes)	Frame Size (amperes)	Continuous Current Ratings (amperes)		
						Range of Trip Ratings†		
						Short-Time-Delay Trip		
	Rated Voltage (volts)	Rated Maximum Voltage (volts)				Minimum Time Band	Intermediate Time Band	Maximum Time Band
1	600	635	2200	14 000	225	100–225	125–225	150–225
2	600	635	2200	22 000	600	175–600	200–600	250–600
3	600	635	2200	42 000	1600	350–1600	400–1600	500–1600
4	600	635	2200	42 000	2000	350–2000	400–2000	400–2000
5	600	635	2200	65 000	3000	2000–3000	2000–3000	2000–3000
6	600	635	2200	85 000	4000	4000	4000	4000
7	480	508	2200	14 000	225	100–225	125–225	150–225
8	480	508	2200	22 000	600	175–600	200–600	250–600
9	480	508	2200	42 000	1600	350–1600	400–1600	500–1600
10	480	508	2200	50 000	2000	350–2000	400–2000	500–2000
11	480	508	2200	65 000	3000	2000–3000	2000–3000	2000–3000
12	480	508	2200	85 000	4000	4000	4000	4000
13	240	254	2200	14 000	225	100–225	125–225	150–225
14	240	254	2200	22 000	600	175–600	200–600	250–600
15	240	254	2200	42 000	1600	350–1600	400–1600	500–1600
16	240	254	2200	50 000	2000	350–2000	400–2000	500–2000
17	240	254	2200	65 000	3000	2000–3000	2000–3000	2000–3000
18	240	254	2200	85 000	4000	4000	4000	4000

From ANSI C37.16-1973. This table may also be found in IEEE JH 2112-1[1].

* Single-phase short-circuit-current ratings are 87 percent of these values.

† The continuous-current-carrying capability of some circuit-breaker–trip-device combinations may be higher than the trip device current rating.

‡ Short-circuit-current ratings for circuit breakers without direct-acting trip devices, opened by a remote relay, are the same as those listed here.

6.2 Trip-Device Characteristics. The overcurrent trip devices considered here are integral parts of their respective types of circuit breakers. By continually monitoring the current flowing through the circuit breaker, they sense any abnormal current conditions. Based on their built-in intelligence in terms of current and time, they release a latch and permit the circuit breaker operating mechanism to open the contacts and interrupt the circuit.

The basic overcurrent trip device characteristics used on molded-case circuit breakers and low-voltage power circuit breakers are long-time delay and instantaneous. The combination of these characteristics provides time delay to override transient overloads, delayed tripping for those low-level short circuits or overloads that persist, and instantaneous tripping for higher level short circuits.

Trip devices of low-voltage power circuit breakers and some new molded-case circuit breakers may be equipped with a short-time-delay characteristic in place of the instantaneous characteristic. The resulting combination of long-time-delay and short-time-delay characteristics provides delayed tripping for all levels of current up to the maximum allowable available short-circuit-current limit of the circuit breaker without instantaneous trip elements.

6.2.1 *Molded-Case Circuit Breakers.* A typical time—current curve is shown in Fig 79 for a 600 A frame size molded-case circuit breaker. This curve applies, as indicated, for trip-device current ratings from 125 through 600 A. To be noted is the long-time-delay portion where delays in the order of seconds and minutes vary inversely with current. At the current level setting of the instantaneous element, the operating time drops abruptly to that required by the circuit breaker to interrupt with no intentional time delay. For a given trip device rating, the long-time-delay portion of the curve generally cannot be varied or modified in any way by adjustment. However, some forms of molded-case circuit breakers are available with an adjustable long-time-delay element. In frame sizes of 100 A and above, which have replaceable trip units, the instantaneous element is generally adjustable up to 10 times the trip-unit rating.

New models of molded-case circuit breakers are available which are equipped with solid-state tripping devices having long time, short time, instantaneous, and ground-fault functions in many combinations.

6.2.2 *Low-Voltage Power Circuit Breakers.* The time-current curves of low-voltage power circuit breakers in the simplest and most basic form, combining long-time-delay and instantaneous characteristics, are somewhat similar to those of the molded-case circuit breakers. This is true even though the long-time-delay elements generally used are of different types, bimetallic for molded-case circuit breakers and mechanical dashpots for low-voltage power circuit breakers. However, some molded-case circuit breakers are available with mechanical dashpots for the long-time-delay element. The differences arise from the adjustability of the trip devices used on low-voltage power circuit breakers. The adjustments normally provided are long-time-delay pickup, long-time-delay operating time, and instantaneous pickup. In addition to these adjustments, most manufacturers make available three different long-time-delay operating bands referred to as minimum, intermediate, and maximum. In some cases these can be selected on the trip device, in others, they must be specified when ordered. A typical time—current curve illustrating one of the three long-time-delay operating bands (with instantaneous curve) is shown in Fig 80.

The second most common combination of trip device characteristics is that of long-time delay and short-time delay. In some cases it is possible to combine a high-set instantaneous characteristic with the two time-delay characteristics. Fig 81 presents these combinations of characteristics. As in the case of the long-time-delay characteristic, the short-time-delay pickup is adjustable, and three different operating bands, minimum. intermediate, and maximum, are usually available.

Throughout the many years that both molded-case circuit breakers and low-voltage power circuit breakers have been available, the basic design of trip device has been the electromechanical type, using a displacement dashpot for the low-voltage power circuit breaker, while the thermal electromechanical type has been in the mainstay

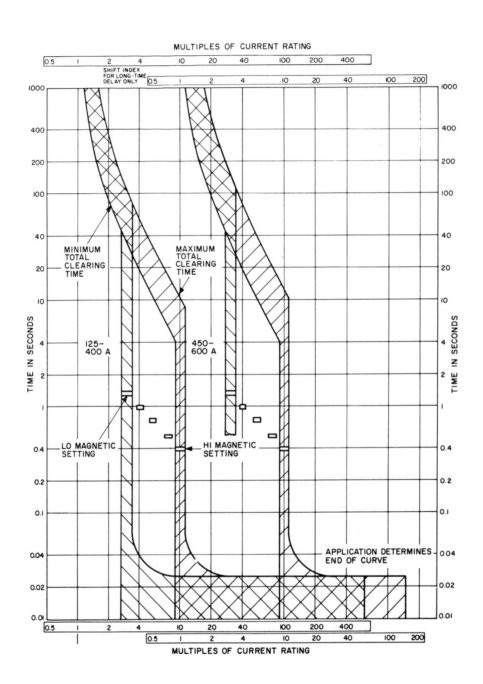

Fig 79
Typical Time—Current Curves of a 600 A Frame Size
Molded-Case Circuit Breaker

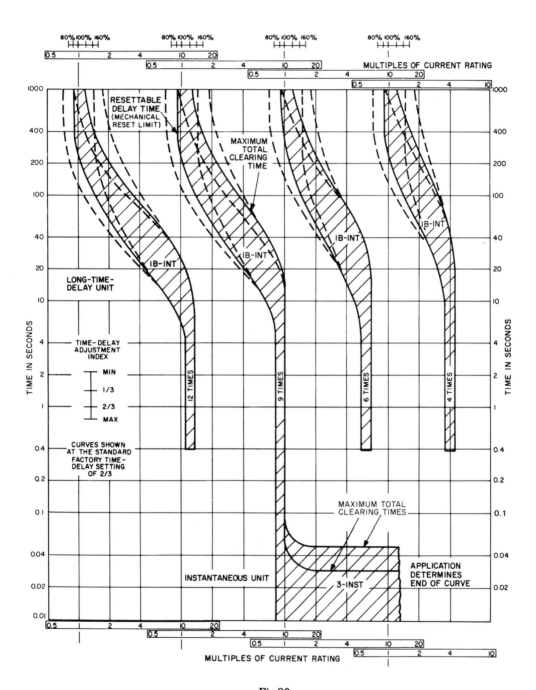

Fig 80
Typical Time—Current Curves for Combinations of Long-Time-Delay and Instantaneous
Characteristics of a Low-Voltage Power Circuit Breaker

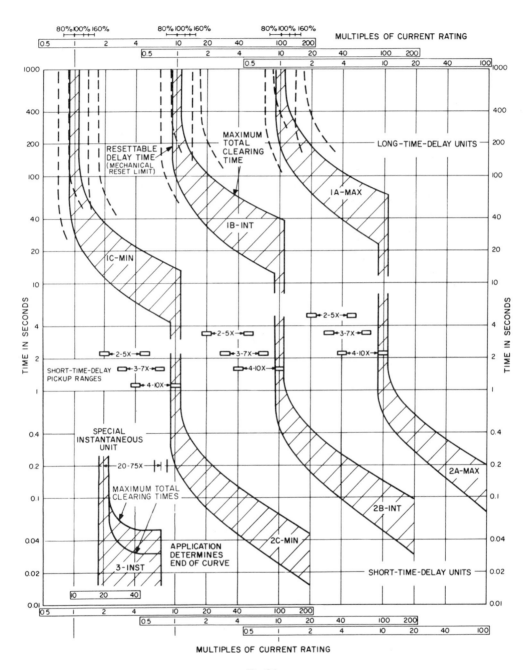

**Fig 81
Typical Time—Current Curves for Combinations of Long-Time-Delay and Short-Time-Delay
Characteristics of a Low-Voltage Power Circuit Breaker**

for molded-case circuit breakers. However, in recent years the direct-acting solid-state trip device has been developed and is available for use on both types of circuit breakers. Figs 82 and 83 illustrate typical characteristics for solid-state trip devices.

6.3 Application. The proper application of a circuit breaker, either molded-case circuit breaker or low-voltage power circuit breaker, involves considerations that go beyond the basic ones of voltage, current, and interrupting ratings. The performance record of a circuit breaker will be influenced by many factors related to the installation environment and even by factors of non-electrical character. Enclosures, service conditions, loads and their characteristics, outgoing conductors, the system supplying power to the circuit breaker, and other protective devices upstream and downstream from the circuit breaker under consideration, even operating and maintenance personnel, must all be taken into account. For purposes of this chapter, application considerations will be limited to those involving just abnormal current conditions.

While protection is the basic application covered in this chapter as it relates to abnormal current conditions, it is well to list certain differences between molded-case circuit breakers and low-voltage power circuit breakers which affect their application.

(1) Low-voltage power circuit breakers are rated to carry 100 percent of their continuous-current rating inside enclosures in a 40°C ambient. Molded-case circuit breakers are basically rated to carry 100 percent of their rated continuous current when tested in the open in a 25°C ambient. Generally, molded-case circuit breakers must be derated for continuous current when used in an enclosure, and they must also have special 40°C ambient calibration compensation for the thermal trip elements. Some molded-case circuit breakers, particularly those of the larger frame sizes, are rated for use in enclosures.

(2) Low-voltage power circuit breakers have short-time current ratings for puposes of selectivity between circuit breakers in series during short-circuit conditions, such that only the one nearest the fault opens. This short-time rating is based on a 30 cycle duration test. Molded-case circuit breakers, in certain cases, have limited short-time ratings for time durations in the neighborhood of 5 cycles.

(3) Low-voltage power circuit breakers are designed to permit routine maintenance, a feature which can contribute to long life. In contrast, most molded-case circuit breakers are sealed and maintenance of the internal mechanism is not possible. Both types allow field adjustment of the tripping boundaries to some degree, although some varieties of molded-case circuit breakers are preset and sealed at the factory and are not field adjustable.

The fundamental rules of applying circuit breakers within voltage and continuous-current ratings are quite well observed. Another very fundamental rule is to apply circuit breakers within their interrupting or short-circuit-current ratings, regardless of the location in the system. Careful attention must be given to this requirement.

Even if the circuit breakers operate perfectly in their switching and interrupting performance, special attention is needed in the selection and settings of the trip devices to obtain the two ultimate goals, protection and coordination.

6.3.1 *Protection.* Referring to Chapter 1, two additional statements bear repeating. First in talking about protection,

"The function of system protection may be defined as the detection and prompt isolation of the affected portion of the system whenever a short circuit or other abnormality occurs which might cause damage to, or adversely affect, the operation of any portion of the system or the load which it supplies." (Section 1.3.)

In speaking of short circuit this was said,

"Treatment of the overall problem of system protection and coordination of electric power systems will be restricted to the selection, application, and coordination of devices and equipments whose primary function is the isolation and removal of short circuits from the system. Short circuits may be phase to ground, phase to phase, phase to phase to ground, three phase, or three phase to ground. Short circuits may range in magnitude from extremely low current faults having high impedance paths to extremely high current faults having very low impedance paths.

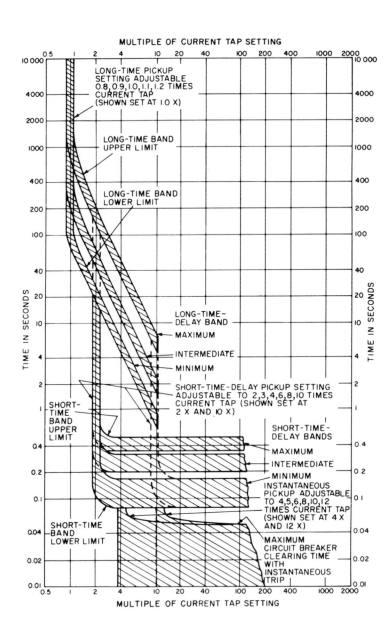

Fig 82
Typical Time—Current Curves for Phase Overcurrent Characteristics of a Solid-State
Trip Device for a Low-Voltage Power Circuit Breaker

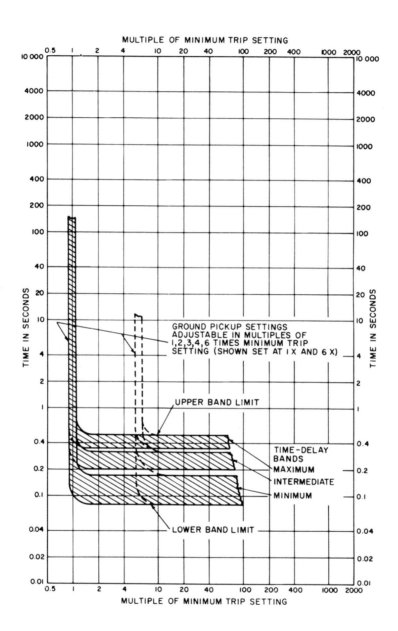

Fig 83
Typical Time—Current Curves for Ground-Overcurrent Characteristics of a Solid-State
Trip Device for a Low-Voltage Power Circuit Breaker

However, all short circuits produce abnormal current flow in one or more phase conductors or in the neutral or grounding circuit. Such disturbances can be detected and safely isolated." (Section 1.4.)

A careful analysis of this paragraph indicates that much has been said and some additional things have been implied.

Of many varieties described, there are two types of overcurrent protection which are emphasized, phase overcurrent and ground fault. At the present state of the art, phase-overcurrent conditions are detected solely on the basis of their magnitudes. Ground-fault conditions of a sufficient magnitude are detectable by phase-overcurrent devices. Those below the minimum sensitivity of phase-overcurrent devices will not be detected. Since ground-fault currents, by their very nature, involve a ground return path, sensitivity to the path of the current becomes a practical way to detect them. Circuit breakers have the advantage that they can be made to provide ground-fault protection that is generally more sensitive than phase-overcurrent protection. Several low-voltage power circuit breakers offer this type of protection and recent papers have discussed the methods used to achieve it [2]−[4].

Perhaps the best example of the expanded versatility in protection made possible by solid-state technology is the ground-overcurrent trip function. This function has become most important because of the ground-fault protection requirements in the National Electrical Code (NFPA No 70) (1975) (ANSI C1-1975), Section 230-95.

Historically, the only means of providing protection against ground-overcurrent damage has been the use of separately mounted overcurrent relays and current transformers. The characteristics of the elements, such as high relay burdens and current transformer saturation, have limited the effectiveness of these schemes.

However, if the direct-acting trip device alone is the only available means for detecting the presence of ground faults, the situation is much worse. Refer to Chapter 8, Sections 8.4 and 8.8.

Conrad and Dalasta [5] discuss the correlation between the extent of the damage and the energy associated with the arc. The energy released at the fault is determined by the product of three quantities, the arcing-fault-current magnitude,

the arc-voltage magnitude, and the fault duration permitted by the overcurrent protective device. The long-time-delay and short-time-delay elements are not sensitive enough to detect an arcing ground fault. The usual high-set instantaneous elements are not usable at all. Since the arc resistance acts to limit the fault current, perhaps to levels below the short-time pickup, the fault is then cleared only after the expiration of the long-time delay with intolerable damage levels.

Electronic technology has provided a basis for design of a ground-fault tripping fuction which combines the often conflicting requirements of protection and security. That is, the system must be sensitive enough to detect very low ground overcurrents, but yet must be immune to nuisance tripping under a host of system conditions, such as motor starting and phase-to-phase faults. These must be detected by the instantaneous or delay elements to ensure coordination in a selective system. Also in order to allow the greatest flexibility of application in a selective system, there must be enough adjustment in pickup and time delay to allow at least two levels of coordination. The characteristic curves of the solid-state system ground-fault tripping functions are shown in Fig 83. These devices are used extensively to provide ground-fault protection in air circuit breakers. Generally, three-level coordination is available and a careful analysis of the system operation is necessary if more levels have to be used, to prevent nuisance tripping.

An important aspect of good protection, which has been stressed in [6], is the simultaneous disconnection of all phases of a polyphase circuit under short-circuit conditions. This is not for the sake of protection of motors against single phasing but rather to prevent a short circuit from being backfed from other energized phases. Circuit breakers satisfy the requirements for this type of performance, the significance of which needs to be appreciated by more system designers.

Several basic rules can be set down which apply to phase-overcurrent protection.

(1) Select continuous-current ratings and pickup settings of long-time-delay characteristics, where adjustable, which are no higher than necessary and which are within the requirements of the NEC.

(2) The amount of time delay provided by the long-time-delay characteristic of low-voltage power circuit breakers should be selected to be no higher than necessary to override transient overcurrents associated with the energizing of load equipment, and to coordinate with downstream protective devices.

(3) Take advantage of the adjustable instantaneous trip elements of trip devices on molded-case circuit breakers and low-voltage power circuit breakers. Set the instantaneous trip elements no higher than necessary to avoid nuisance tripping. Be sure that instantaneous trip settings do not exceed the available short-circuit current at the location of the circuit breaker on the system. This point is frequently overlooked, particularly in service entrance applications.

6.3.2 *Coordination.* When protection is being considered, the performance of a circuit breaker with respect to the connected conductors and load is of primary concern. In contrast, the concern in coordination is with the performance of a circuit breaker with respect to other upstream and downstream circuit breakers and protective devices. The objective in coordinating protective devices is to make them selective in their operation with respect to each other. In so doing, the effects of short circuits on a system are reduced to a minimum by disconnecting only the affected part of the system. Stated in another way, only the circuit breaker nearest the short circuit should open leaving the rest of the system intact and able to supply power to the unaffected parts.

Normally coordination is demonstrated by plotting the time—current curves of the circuit breakers involved and by making sure that no overlapping occurs between the curves of adjacent circuit breakers. Often selectivity is possible only when circuit breakers with delayed trip devices are used in all circuit positions except the one closest to the load. This is particularly true when there is little or no circuit impedance between successive circuit breakers. This condition exists in a main switchboard or load center unit substation between the main and feeder circuit breakers. Here, to be selective for all levels of possible short-circuit current beyond the load terminals of the feeder circuit breakers requires that the main circuit breaker be equipped with a combination of long-time-delay and short-time-delay trip characteristics. Moving downstream, on many feeder circuits there is sufficient impedance to appreciably lower the available short-circuit current at the next downstream circuit breaker. Should the available short-circuit current at this next circuit breaker be less than the setting of an instantaneous trip on the feeder circuit breaker, selectivity is possible.

It should be recognized that the use of excessively high instantaneous trip settings in combination with only long-time delay to obtain selectivity with downstream protective devices is poor practice from a protection point of view. Combinations of long-time and short-time delay with a high-set instantaneous trip are sometimes useful in providing both good protection and selectivity.

A real basis for judging selectivity between two molded-case circuit breakers in series has been given in [7]. Molded-case circuit breakers always have instantaneous trip elements. It is pointed out in [7] that if the fault current being interrupted by the downstream circuit breaker flows through the upstream circuit breaker for a period of time equal to or greater than its unlatching (resettable) time, the upstream circuit breaker will trip. under these conditions the circuit breakers will not be selective. However, if because of impedance between the circuit breakers the maximum current that can flow in a short circuit at the load terminals of the downstream circuit breaker is insufficient to unlatch the upstream circuit breaker, selectivity will exist.

6.4 Conclusions. Low-voltage circuit breakers have been providing better protection in several ways. Improved trip devices of the solid-state type make it possible to give better protection and to coordinate more easily and more completely. Ground-fault protection as an integral part of overcurrent trip devices without external current transformers and relays is also available.

6.5 Standards References. The following standards publications were used as references in preparing this chapter.

ANSI C37.16-1973, Preferred Ratings, Related Requirements, and Application Recommendations for Low-Voltage Power Circuit Breakers and AC Power Circuit Protectors

ANSI C37.17-1972, Trip Devices for AC and General Purpose DC Low-Voltage Power Circuit Breakers

ANSI C37.100-1972, Definitions for Power Switchgear

IEEE Std 20-1973, Low-Voltage AC Power Circuit Breakers Used in Enclosures (ANSI C37.13-1973)

NEMA AB 1-1969, Molded-Case Circuit Breakers

NEMA SG 3-1971, Low-Voltage Power Circuit Breakers

NFPA No 70, National Electrical Code (1975) (ANSI C1-1975)

6.6 References

[1] IEEE Committee Report. Protection Fundamentals for Low-Voltage Electrical Distribution Systems in Commercial Buildings. IEEE JH 2112-1, 1974.

[2] STEEN, F.L, REIFSCHNEIDER, P.J., and TOLSON, J.J. A New Approach to Solid-State Overcurrent Trip Devices for Low-Voltage Power Circuit Breakers. Paper 31 PP 67, presented at the IEEE Winter Power Meeting, New York, NY, Jan 29-Feb 3, 1967.

[3] LAUBACH, W.E., WALDRON, J.E., and DAVIS, L.A. New Versatility in Solid-State Trip Devices for Low-Voltage Power Circuit Breakers. *Conference Record, 1972 7th Annual Meeting of the IEEE Industry Applications Society*, IEEE 72 CHO 685-8-IA, pp. 187-195.

[4] BAILEY, B.G. Clearing Low-Voltage Ground Faults with Solid-State Trips. *IEEE Transactions on Industry and General Applications*, vol IGA-3, Jan/Feb 1967, pp 60-65.

[5] CONRAD, R.R., and DALASTRA, D. A New Ground-Fault Protection System for Electrical Distribution Circuits. *IEEE Transactions on Industry and General Applications*, vol IGA-3, May/Jun 1967, pp 217-227.

[6] KAUFMANN, R.H. Application Limitations of Single-Pole Interrupters in Polyphase Industrial and Commercial Building Power Systems. *IEEE Transactions (Applications and Industry)*, vol 82, Nov 1963, pp 363-368.

[7] VALVODA, F.R. Selective Systems for Overcurrent Protection. *Actual Specifying Engineer*, Nov 1966.

7. Coordination

7.1 General Discussion. The coordination study of an electric power system consists of an organized time—current study of all devices in series from the utilization device to the source. This study is a comparison of the time it takes the individual devices to operate when certain levels of normal or abnormal current pass through the protective devices.

A preliminary coordination study should be made in the early planning stages of a new system. Such a study may indicate that sizes of transformers should be modified or cable sizes changed. This tentative study should be confirmed by a final study after exact equipment characteristics are determined.

A coordination study or revision of a previous study should be made for an existing plant when new loads are added to the system or when existing equipment is replaced with higher rated equipment. A coordination study should also be made when the available short-circuit current of the source to a plant is increased. This study determines settings or ratings necessary to assure coordination after system changes have been made.

A coordination study definitely should be made for an existing plant when a fault on the periphery of the system shuts down a major portion of the system. Such a study may indicate a need to change or replace devices.

The objective of a coordination study is to determine the characteristics, ratings, and settings of overcurrent protective devices which will ensure that the minimum unfaulted load is interrupted when the protective devices isolate a fault or overload anywhere in the system. At the same time, the devices and settings selected must provide satisfactory protection against overloads on the equipment, and interrupt short circuits as rapidly as possible.

The coordination study provides data useful for the selection of instrument transformer ratios, protective relay characteristics and settings, fuse ratings, low-voltage circuit breaker ratings, characteristics, and settings. It also provides other information pertinent to the provision of optimum protection and selectivity or coordination of these devices.

7.2 Primary Considerations

7.2.1 *Short-Circuit Currents.* In order to obtain complete coordination of the protective equipment applied, it may be necessary to obtain some or all of the following information on short-circuit currents for each local bus.

(1) Maximum and minimum 0 to 3 cycle (momentary) total rms short-circuit current

(2) Maximum and minimum 3 cycle to 1 s (interrupting duty) total rms short-circuit current

(3) Maximum and mimimum ground-fault currents

These short-circuit-current values are obtained as described in Chapter 2.

The maximum and minimum 0 to 3 cycle (momentary) currents are used to determine the maximum and mimimum currents to which instantaneous and direct-acting trip devices respond, and to verify the capability of the apparatus applied such as circuit breakers, fuses, switches, and reactor and bus bracings.

The maximum 3 cycle to 1 s (interrupting) current at maximum generation will verify the ratings of circuit breakers, fuses, and cables. This is also the value of current at which the circuit protection coordination interval is established. The maximum 3 cycle to 1 s (interrupting) current at minimum generation is needed to determine whether the circuit-protection sensitivity of the circuits is adequate.

7.2.2 *Coordination Time Intervals.* When plotting coordination curves, certain time intervals must be maintained between the curves of various protective devices in order to ensure correct sequential operation of the devices. These intervals are required because relays have overtravel, fuses have damage characteristics, and circuit breakers have certain speeds of operation. Sometimes these intervals are called margins.

When coordinating inverse time overcurrent relays, the time interval is usually 0.3–0.4 s. This interval is considered between relay curves either at the instantaneous setting of the load side feeder circuit breaker relay or the maximum short-circuit current which can flow through both devices simultaneously, whichever is the lower value of current. The interval consists of the following components:

Circuit breaker opening time
(5 cycles)   0.08 s
Overtravel   0.10 s
Safety factor   0.12–0.22 s

This margin may be decreased if field tests of relays and circuit breakers indicate that the system still coordinates with the decreased margins. The overtravel of very inverse and extremely inverse time overcurrent relays is somewhat less than for inverse relays, allowing a decrease in time interval for carefully tested systems to 0.3 s. When solid-state relays are used, overtravel is eliminated and the time may be reduced by the amount included for overtravel. For systems using induction disk relays, a decrease of the time interval may be made by employing an overcurrent relay with a special high-dropout instantaneous element set at approximately the same pickup as the time element with its contact wired

in series with the main relay contact. This eliminates overtravel in the relay so equipped. The time interval often used on carefully calibrated systems with high-dropout instantaneous relays is 0.25 s. The minimum time interval using a high-dropout instantaneous relay could be 0.15 s (that is, 0.03 s instantaneous reset + 0.05 s vacuum circuit breaker opening time + 0.07 s safety factor).

When coordinating relays with downstream fuses the relay overtravel and circuit breaker opening time do not exist for the fuse. The margin for overtravel is plotted beneath the relay curve and since some safety factor is desirable above the total clearing time of the fuse, the same time margin is needed as for relay-to-relay coordination. However below 1 s, some reduction of margin is accepted. The same margin is used between a downstream relayed circuit breaker and the damage curve of the fuse.

When coordinating direct-acting-trip low-voltage power circuit breakers with source-side fuses at the same voltage level, a 10 percent current margin is sometimes used. This allows for possible fuse damage below the average melting time characteristics. The published minimum melting time–current curve should be corrected for ambient temperature or preloading if the fuse manufacturer provides the data necessary to perform this correction. However, if the fuse is preloaded to less than 100 percent of its current rating and the ambient temperature is lower than about $50°C$, the correction to the minimum melting time–current curve of the fuse is usually less than 20 percent in time. Since the characteristic curves are relatively steep at the point where the margin is measured, the normal current margin applied probably is sufficient to allow coordination without making a fuse characteristic correction also.

When low-voltage circuit breakers equipped with direct-acting trip units are coordinated with relayed circuit breakers, the coordination time interval is usually regarded as 0.4 s. This interval may be decreased to a shorter time as explained previously for relay-to-relay coordination. The time margin between the fuse total clearing curve and the upstream relay curve could be as low as 0.1 s where clearing times below 1 s are involved.

When coordinating circuit breakers equipped with direct-acting trip units, the characteristic

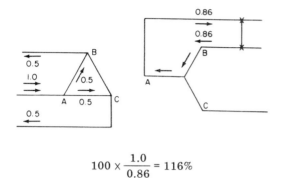

$$100 \times \frac{1.0}{0.86} = 116\%$$

Fig 84
Currents in Delta—Wye Transformer for
Secondary Phase-to-Phase Fault

curves should not overlap. In general, only a slight separation is planned between the different characteristic curves. This lack of a specified time margin is explained by the incorporation of all the variables plus the circuit breaker operating times for these devices within the band of the device characteristic curve.

7.2.3 *Delta—Wye Transformers.* When the source-side medium-voltage fuse is applied on the high side of a delta—wye transformer, an additional 16 percent current margin over margins mentioned in Section 7.2.2 is used between the minimum melting time of the fuse and the circuit breaker characteristic. This helps maintain selectivity for phase-to-phase faults since the per-unit primary current in one line for this type of fault is 16 percent greater than the per-unit secondary current. This is illustrated in Fig 84.

7.2.4 *Cable Ampacity.* Ampacity is the current-carrying capacity of a conductor, expressed in amperes. The current capacity is the maximum allowable current which may flow through a conductor without damaging the conductor or its insulation. The maximum continuous ampacities of 600 V copper and aluminum conductors are given in Tables 310-16 to 310-19 of NFPA No 70, National Electrical Code (1975) (ANSI C1-1975). As stated in the tables, these ampacities are based on the fact that there are no more than three conductors in a conduit or cable and that the ambient temperature is not greater than $30°C$ or $86°F$. If these conditions are exceeded, then ampacity reduction factors as given in Sec-

tion 310-15 of the NEC must be applied. The conductors shall be protected in accordance with the ampacities as determined per the above information, with the noted exceptions in the NEC, Sections 240-3 and 240-4. For high-voltage cable ampacities, refer to Tables 310-39 through 310-50.

Another important factor in determining the size of the circuit cable is the maximum short-circuit current available at the extremity of the cable circuit. The conductor insulation should not be damaged by the high conductor temperature resulting from current flowing to a fault beyond the cable termination. As a guide in preventing insulation damage, curves of conductor size and short-circuit current based on temperatures which damage insulation are available from cable manufacturers. In coordinating system protection, the cable should withstand the maximum through short-circuit current for a time equivalent to the tripping time of the back-up relay, not the primary relay protection. Many times this will determine the minimum conductor applicable to a particular power system.

7.2.5 *National Electrical Code.* The NEC is primarily an installation code. It does have many application requirements, some of which affect system selectivity. Three areas of primary concern are overcurrent protection, motor circuits, and transformer protection.

Interpretation and enforcement of the NEC is the prerogative of the enforcing authority. The Federal Occupational Safety and Health Act makes compliance with the NEC mandatory. Comments made on code articles in this publication are not to be construed as official interpretations of the applicability or meaning of the NEC, nor are they to be construed as covering all implications of any article.

The comments given in this publication pertain to the 1975 NEC. Since the NEC is revised every three years, the latest issue should always be consulted.

(1) *Overcurrent Protection — Article 240.* This article covers overcurrent protection for conductors and cross references other articles containing specific protection requirements for motors, generators, transformers, capacitors, and other equipment. The articles on branch circuits, feeders, busways, cable bus, and services contain pertinent overcurrent protection requirements.

The article states that "Conductors shall be protected in accordance with their ampacities ... except ... where standard ampere ratings of ... nonadjustable circuit breakers do not correspond with the allowable ampacities of conductors, the next higher standard rating may be used, only where the rating is 800 amperes or less."

Selectivity of fault protective devices cannot be maintained without giving proper consideration to all the circuit parameters involved. These include not only load and cable size but also settings or ratings and characteristics of downstream protective devices. To meet the requirements of this article and maintain system selectivity, many applications require installation of larger cable than originally anticipated or installation of supplementary overcurrent devices. The alternative to these remedies is a compromise of system selectivity.

(2) *Motor Circuits — Article 430.* The provisions of this article establish methods for selecting cable and protective device settings for motor branch circuits and motor feeder circuits. Sections 430-22 through 430-28 and 430-51 through 430-58 concern selection of cable size and specify overcurrent devices intended to protect the motor branch circuit conductors. The provisions of these sections are in addition to or amendatory of the provisions of Article 240.

(3) *Transformer Protection — Article 450.* The 1975 NEC contains the following requirement in Section 450-3: "A transformer 600 V or less having an overcurrent device on the secondary side, rated or set at not more than 125 percent of the rated secondary current of the transformer, shall not be required to have an individual overcurrent device on the primary side if the primary feeder overcurrent device is rated or set at a current value of not more than 250 percent of the rated primary current of the transformer."

Without a main secondary device this article states that the primary device shall be rated or set at no more than 125 percent of the transformer full-load current, or the next higher standard rating of the primary device.

Applications of transformers rated 600 V or less without main secondary overcurrent protective devices require particular attention since some devices meeting the requirements of this article may operate on magnetizing inrush. Selec-

tivity with secondary devices may be difficult to achieve.

Transformers 600 V or less equipped with coordinated thermal overload protection, which is arranged to interrupt the primary current, are not required to have an individual overcurrent device in the primary connection provided the primary feeder overcurrent protective device is rated or set to open at 600 percent of transformer full-load current for transformers of 6 percent or less impedance (400 percent for impedances from 6 to 10 percent).

For transformers over 600 V the requirements of Section 450-3 differ depending on whether fuses or circuit breakers are used. Section 450-3 should be consulted for specific details.

7.2.6 Pickup. The term "pickup" has acquired several meanings. For many devices, pickup is defined as that minimum current which starts an action. It is accurately used when describing a relay characteristic. It is also used in describing the performance of a low-voltage power circuit breaker. The term does not apply accurately to the thermal trip or a molded-case circuit breaker, which deflects as a function of stored heat.

The pickup current of an overcurrent protective relay is the minimum value of current which will cause the relay to close its contacts. For an induction disk time overcurrent relay, pickup is the minimum current which will cause the disk to start to move and ultimately close its contacts. For solenoid-actuated devices with time-delay mechanisms, this same definition applies. For solenoid-actuated devices without time-delay mechanisms, the time to close the contacts is extremely short. Taps or current settings of these relays usually correspond to pickup current.

For low-voltage power circuit breakers, pickup is defined as that calibrated value of minimum current, subject to certain tolerances, which will cause a trip device to ultimately close its armature, either unlatching the circuit breaker or closing an alarm contact. A trip device with a long-time delay, short-time delay, and an instantaneous characteristic will have three pickups. All these pickups are given in terms of multiples or percentages of trip-device rating.

If the long-time-delay element is set at 100 percent of the trip-device rating, pickup is equal to the trip-device rating, even though the minimum current to actuate the device is ± 10 percent of this setting. If the long-time-delay element is set on 80 percent of the trip-device rating, the pickup is equal to 80 percent of the trip-device rating, even though the minimum current to actuate the device is ± 10 percent of the 80 percent setting.

If the short-time-delay element is set on five times the trip-device rating, no matter what the settings of the long-time-delay element or the instantaneous element, the short-time-delay pickup will be a value of current equal to 500 percent of the trip-device rating with a ± 10 percent tolerance.

If the instantaneous element is set at nine times the trip-device rating, regardless of the settings of the long- or short-time-delay elements, the instantaneous pickup is a current equal to 900 percent of the trip-device rating with a ± 10 percent tolerance.

For molded-case circuit breakers with thermal trip elements, tripping times, not pickups, are discussed, since a properly calibrated molded-case circuit breaker carries 100 percent of its rating at $25°C$ in open air. The instantaneous magnetic setting could be called a pickup in the same way as that for low-voltage power circuit breakers.

Finally it should be noted that it is usually easier and less confusing, when making coordination studies particularly, to think in terms of current and time setting.

7.2.7 *Current-Transformer Saturation.* The function of a current transformer is to produce a current which is applicable to standard protective relays and which is a representation of the primary current in known proportionality and phase relationship.

Current transformers are designed with standards in mind. The standard values have been selected to give satisfactory results under the many varied conditions of metering and relaying. In most applications, instruments and relays are initiated from the same set of current transformers and perfectly satisfactory operation is obtained with standard-accuracy current transformers. An analysis of the performance of most standard current transformers will indicate that current transformers which are far from perfect will still be adequate for the particular application involved.

The major criterion for the selection of the

current-transformer ratio is almost invariably the maximum load current. A second criterion for the transformer ratio is determined by the maximum interrupting short-circuit current. The result of the maximum fault current divided by the short-time thermal rating of the current-transformer secondary apparatus should be checked against the current-transformer ratio. If the transformer ratio is lower, which means that the secondary apparatus may be damaged, then a more refined calculation applying the current-transformer saturation curve is necessary (see Chapter 4).

When checking coordination, the effect of transformer saturation is to slow the induction disk relay operation. When the current transformer becomes saturated due to a high burden or many times full-load current, the actual secondary relay current is less than it should be, and the relay operates more slowly.

For practical purposes, the maximum secondary current available with negligible saturation is represented by the secondary exciting current at the point of intersection of the unsaturated burden line with the current-transformer saturation curve. The unsaturated burden line is actually a $45°$ line initiated at the relay pickup point on the current-transformer saturation curve and drawn up and to the right until it crosses the plotted saturation curve. (Fig 38 of Chapter 4 shows the saturation curve of a typical current transformer.)

Actually, a burden applies only to a particular value of secondary current. This is because most of the equipment applied has a magnetic circuit in which the burden decreases due to saturation as the current increases. Thus the impedances of the applied apparatus should be known for several values of overcurrent so that the impedance values can be approximated for a particular value of current.

Instantaneous elements should be set below the current-transformer saturation point so that they will not be affected by any saturation condition.

In most industrial systems, current-transformer saturation is a problem only on circuits with relatively low-ratio current transformers. However, in most cases these circuits feed utilization equipment; relays with instantaneous settings which are unaffected by saturation can be applied. As one progresses back toward the source, the current-transformer ratios get larger, the transformers have more turns, developing higher voltages, and therefore they are less likely to get saturated when normal burdens are applied.

Differential relaying is not usually susceptible to saturation problems if the current transformers are closely matched or the proper ratio-matching taps on the relays are selected.

Usually current-transformer saturation problems occur only between the two lowest setting time element relays. Since the minimum coordination interval is at the theoretical maximum fault-current point, the overall problem is more or less minimized.

Saturation of current transformers due to the dc component of an asymmetrical fault current can cause a delay in the operation of some instantaneous relays. It can also cause false tripping of residually connected ground relays if time delay is not used.

7.2.8 *How to Read Curves.* A basic understanding of time—current characteristics is essential to any study. On an ordinary coordination curve, time 0 is considered as the time at which the fault occurs, and all times shown on the curve are the elapsed time from that point. The curves that are drawn are response times since, for a radial system, all the devices between the fault and the source experience the same current until one of them interrupts the circuit. After interruption, relay overtravel as well as circuit breaker and relay reset times are considered in order to determine if any device which has started to operate will continue to operate under reduced current and trip a backup protective device. These overtravel and reset times are taken into account by shifting curves or allowing time margins between curves.

A coordination curve is arranged so that the region below and to the left of the curve represents an area of no operation. The curves represent a locus of a family of paired coordinates (current and time) which indicate how long a period of time is required for device operation at a selected value of current. Protective relay curves are usually represented by a single line only. Circuit breaker tripping curves which include the circuit breaker operating time as well as the trip device time are represnted as bands. The bands represent the limits of maximum and mini-

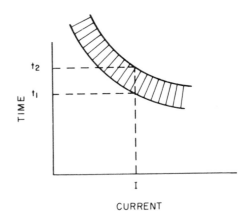

Fig 85
Time—Current Curve Band
Including Resettable Allowance
(Impulse Characteristic)

mum times at selected currents during which circuit interruption is expected. The region above and to the right of the curve or band represents an area of operation.

Fig 85 shows a time—current curve represented as a band. Time t_2 is the maximum time from the initiation of the current flow I within which operation of the device and circuit breaker is assured.

Time t_1 is the time from initiation of the current flow I within which the current must be normalized to prevent the device under consideration from operating due to the impulse characteristic of the trip device.

Reading current along the abscissa, the time or range of times in which any device is expected to operate corresponds to the ordinate or ordinates of the curve plotted. Usually circuit breaker curves begin at a point of low current close to the trip device rating or setting and an operating time of 1000 s; relay curves begin at a point close to 1½ times pickup and the corresponding time for this point. Curves usually end at the maximum short-circuit current to which the device under consideration can be subjected. A single curve can be drawn for any device under any specified condition, although most devices (except relays) plot an envelope within which operation takes place. This envelope takes into consideration most of the variables which affect operation.

Some of these variables are ambient temperature, manufacturing tolerances, and resettable time delay.

7.3 Initial Planning. There are four steps to follow in planning a coordination study.

(1) Develop a one-line diagram including the data indicated in Section 7.4. Adhering to the principles of protection outlined in other chapters of this publication will minimize the number of modifications of the one-line diagram necessary to achieve a coordinated system.

(2) Determine the level of short-circuit current at each location in the system. Section 7.2.1 gives the fault currents necessary for a study.

(3) Determine the protection requirements of various elements of the system and the load flow requirements of the system.

(4) Assemble the characteristics of the protective devices involved in the system.

Section 7.5 indicates the procedure to follow and gives examples of a coordination study.

7.4 Data Required for a Coordination Study. The first requisite for a coordination study is a one-line diagram of the system or portion of the system involved in the study. This one-line diagram should show the following data.

(1) Apparent power and voltage ratings as well as the impedance and connections of all transformers

(2) Normal and emergency switching conditions

(3) Nameplate ratings and subtransient reactance of all major motors and generators as well as transient reactances of synchronous motors and generators, plus synchronous reactances of generators

(4) Conductor sizes, types, and configurations

(5) Current transformer ratios

(6) Relay, direct-acting trip, and fuse ratings, characteristics, and ranges of adjustment

The second requirement is a complete short-circuit-current study as described in Section 7.2.1 for both first-cycle and interrupting duties. This study should include maximum and mimimum expected duties as well as available short-circuit-current data from all sources.

The third requirement is the time—current characteristics af all the devices under consideration, and the fourth requirement is the expected maximum loading on any circuit considered. Any limiting devices such as utility settings on relays must be noted.

It is usually presumed that continuous current, close and latch, and interrupting ratings have been evaluated by a short-circuit-current study. These items should be noted during a coordination study if they have not been previously noted.

7.5 Procedure. The principle of using overlays for making coordination curves removes much of the tediousness from coordination studies. Once a specific current scale has been selected, the proper multipliers for the various voltage levels considered in the study are calculated. Protective device curves for the various devices are then placed on a smooth bright surface such as a white sheet of paper, a window pane, or a glass-topped box with a lamp in it. The sheet of log—log paper on which the study is being made is placed on top of the device characteristic curve, the current scale of the study lined up with that of the device characteristic. The curves for all the various settings and ratings of devices being studied may then be traced or examined.

7.5.1 *Selection of Proper Current Scale.* Considering a large system or one with more than one voltage transformation, the characteristic curve of the smallest device is plotted as far to the left of the paper as possible so that the curves are not crowded to the right of the paper. The maximum short-circuit level on the system is the limit of the curves to the right, unless it seems desirable to observe the possible behavior of a device above the level of short-circuit current on the system under study. A minimum number of trip characteristics should be plotted on one sheet of paper.

Indexing the various curves to a common scale sometimes confuses even a competent plant engineer. The manipulation is explained below.

Consider a system on which a 750 kVA transformer with a 4160 V delta primary and a 480 V wye secondary is the largest device. Assume that this transformer is equipped with a primary circuit breaker and a main secondary circuit breaker supplying some feeder circuit breakers. On this system, full-load current of the transformer at 480 V is $(750 \times 10^3)/(480 \times \sqrt{3}) = 902$ A. When 902 A is flowing in the secondary of the transformer, the current in the primary of the transformer will be this same value of current (902 A) multiplied by the turns ratio or voltage ratio of the transformer $(480/4160 = 0.115)$. In the case under consideration, the primary current will be $902 \times 0.115 = 104$ A. If we establish the full-load current to be 1 per unit, then 902 A at 480 V = 1 per unit = 104 A at 4160 V. As far as the time–current coordination curve is concerned, both 104 A at 4160 V and 902 A at 480 V represent the same value of circuit current: full load of the 750 kVA transformer and 1 per unit

current. Plotting current on the time–current plot, 902 A at 480 V is the same as plotting 104 A at 4160 V. This type of manipulation permits the study of devices on several different system voltage levels on one coordination curve if the proper current scales are selected for the plot.

7.5.2 *Example of Step-by-Step Phase Coordination Study*

(1) *One-Line Diagram.* Draw the one-line diagram of the portion of the system to be studied with the ratings of all known devices shown (Fig 86).

(2) *Short-Circuit-Current Study.* Calculate the short-circuit-current values available at different points in the system:

34.4 kV system	500 MVA (from utility or system study)
4160 V bus	55.5 MVA (from system only with no motor contribution, by calculation)
480 V bus	12 800 A, symmetrical, from 750 kVA transformer
480 V substation feeder	12 800 A, symmetrical, from 750 kVA transformer plus 3600 A, symmetrical, motor contribution from 480 V system
480 V, 100 A panelboard	11 000 A, symmetrical (by calculation)

(3) *Protection Points.* Determine the protection points desired for certain large system components.

3750 kVA Transformer	750 kVA Transformer
(a) *ANSI Point** $16.6 \times I_{fl} \times 0.58 = I_{ANSI}$ for 4 s $16.6 \times 520 \times 0.58 = 5000$ A at 4160 V $16.6 \times 63 \times 0.58 = 606$ A at 34.4 kV	$17.6 \times I_{fl} \times 0.58 = I_{ANSI}$ for 3.75 s $17.6 \times 902 \times 0.58 = 9300$ A at 480 V $17.6 \times 104 \times 0.58 = 1060$ A at 4160 V
(b) *Inrush Point* $12 \times I_{fl} = I_{inrush}$ for 0.1 s $12 \times 520 = 6250$ A at 4160 V $12 \times 63 = 755$ A at 34.4 kV	$8 \times I_{fl} = I_{inrush}$ for 0.1 s $8 \times 902 = 7216$ A at 480 V $8 \times 104 = 832$ A at 4160 V
(c) *6 Times Full Load (NEC Rule)* $6 \times 520 = 3120$ A at 4160 V $6 \times 63 = 378$ A at 34.4 kV	$6 \times 902 = 5472$ A at 480 V $6 \times 104 = 624$ A at 4160 V

*When a delta–wye transformer is involved in a system, a line-to-ground fault producing a 100 percent fault current in the secondary winding will produce only 58 percent fault current $(1/\sqrt{3})$ in each of two phases of the incoming line to the primary of the transformer. This means that the indicated current of the ANSI point must be decreased to 58 percent of the value used for three-phase faults. In ungrounded or impedance grounded systems where line-to-ground faults are limited to low values, the ANSI point is decreased to 87 percent of the 100 percent three-phase fault point.

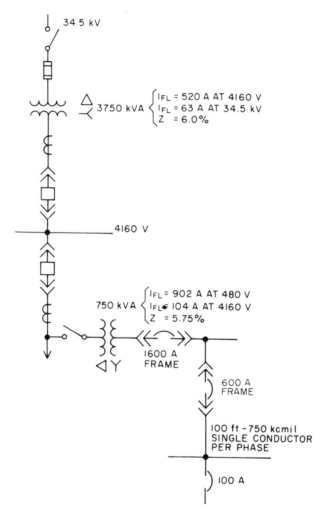

34.5 kV

I_{FL} = 520 A AT 4160 V
3750 kVA $\{$ I_{FL} = 63 A AT 34.5 kV
Z = 6.0%

4160 V

I_{FL} = 902 A AT 480 V
750 kVA $\{$ I_{FL} = 104 A AT 4160 V
Z = 5.75%

1600 A
FRAME

600 A
FRAME

100 ft – 750 kcmil
SINGLE CONDUCTOR
PER PHASE

100 A

Fig 86
One-Line Diagram

Table 27
Range of Currents at Various System Voltages

| System | 34.4 kV System | | 4160 V System | 480 V System | |
	Full-Load Current for 3750 kVA Transformer	500 MVA Short-Circuit Capacity	55.5 MVA Short-Circuit Capacity	100 A Load	16 400 A Short-Circuit Current
34.4 kV	63 A	8 400 A	930 A	1.4 A	229 A
4160 V	520 A	69 200 A	7 700 A	11.5 A	1 890 A
480 V	4500 A	600 200 A	66 500 A	100 A	16 400 A

NOTE: Since the cable size determines the current transformer size and may materially constrict coordination, leave selection of the cable size until later, particularly on a medium-voltage system.

(4) *Scale Selection*

(a) Examine the range of currents to be depicted at different voltages (Table 27). The range of currents shown in the table extends over four cycles of log paper to completely depict all the devices under consideration.

(b) Select a scale which will minimize multiplications and manipulations on devices where a range of settings is available. Since the load-end device is fixed, settings will be selected for two devices at 480 V and two at 4160 V in addition to determining cable sizes. Since relays have current transformer multipliers, an even-digit scale at 4160 V appears to offer the easiest working scale. Using a multiplier of 10 for 4160 V currents, multipliers of 87 for 480 V currents and of 1.21 for 34 400 V currents follow.

(5) *Fixed Points.* Plot the following on log–log paper (Fig 87):

(a) ANSI, inrush, and 6 times full-load points for transformers

(b) Short-circuit currents

(c) 100 A low-voltage circuit breaker (load device)

(6) *High-Voltage Fuse.* According to published tables, a standard-speed 100E fuse will protect the 3750 kVA transformer. However, an examination of the curves plotted shows that the coordination will be close for fitting all the devices necessary between this rating fuse and the largest load device. Hence a slow-speed characteristic is selected, since it will protect the transformer according to the established criteria. A larger fuse would also protect the transformer, but a tentative selection of the smaller fuse provides the transformer with better protection (Fig 88).

(7) *Low-Voltage Circuit Breakers.* By examination of the ANSI and inrush points, the limits of the curve for relay protection of the 750 kVA transformer can be determined. Keeping in mind that a low pickup for this relay is desirable for good cable protection, it is good practice to keep the circuit breaker characteristics as far to the left as possible.

A 750 kcmil cable has an ampacity of about 500 A; hence a trip device set at 500 A adequately protects this cable. Either a short-time-delay trip element or a low-set intantaneous element prevents conflict with 4160 V overcurrent relay setting. A short-time-delay trip device is selected to be selective with the downstream molded-case circuit breaker. Select a 600 A medium-time trip element set at 80 percent (480 A) and a short-time trip element set at 4 times (2400 A) with a minimum time characteristic (Fig 89).

It becomes immediately apparent that there are difficulties to be met squeezing all the relays in between the settings selected. However, the next circuit breaker to be selected is the 750 kVA transformer secondary circuit breaker. For a 902 A full load, a 1200 A trip is selected with the maximum time characteristic on both long- and short-time-delay elements. Set the short-time setting at 3 times (3600 A).

While in this example liberal curve separation is effected for clarity, in actual practice many engi-

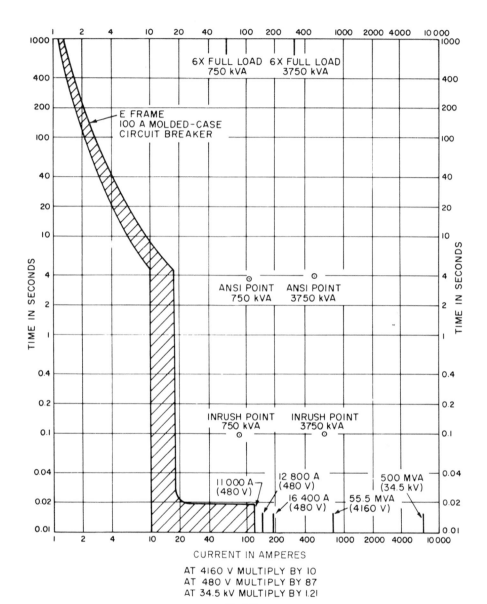

**Fig 87
Plot Showing Fixed Points, Maximum Short-Circuit Currents,
and 100 A Circuit Breaker Characteristics**

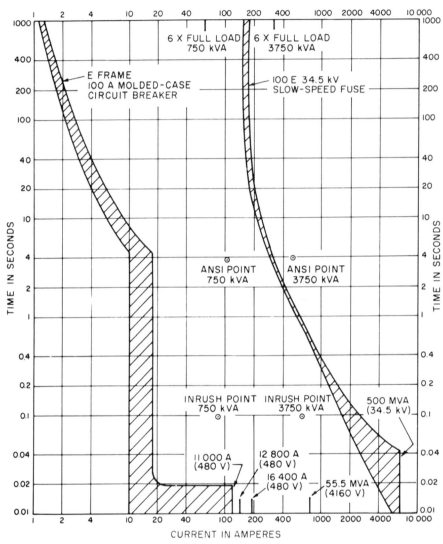

Fig 88
Selection of High-Voltage Fuse

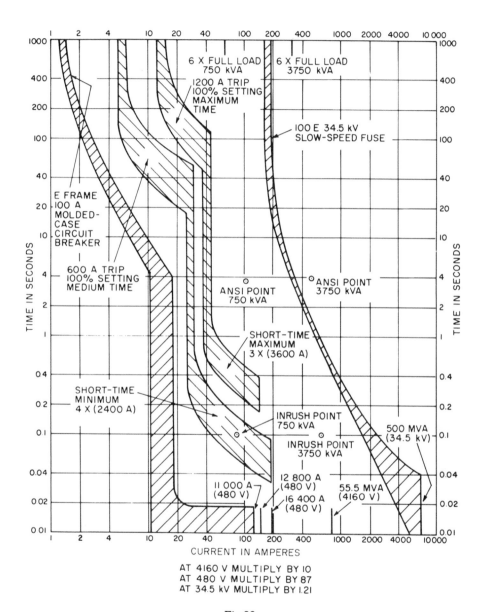

**Fig 89
Selection of Main and Feeder Low-Voltage Circuit Breakers**

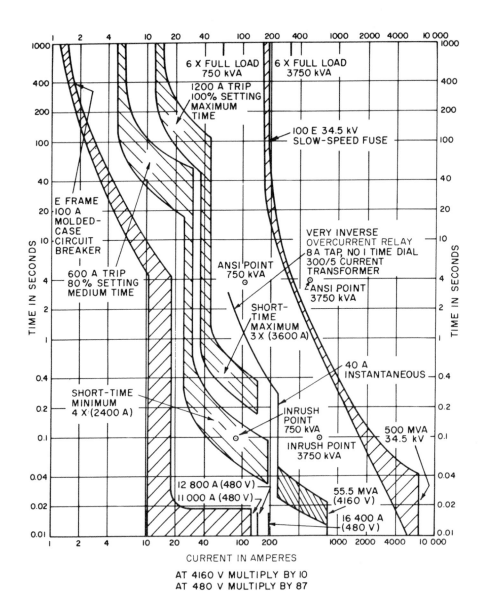

Fig 90
Selection of Overcurrent Relay Curve and Instantaneous Setting

neers prefer to use faster bands more closely stacked in order to provide faster protection.

(8) *Medium-Voltage Feeder Relays.* Allowing 16 percent current margin between the short-time setting of the main secondary circuit breaker (3600 A at 480 V), select a pickup for the medium-voltage feeder overcurrent relay. This should be less than 624 A at 4160 V and more than 3600 × (480/4160) × 1.16 = 480 A. Since this setting will also protect the cable supplying the substation, the lowest possible pickup is selected. With a 300/5 current transformer, 480 A is 480 × 5/300 = 8 A, a standard tap on induction relays. The 8 A tap allows for addition of future load. Tap settings of 5 A or lower are often desirable and can be obtained by closer stacking of curves and tolerating some overlap with the transformer secondary circuit breaker curves. Also, an ammeter with a 300 A scale reads 1/3 scale on full load of the 750 kVA transformer.

Trying to select a characteristic of this relay, it is better to be tentative since there may be trouble with the main 4160 V circuit breaker relays. Hedging a little, pick a very inverse instead of the inverse characteristic sometimes recommended. The 16 percent current margin is to be maintained at all times, hence a time dial of between 3/4 and 1 is required.

The instantaneous element is set above the available asymmetrical short-circuit current on the 480 V bus so that it does not trip for 480 V faults. This value is 12 800 × (480/4160) × (5/300) × 1.6 = 39.4 A (Fig 90).

Notice that a 4160 V feeder current transformer has been selected and a pickup value determined for the overcurrent relay protecting the cable to the substation. The pickup of the relay is no greater than the protected cable ampacity. The cable ampacity is more than 480 A. One 750 kcmil cable per phase meets the requirements of coordination and protection and allows for future expansion. A smaller cable could be used, relying on the setting of the low-voltage circuit breaker for overcurrent protection, but there would be no assurance that at some future date this protection would not be compromised. Such a compromise could take place by the addition to this circuit of medium-voltage fused starters or unit substations without main circuit breakers.

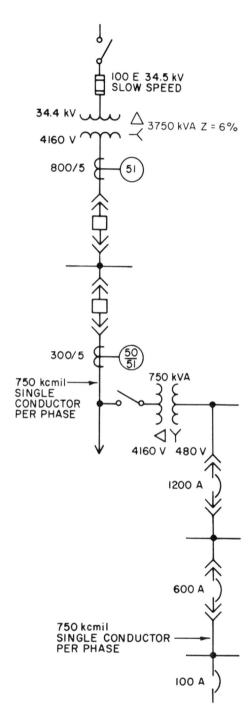

Fig 91(a)
One-Line Diagram for
Coordination Study of Fig 91(b)

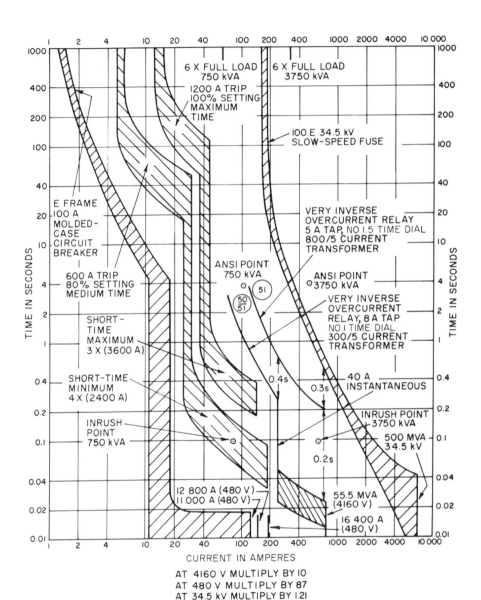

Fig 91(b)
Selection of Main Overcurrent Relay Curve

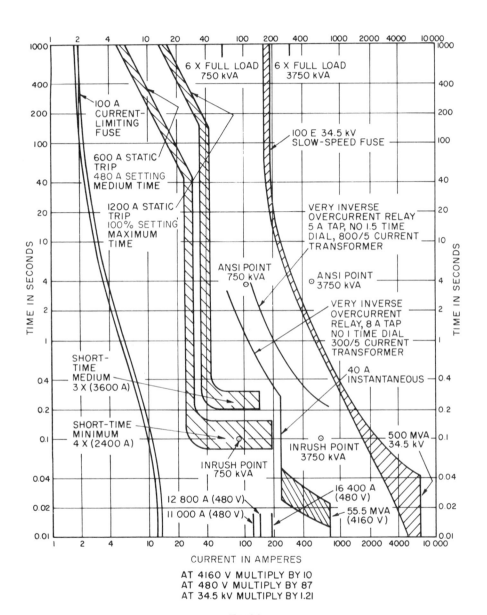

AT 4160 V MULTIPLY BY 10
AT 480 V MULTIPLY BY 87
AT 34.5 kV MULTIPLY BY 1.21

Fig 92
Fig 91 Replotted Using Solid-State Trip Device Characteristics

(9) *Medium-Voltage Main Relays*. Select a pickup for the 3750 kVA transformer secondary circuit breaker relays no lower than 125 percent of full load (520 × 125 = 650 A) and no higher than 250 percent full load (520 × 2.50 = 1300 A). A good selection is 800 A with an 800/5 current transformer. Do not put an instantaneous attachment on this relay since it is in series with the feeder circuit breaker relay for feeder faults.

Select a time dial setting such that 0.3—0.4 s is obtained between this relay and the feeder relay at the instantaneous setting of the feeder relay (40 × 300/5 = 2400 A). The setting selected allows only 0.2 s at the theoretical 100 percent fault current point between the main secondary circuit breaker and the feeder circuit breaker. This compromise is usually acceptable because of the desirability of maintaining the margin between the transformer primary fuse and main secondary circuit breaker setting. A margin of 0.2—0.4 s should exist between the primary fuse minimum melting time curve and 3750 kVA transformer main secondary circuit breaker relay characteristic at the maximum 4160 V value of short-circuit current. This value is 55.5 MVA or 7600 A (Fig 91).

Fig 92 illustrates the same system equipment with static trip devices on the circuit breakers and a downstream fuse as the final load device. As previously mentioned, the curves can probably be stacked closer to obtain faster tripping for abnormal currents.

7.6 Coordination Trends. Work is in progress in various IEEE committees concerning the extension of time—current characteristics to much shorter times than the examples shown (extending the time coordinates to 0.001 s).

7.7 Standards References. The following standards publications were used as references in preparing this chapter.

IEEE Std 141-1969, Electric Power Distribution fo Industrial Plants

IEEE Std 241-1974, Electric Power Systems in Commercial Buildings

NFPA No 70, National Electrical Code (1975) (ANSI C1-1975)

7.8 References

[1] BEEMAN, D.L., Ed. *Industrial Power Systems Handbook*. New York: McGraw-Hill, 1955.

8. Ground-Fault Protection

8.1 General Discussion. In recent years there has been an increasing interest in the use of ground-fault protection in electric distribution circuits. This interest has been intensified by the requirement of the National Electrical Code, NFPA No 70 (1975) (ANSI C1-1975) (NEC) for ground-fault protection on certain service entrance equipments and the incorporation of the NEC in the OSHA standards. This is evident when one inspects today's electrical indoor distribution, construction, and consulting engineering press and notes the number of feature articles dealing with this subject. These articles and the unusual interest in ground-fault protection have been brought about by a disturbing number of electric failures. One editor [1] reports the cost of arcing faults as follows: "One five-year estimate places the figure between $1 billion and $3 billion annually for equipment loss, production downtime, and personal liability ." This chapter explores the need for better ground-fault protection, pinpoints the areas where that need exists, and discusses the solutions which are being applied today.

Distribution circuits which are solidly grounded or grounded through low impedance require fast clearing of ground faults. This is especially true in low-voltage grounded wye circuits which are connected to busways or long runs of metallic conduit. The problem involves sensitivity in detecting low ground-fault currents as well as coordination between main and feeder circuit protective devices. Fault clearing must be extremely fast where arcing is present.

The appeal of effective ground-fault protection is based on four factors.

(1) The majority of electric faults involve ground. Even those which are initiated phase to phase will spread quickly to any adjacent metallic housing, conduit, or tray which provides a return path to the system grounding point. Ungrounded systems are also subject to ground faults, and require careful attention to ground detection and ground-fault protection.

(2) The ground-fault protective sensitivity can be relatively independent of continuous load current values, and thereby have lower pickup settings than phase protective devices.

(3) Since ground-fault currents are not transferred through system power transformers which are connected delta—wye or delta—delta, the ground-fault protection for each system voltage level is independent of the protection at other voltage levels. This permits much faster relaying than can be afforded by phase-protective devices which require coordination using pickup values and time delays which extend from the load to the source generators, often resulting in considerable time delay at some points in the system.

(4) Arcing ground faults which are not promptly detected and cleared can be extremely destructive. A relatively small investment can provide very valuable protection.

Much of the present emphasis on ground-fault protection centers in low-voltage circuits, 600 V or less. Low-voltage circuit protective devices have usually involved fused switches or circuit breakers with integrally mounted series tripping devices. These protective elements are termed overload or fault overcurrent devices because they carry the current in each phase and clear the circuit only when the current reaches a magnitude greater than full-load current. To accommodate inrush currents such as motor starting or transformer magnetizing inrush, phase overcurrent devices are designed with inverse characteristics which are rather slow at overcurrent values up to about five times rating. For example, a 1600 A low-voltage circuit breaker with conventional phase protection will clear a 3200 A fault in about 100 s, although it can be adjusted in a range of roughly 30–200 s at this fault value. A 1600 A fuse may require 10 min or more to clear the same 3200 A fault. These low values of fault currents are associated predominately with faults to ground and have generally received little attention in the design of low-voltage systems until the occurrence of many serious electric failures in recent years. In contrast, on grounded systems 2400 V and above it has long been standard practice to apply some form of ground-fault protection.

The action initiated by ground-fault sensing devices will vary depending on the installation. In some cases, such as services to dwellings, it may be necessary to immediately disconnect the faulted circuit to prevent loss of life or property. However, the opening of some circuits in critical applications may in itself endanger life or property. Therefore, each particular application should be studied carefully before selecting the action to be initiated by the ground-fault protective devices.

8.2 Types of Systems Relative to Ground-Fault Protection. A comprehensive discussion of grounded and ungrounded systems is given in Chapter 1 of IEEE Std 142-1972, Grounding of Industrial and Commercial Power Systems (ANSI C114.1-1973), known as the IEEE Green Book. When considering the choice of grounding, it is important to determine the types of ground-fault protection available and their effect on system performance, operation, and safety.

An ungrounded system has no intentional connection to ground except through potential indicating or measuring devices, or through surge protective devices. While it is called "unground-

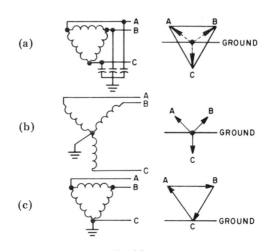

Fig 93
Voltages to Ground under Steady-State Conditions (a) Ungrounded System, Showing System Capacitance to Ground (b) Grounded Wye-Connected System (c) Grounded Delta-Connected System

ed," it is actually coupled to ground through the distributed capacitance of its phase windings and conductors.

A grounded system is intentionally grounded by connecting its neutral or one conductor to ground, either solidly or through a current-limiting impedance. Various degrees of grounding are used ranging from solid to high impedance, usually resistance.

Fig 93 shows ungrounded and grounded systems and their voltage relationships. The term "solidly grounded" and "direct grounded" have the same meaning, that is, no intentional impedance is inserted in the neutral-to-ground connection.

8.2.1 *Classification of System Grounding.* The types of system grounding normally used in industrial and commercial power systems are

(1) Solid grounding
(2) Low-resistance grounding
(3) High-resistance grounding
(4) Ungrounded

Each type of grounding has advantages and disadvantages, and there is no general acceptance of any one method. Factors which influence the choice include

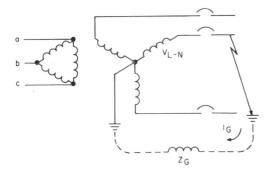

$$I_G = \frac{3V_{L-N}}{Z_1 + Z_2 + Z_0 + 3Z_G}$$

$$\approx \frac{V_{L-N}}{Z_1 + Z_G}$$

$$I_G(\max) \approx \frac{V_{L-N}}{Z_1} \approx I_{3\phi}\text{ fault}$$

$I_G(\min)$ can be very low

Fig 94
Direct or Solid Grounding;
Use Ground Relays to Trip

(1) Voltage level of power system
(2) Transient overvoltage possibilities
(3) Type of equipment on the system
(4) Required continuity of service
(5) Caliber and training of operating and maintenance personnel
(6) Methods used on existing systems
(7) Availability of convenient grounding point
(8) National Electrical Code
(9) Cost of equipment, including protective devices and maintenance
(10) Safety, including fire and shock hazard
(11) Tolerable fault damage levels
(12) Effect of voltage dips during faults

There are many factors involved in selecting grounding methods for the different voltage levels found in power distribution systems. IEEE Std 142-1972 discusses many of these factors in detail while the following discussion mentions only the reasons which relate to ground-fault protection.

8.2.2 Solid Grounding (Fig 94). Most industrial and commercial power systems are supplied from electric utility systems which are solidly grounded. If the user must immediately convert to lower voltage, the power transformers typically have a delta-connected primary and a wye-connected secondary, which can again be connected solidly to ground. This results in a system which can be conveniently protected against overvoltages and ground faults. The system has flexibility since the neutral can be carried with the phase conductors which permits connecting loads from phase to phase and from phase to neutral.

(1) *Systems above 600 V.* Ground relaying of medium-voltage and high-voltage systems which are solidly grounded has been successfully accomplished for many years using residually connected ground relays. The circuit breakers normally have current transformers to provide the signal for the phase overcurrent relays, and the ground overcurrent relay is connected in the wye point ("residual") to provide increased sensitivity for ground faults. Ground-fault magnitudes usually are comparable to phase-fault magnitudes and are therefore easily detected, unless they occur in equipment windings near the neutral point.

The disadvantages include the fact that ground-fault currents are often of high magnitude, and thus are very destructive unless interrupted within a few cycles. For this reason most industrial applications use a grounding resistor in the connection from neutral to ground to limit the ground-fault current to a few hundred amperes. This results in a low-resistance grounded system as shown in Fig 95.

Another disadvantage is the fact that a phase-to-ground fault must be interrupted immediately, thus de-energizing the affected circuit. This forced outage is unacceptable in some applications where continuity of service is very important. The alternative is to use the high-resistance or ungrounded system shown in Figs 96 and 97.

(2) *Systems 600 V and below.* All 208Y/120 V systems are solidly grounded so that loads can be connected from line to neutral to provide 120 V service. Similarly, all 480Y/277 V systems which are to serve 277 V lighting must also be solidly grounded. This results in most commercial building and many industrial plant 480 V systems being solidly grounded, as the lighting loads and motor loads can be served from the same system. Even where 277 V lighting is not used, many industrial plant 480 V systems are solidly grounded to limit overvoltages and to facilitate clearing ground faults.

While higher voltage systems normally use relays, and the ground relay with increased sensitivity is easily provided, low-voltage systems usually use circuit breakers with integrally mounted trip devices in the phases. Until recently it was not felt that the additional cost of supplementary ground-fault relaying was justified. However, it is now recognized that even solidly grounded low-voltage systems can experience ground faults with a relatively low fault current level for the reasons explained in Section 8.3. Because of this, sensitive ground-fault relays and trip devices have been developed for use with low-voltage circuit breakers for proper protection of solidly grounded low-voltage systems.

One disadvantage of the solidly grounded 480 V system involves the high magnitude of ground-fault currents which can occur, and the destructive nature of arcing ground faults. However, if these are promptly interrupted, the damage is kept to acceptable levels as explained in Section 8.3. (While it is possible to resistance-ground low-voltage systems, resistance grounding restricts the use of line-to-neutral loads. Some systems which do not need the neutral voltage for lighting loads are resistance-grounded, but this is always considered high-resistance grounding when applied to low-voltage systems. Low-resistance grounding is not used as the extra cost of the resistor is not justified by the advantages.)

Another disadvantage of solidly grounded 480 V systems is that the ground fault will cause immediate forced outages. If this cannot be tolerated, then either the high-resistance grounded system (without ground-fault tripping) or ungrounded systems are used to delay the required outage for repairs.

8.2.3 *Low-Resistance Grounding* (Fig 95). The low-resistance grounded system is similar to the solidly grounded system in that transient overvoltages are not a problem. The resistor limits ground-fault-current magnitudes to reduce the rate of damage during ground faults. The magnitude of the grounding resistance is selected to allow sufficient current for ground-fault relays to detect and clear the faulted circuit. This type of grounding is used mainly in the 2.4–13.8 kV systems which often have motors directly connected.

The value of resistance depends on the type of relaying and the amount of motor windings which can be protected. Ground faults in wye-connected motors have reduced driving voltage as the neutral of the motor winding is approached; thus ground-fault-current magnitudes are reduced.

8.2.4 *High-Resistance Grounding* (Fig 96). High-resistance grounding limits ground-fault currents to very small values. The fault-current magnitude is predictable for practically all fault locations, since the grounding resistance value inserted in the neutral is very large compared to the remainder of the ground-fault path impedance. This insures essentially a fixed ground-fault current of sufficient magnitude for relaying purposes and control of overvoltages, but low enough so that immediate tripping is not always necessary, particularly on low-voltage circuits. Thus a load process need not be interrupted, but allowed to operate with the ground fault until a more favorable moment for circuit outage arrives. Sensitive relaying is required with high-resistance grounding because of the low ground-fault currents involved. High-resistance grounding schemes can be used at any voltage.

It is important to allow sufficient ground-fault current to flow to compensate for the capacitive charging current of the system which otherwise could trigger transient overvoltages during switching or other changing circuit conditions. This current should not be less than the measured or calculated capacitive charging current of the existing and projected system.

8.2.5 *Ungrounded Systems* (Fig 97). Ungrounded systems employ ground detectors to indicate a ground fault. The system will operate with the ground fault acting as the system ground point. The ground-fault current which flows is

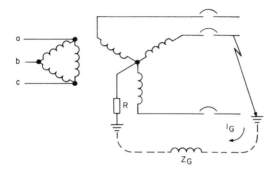

$$I_G \approx \frac{V_{L-N}}{R}$$

Select I_G = 600 A; then

$$R = \frac{2400}{600} = 4\ \Omega$$

for a 4160 V system.

**Fig 95
Low-Resistance Grounding;
Use Ground Relays to Trip**

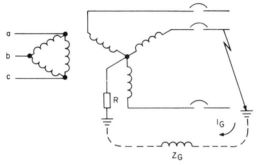

$$I_G \approx \frac{V_{L-N}}{R}$$

Select $I_G \approx$ 5—10 A; then

$$R = \frac{277}{I_G} \approx 28\text{—}56\ \Omega$$

for a 480 V system.

**Fig 96
High-Resistance Grounding;
Use Ground Relays to Trip or Alarm**

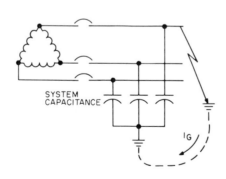

I_G is small; calculate or measure.
Approximate values:
 480V system $I_G < 1.5A$
 4160V system $I_G < 5A$

**Fig 97
Ungrounded System;
Use Bus Ground Detector to Alarm
(Watch for Possible Overvoltage Problem)**

the capacitive charging current of the system. By continuing to operate with this ground fault, the system is subjected to possible overvoltages which can result in multiple ground faults, several circuit outages, and coincident damaged equipment. These overvoltages are caused by resonance between the capacitance to ground and the varying inductances of magnetic elements in the system. An ungrounded system experiencing overvoltage problems is often converted to a high-resistance grounded system by addition of neutral resistance. If the system neutral is not available, it can be formed by utilizing grounding transformers. The selection of suitable grounding transformers and neutral resistance is covered in IEEE Std 142-1972.

Since the ground-fault current is very low, it is not practical to provide ground-fault relays of sufficient sensitivity to indicate the faulted circuit. Thus the faulted equipment can be very difficult to locate. Often one feeder at a time must be opened until the ground-detection lights reveal that the faulted feeder has been opened.

8.3 Ground Faults — Nature, Magnitudes, and Damage.

Ground faults on electric systems can (1) originate in many ways, (2) have a wide range of magnitudes, and (3) cause varying amounts of damage. The most serious faults from the standpoint of rate of eroded material are arcing faults, both phase to phase and phase to ground.

8.3.1 *Origin of Ground Faults.* Ground faults originate from insulation breakdown from causes which can be classified roughly as follows:

(1) Reduced insulation due to moisture, atmospheric contamination, foreign objects, insulation deterioration, etc

(2) Physical damage to insulation system due to mechanical stresses, insulation punctures, etc

(3) Excessive transient or steady-state voltage stresses on insulation

Good installation and maintenance practices ensuring adequate connections and the integrity of the insulation of the equipment have a significant effect in reducing the probability of ground faults. However, insulation breakdowns to ground can occur at any point in the system where the exposed phase conductors are in close proximity to a grounded reference. The contact

between the phase conductor and ground is usually not a firm metallic contact, but rather usually includes an arcing path in air or across an insulating surface, or a combination of both. In addition to these arcing ground faults certain bolted-type faults occur, usually during installation or maintenance when there is an inadvertent firm metallic connection from phase to ground.

8.3.2 *Magnitude of Ground-Fault Currents.* Ground-fault-current magnitudes can vary greatly. Using the method of symmetrical components (see Chapter 2), the single line-to-ground fault current I_{GF} is calculated by the formula

$$I_{GF} = \frac{3V_{L-N}}{Z_1 + Z_2 + Z_0 + 3Z_G}$$

where Z_1 and Z_2 are the positive and negative impedances, and $Z_0 + 3Z_G$ is the zero-sequence impedance. The term Z_G is the sum of the impedances of the fault arc, the grounding circuit, and the intentional neutral impedance, when present.

To illustrate how ground-fault currents can vary greatly in magnitude, consider a solidly grounded system with a bolted ground fault very close to the generator terminals. In this example Z_G could approach zero, and if we assume that $Z_1 = Z_2 = Z_0$, then

$$I_{GF} = \frac{V_{L-N}}{Z_1}$$

which is actually the formula for a bolted three-phase fault. In fact, with many generators, since Z_0 is smaller than Z_1, it is necessary to add an intentional neutral impedance Z_N to reduce the bolted ground-fault current to the magnitude of the bolted three-phase fault current.

For a ground fault in a high-resistance grounded system, the neutral resistance R_N is usually very large compared to Z_1, Z_2, Z_0, and the remainder of Z_G. Then I_{GF} is approximately equal to E_{L-N} divided by R_N.

In a high-resistance grounded 480 V system with a neutral resistance of 20 Ω the ground-fault current will be

$$I_{GF} = \frac{480/\sqrt{3}}{20} = 14 \text{ A}$$

This is true, since the fault-arc impedances and the ground-circuit impedances are negligible when compared to 20 Ω.

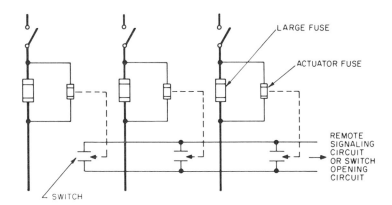

Fig 98
Anti-Single-Phasing Provisions for
Fusible Switches

Precise calculation of ground-fault-current magnitudes in solidly grounded systems is much more difficult than the previous example. The reason for this is that the circuit impedances including the fault arc impedance which were negligible in the high-resistance example play an important part in reducing ground-fault-current magnitudes. This applies even in most cases where a sizable grounding conductor is carried along with phase conductors. Fortunately, however, it is seldom necessary to make exact fault calculations to apply ground-fault protection as most ground-fault relays cause tripping for all ground-fault values above a minimum operating current.

The two main setting characteristics which need to be determined for ground-fault relays are (1) minimum operating current, and (2) speed of operation. Selection of the minimum operating current (pickup) setting is based primarily on the characteristics of the circuit being protected. If the circuit serves an individual load such as a motor, transformer, or a heater circuit, then the pickup setting can be quite low, such as 5—10 A. If the protected circuit feeds multiple loads, each with individual overcurrent protection, for example, a panelboard, feeder duct, motor control center, etc, the pickup settings will be higher. These higher settings in the order of 200—1200 A are selected to allow the branch phase overcurrent devices to clear low-magnitude ground faults in their respective circuits.

8.3.3 *Damage Due to Arcing Faults.* The arcing fault causes a large amount of energy to be released in the arcing area. The ionized products of the arc spread rapidly. Vaporization at both arc terminals occurs, and the erosion at the electrodes is concentrated when the arc does not travel. While there is a tendency of the arc to travel away from the source, this does not necessarily occur at low levels of fault current or at higher levels of current in circuits with insulated conductors. An arcing fault, if allowed to persist, (1) is dangerous to personnel, (2) is a serious fire hazard, (3) causes considerable damage, (4) often results in an extended power outage, and (5) can create transient overvoltages.

Single-pole interrupters such as fuses without anti-single-phasing provisions are especially ineffective against arcing faults [2]. The reason for this is that often an initial ground fault or phase-to-phase fault can cause one or two fuses to clear faulted phases, but the fault continues to be fed through the load impedances from the uncleared phases. The fault is thus of a diminished magnitude and may never be cleared by the remaining fuses until other circuits are involved. This sequence has caused extended outages of many physically adjacent circuits. In many applications, the probability of extended damage can be reduced by careful design including compatible ground-fault protection.

Fusible switches can be equipped with anti-single-phasing provisions by installing small actuator-type fuses in parallel with the large fuses in the switch. These fuses will close a contact to actuate a signal or switch-opening circuit to open all three poles of the switch. Fig 98 shows this particular scheme, which can also be used in conjunction with ground-fault protective relaying.

Fused circuit breakers and service protectors have anti-single-phasing devices incorporated in their basic designs.

The basic need for ground-fault protection in low-voltage grounded systems is illustrated in Fig 99. Shown is a 1000 kVA transformer with a 1600 A main circuit breaker with typical long-time and short-time characteristics and optionally with a fused switch.

A 1500 A ground fault (point I) on the 480Y/277 V grounded neutral system would not be detected by the main circuit breaker or fuse. A ground relay set at 0.2 s time delay would cause the circuit breaker to clear the fault in about 0.33 s. A 4000 A ground fault (point II) could persist for about 33 s, even if the circuit breaker minimum long-time band were used. The fuse would require up to 5 min to clear this fault. The ground relayed circuit breaker would clear the fault in about 0.25 s. An 8000 A ground fault (point III) would be cleared within about 0.2–0.4 s by the circuit breaker short-time device, assuming it is present; otherwise between 8–20 s would elapse before the long-time device clears the fault.

Arc energies for these assumed faults are tabulated in Table 28. Arc voltages are assumed to be 100 V. Since the arc voltage tends to have a flat top characteristic (nonlinear arc resistance), the energy of the arc in watt-seconds can be estimated by obtaining the product of the current in rms amperes, the arc voltage in volts, and the clearing time in seconds. Approximate calculation of the energy required to erode a certain amount of electrode material shows that 50 kW·s of energy divided equally between conductor and enclosure will vaporize about $1/8$ in^3 of aluminum or $1/20$ in^3 of copper. The calculation assumes that most of the arc energy goes into the electrodes while the energy lost to the surrounding air is neglected. Comparisons were made from several arcing fault tests [3],[4], and good correlation was obtained between calculated energy from test data and measured conductor material eroded.

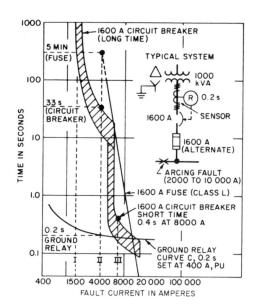

Fig 99
Time—Current Plot Showing Slow
Protection Provided by Phase Devices
for Low-Magnitude Arcing Ground Faults

For the assumed 8000 A fault, even though the current values are the calculated result using all source, circuit, and arc impedances, the actual rms current values passing through the circuit breaker can be considerably lower. This is because of the spasmodic nature of the fault caused by (1) arc elongating blow-out effects, (2) physical flexing of cables and some bus structures due to mechanical stresses, (3) self-clearing attempts and arc reignition, and (4) shifting of the arc terminals from point to point on the grounded enclosures (as well as on the faulted conductors for noninsulated construction). All these effects tend to reduce the rms value of arcing fault currents. Therefore, a ground fault which would normally produce 8000 A under stabilized conditions might well result in an effective value of only 4000 A, and would have the arc energies associated with point II in Table 28.

8.3.4 *Selection of Protective Device Settings.* Maximum protection against ground faults can be obtained by applying ground protection on every feeder circuit from source to load. The

Table 28
Arc Energies for Assumed Faults of Fig 99

Fig 99 Points	Fault (A rms)	Main Device	Clearing Time (seconds)	Arc Energy (kW · s)
I	1500	Relay	0.33	50
		Circuit breaker	∞	∞
		Fuse	∞	∞
II	4000	Relay	0.25	100
		Circuit breaker	33	13 200
		Fuse	300	120 000
III	8000	Relay	0.25	200
		Circuit breaker	0.4	320
		Fuse	10	8000

minimum operating current for all series devices is set at about the same pickup setting, but the time curves are selected so that each circuit protective device is opened progressively faster, moving from the source to the load. The load switching device can be opened instantaneously upon occurrence of a ground fault.

The delay required between devices is determined by the addition of (1) the trip operating time of the feeder circuit breaker, (2) the arcing time of the feeder circuit breaker, and (3) a margin of safety. The trip operating time of modern circuit breakers, molded-case type, service protectors, large air-type, and power circuit breakers used in indoor and in-plant distribution, is usually two cycles or less. Shunt tripped switches may take somewhat longer.

Recent developments in bolted-pressure fused switch design has made possible the use of most of the ground-fault protection schemes presently available. Now available are stored energy switching mechanisms which can be shunt tripped to cause the switch to open very quickly. In order to avoid the situation that occurs when the switch might open on ground faults above its nominal interrupting rating but before the fuses can operate, Underwriters Laboratories, Inc. (UL) approved switches must incorporate one of two features: (1) a lock-out feature activated by a fault detector on each phase which allows the fuses to clear on faults above the interrupting rating of the switch (usually 6—7.5 times switch rating), or (2) the switch interrupting rating must

equal 12 times the switch continuous current rating.

Either of these two features will qualify the bolted-pressure fused switch to be used with ground-fault protection provided either the withstand rating [feature (1)] or the interrupting rating [feature (2)] of the fused switch is above the available short-circuit current at the point of application.

From the standpoint of damage alone, speed of clearing is paramount. However, there are situations where some delay is desirable. This is primarily to obtain coordination between main and feeder circuits and branch currents. Consider a typical 480Y/277 V application consisting of a 3000 A main, an 800 A feeder, and a 100 A branch. If the branch circuits do not have ground-fault protection, then the feeder ground-fault protection must be set with a time delay to allow the branch circuit instantaneous units to clear moderately high-magnitude ground-fault currents without tripping the feeder circuit breaker. When full coordination is essential, it is desirable to set the feeder ground-fault pickup equal to the instantaneous setting of the branch circuits. While infrequent loss of coordination is often acceptable between feeders and branch circuits, it is recommended that full coordination be maintained between main and feeder circuit breakers.

Another reason for not clearing ground faults instantaneously on main or large feeder circuits is the threat of a trip where the power outage itself is of greater consequence than a slight amount of damage. In long circuits it is sometimes difficult to locate ground faults cleared instantaneously, that is, within a few cycles. In the absence of fault-locating equipment, the operator may choose to reclose on the fault a few times until the fault location is evident. This action is likely to create a higher magnitude fault. Slightly slower clearing of the original fault is preferable to this procedure. Also, the operator, by habit, may make the error of closing into a high-magnitude fault with disastrous results. The foregoing does not apply to outdoor overhead circuits where faults can be transient and fault reclosing has a long and successful record. Also, it is very important to always use as low a setting as possible for the phase instantaneous devices in order to assure that all high-magnitude faults are cleared quickly.

In summary, the sensitivity (minimum operating current setting) of ground-fault protection in solidly grounded low-voltage systems is determined by the following considerations.

(1) When the ground-fault protection is used on devices protecting individual loads such as motors, instantaneous devices with the lowest available settings can be used providing the devices will not cause false tripping from inrush currents.

(2) For mains and feeders, the setting for ground-fault protective devices is normally in the range of 10 to 100 percent of the circuit trip rating or fuse rating. If downstream devices do not have sensitive ground-fault protection, then the circuit ground-fault protection may have to be set higher than the downstream phase protective device instantaneous setting to ensure full coordination.

For full protection the setting should be somewhat lower than the minimum estimated ground-fault current in the zone of protection for which the circuit protective device in responsible.

8.4 Frequently Used Ground-Fault Protective Schemes. While ground-fault protective schemes may be elaborately developed, depending on the ingenuity of the relaying engineer, nearly all schemes in common practice fall into one of these broad classes: (1) residually connected overcurrent relays, (2) core balance (window)

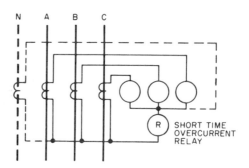

Fig 100
Residually Connected Ground Relay

current transformer magnetically enclosing the neutral along with the phase conductors, (3) differential relaying with a current transformer connected into the transformer neutral-to-ground connection, and (4) detection of ground-return current in the equipment grounding circuit.

8.4.1 *Residual Connection.* A residually connected ground relay is widely used to protect medium-voltage systems. The actual ground current is measured by current transformers which are interconnected in such a way that the ground relay responds to a current proportional to the ground-fault current. This scheme, using individual relays and current transformers, is not often applied to low-voltage systems. However, there are available low-voltage circuit breakers with three current transformers built into them and connected residually with the solid-state trip devices of the circuit breakers to provide ground-fault protection.

The basic residual scheme is shown in Fig 100. Each phase relay is connected in the output circuit of its respective current transformer while a ground relay connected in the common or residual circuit will measure the ground-fault current. In three-phase three-wire systems such as shown in Fig 100 no current flows in the residual leg under normal conditions, since the net effect of the three current transformers is zero. This is true for phase-to-phase short circuits also. When a ground fault occurs, current bypasses the phase conductors and their current transformers, the net flux is not zero, and current flows in the residual leg, operating its relay.

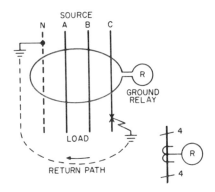

Fig 101
Core-Balance Current Transformer
Encircles All Phase and Neutral Conductors

Residually connected relays cannot have sensitive settings because of unequal saturation of the current transformers. If sensitive ground-fault protection is needed, use the core-balance method, Section 8.4.2.

On four-wire circuits a fourth current transformer should be connected in the neutral circuit as shown. The neutral conductor carries both 60 Hz single-phase load unbalance current as well as 180 Hz harmonic currents caused by the nonlinear inductance of single-phase loads such as fluorescent lighting. Without the neutral-conductor current transformer the current in that conductor will appear to the ground relay as ground-fault current, and the ground relay would have to be desensitized sufficiently to prevent tripping under load conditions.

8.4.2 *Core Balance.* The core-balance current transformer is the basis of several low-voltage ground-fault protective systems introduced in recent years. (The core-balance current transformer is frequently called a "window" current transformer, but the term "core balance" is preferable since it more specifically describes the function of the current transformer.) The principle of the core-balance current transformer circuit is shown in Fig 101. The phase and neutral conductors all pass through the same opening in the current transformer and are surrounded by the same magnetic core. Core-balance current transformers are available in several convenient shapes and sizes, including rectangular designs for use over bus bars.

Under normal conditions, that is, balanced, unbalanced, or single-phase load currents or short circuits not involving ground, all current flows out and returns through the current transformer. The net flux produced in the current transformer core will be zero and no current will flow in the ground relay. When a ground fault occurs, the ground-fault current returns through the equipment grounding circuit conductor (and possibly other ground paths) bypassing the current transformer. The flux produced in the current transformer core is proportional to the ground-fault current, and a proportional current flows in the relay circuit. Relays connected to core-balance current transformers can be made quite sensitive, detecting even currents of milliamperes. However, care is necessary to prevent false tripping during normal inrush conditions or through faults not involving ground.

By properly matching the current transformer and relay, ground-fault detection can be made as sensitive as the application requires. The relays are fast to limit damage and may be adjustable (for current, time, or both) in order to obtain selectivity. Many ground protective systems now have solid-state relays specially designed to operate with core-balance current transformers. The relays in turn trip the circuit protective device. Power circuit breakers, molded-case circuit breakers with shunt trips, or electrically operated fused switchgear can be used. The latter includes service protectors, which use circuit breaker contacts and mechanism but depend on current-limiting fuses to cope with high available short-circuit currents. Fused contactors and combination motor starters may be used where the device interrupting capability equals or exceeds the available ground-fault current.

8.4.3 *Ground Differential.* Ground differential relaying is effective for main bus protection since it has inherent selectivity. With the differential scheme (Fig 102), core-balance current transformers are installed on each of the outgoing feeders and another smaller current transformer is placed in the transformer neutral connection to ground. This arrangement can be made sensitive to low ground-fault currents without incurring tripping for ground faults beyond the feeder current transformers. All current transformers must be very carefully matched to prevent improper tripping for

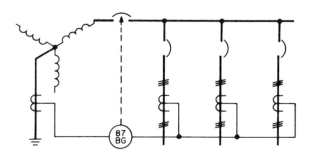

Fig 102
Ground Differential Scheme

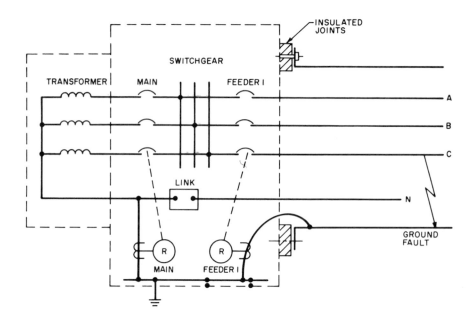

Fig 103
Ground-Return Relay Scheme

high-magnitude faults occurring outside the differential zone.

Bus differential protection protects only the zone between current transformers and does not provide backup protection against feeder faults.

8.4.4 *Ground Return.* Ground return relaying is illustrated in Fig 103. The ground-fault current returns through the current transformer in the neutral-bus to ground-bus connection. For feeder circuits an insulating segment may be intro-duced in busway or conduit, as shown in Fig 103, and a bonding jumper connected across the insu-lator to carry the ground-fault current. A current transformer enclosing this jumper will then de-tect a ground fault. This method is not recom-mended for feeder circuits due to the likelihood of multiple ground current return paths and the difficulty of maintaining an insulated joint.

8.5 Typical Applications. The application of ground-fault protection to typical power distribution systems is illustrated by Figs 104—114 for the following types of power distribution systems:

The one-line diagrams show the locations for the ground-fault sensing devices as well as the locations for the protective devices. Additional considerations in the application of ground-fault protection follow.

(1) A common economy in system design is to use the simple radial system without transformer secondary main circuit breakers. This results in a particular hazard when low-magnitude ground faults are considered, as there is no protective device to open should a ground fault occur between the transformer secondary winding and the feeder circuit breaker. Some systems are designed so that a transformer primary protective device can be opened, which is ideal. However, this is not practical in many systems, and the use of secondary mains should be considered.

(2) Even with a secondary main circuit breaker, the zone from the transformer secondary to the main circuit breaker is not protected. Even though the ground-fault sensing device may detect the fault and trip the secondary main, the fault is not removed. Thus to minimize the possibility of trouble in this zone, it should be kept small by locating the main circuit breaker as close to the transformer as possible, and it should be designed with extra care to reduce the possibility of faults.

(3) The use of sensitive ground-fault protection makes the coordination of protective devices extremely important. The first consideration is where to apply the ground-fault protection.

In any evaluation of whether or not to apply certain protective devices, one eventually arrives at a comparison of cost of application versus probable consequence of omission. In considering the application of ground-fault protection, there are several alternatives, each varying in cost of application. This discussion will consider two basic approaches, (1) ground-fault protection on the mains only, and (2) ground-fault protection on mains and feeders. In this short treatment, protection versus coordination will be explored in only the most fundamental manner.

8.5.1 *Ground-Fault Protection on Mains Only.* An example of this approach is shown in Fig 104. Here there is a 3000 A main with long-time and short-time trip devices, a 1200 A feeder with long-time and instantaneous trip devices, and a molded-case circuit breaker in a branch circuit with thermal and instantaneous trip devices. The ground-fault protection on the main will coordinate with both instantaneous trip devices if given about 0.2 s time delay with a relatively flat characteristic.

The problem arises, where do we set the minimum ground pickup? For full coordination with all feeders, the setting would have to be above 6000 A (above the instantaneous setting of the largest feeder). Obviously, this is too high. For excellent protection of circuit ground faults, the pickup setting should be about 200 A. This, however, produces loss of coordination for ground faults at A of magnitude between 200 A and 1000 A and loss of coordination for faults at B of magnitude between 200 A and 6000 A. Thus, while the 200 A setting on the main will provide excellent arcing fault protection, we can expect the main circuit breaker to trip for certain feeder faults where heretofore we were accustomed to having them handled by the feeder or branch circuit breakers. In short, we have lost a rather substantial degree of coordination. In some applications, this loss of coordination can be tolerated.

Under the circumstances, the best setting is approximately a 1200 A minimum pickup. Here

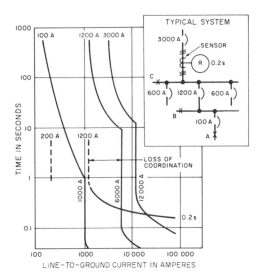

Fig 104
Ground Relays on Main Circuit Only

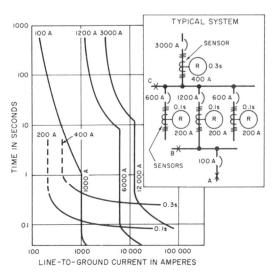

Fig 105
Ground Relays in Both Main and Feeder Circuits

we have protection against the most severe arcing faults and we have only lost coordination on faults between 1200 A and 6000 A. The above scheme is fairly common, but it is still clearly a compromise, which should be noted.

8.5.2 *Ground-Fault Protection on Mains and Feeders.* An example of this approach is shown by Fig 105. Here we have included ground-fault protection on the 3000 A main and also on all feeders above roughly 400—800 A. This application shows a 200 A minimum pickup with a time delay of 0.1 s on each feeder in addition to a 400 A minimum pickup and a 0.3 s time delay on the main.

In this example the main circuit breaker is fully coordinated with each feeder circuit breaker. Also, both main and feeders have sufficiently low settings to provide excellent arcing fault protection. There is some loss of coordination between the feeder and branch devices, but this is felt to be acceptable in most applications.

8.5.3 *Ground-Fault Protection, Mains Only — Fused System.* Fig 106 shows a situation similar to that of Fig 104, involving a fused system with 1200 A ground pickup. This setting will coordinate with the 200 A fuses and branch circuit lighting circuit breakers. However, coordination of the 1200 A pickup with the 800 A feeder fuses is sacrificed.

8.6 Special Applications

8.6.1 *Ground-Fault Detecting Schemes.* The conventional method of ground-fault detection used on ungrounded wye or delta three-wire systems utilizes three-potential transformers supplying (1) ground detector lamps, (2) ground detecting voltmeters, or (3) ground alarm relays. However, it should be noted that the presence of potential transformers connected to ground from each phase may in itself be the cause of dangerous overvoltages.

To reduce the probability of transient overvoltages, high-resistance grounding is often applied as described in detail in IEEE Std 142-1972. When high-resistance grounding is used, the basic reason is to eliminate a trip operation when the first ground fault occurs. Operation with a ground fault on the system entails a substantial hazard and it becomes important to (1) locate the fault as soon as system operation allows, and (2) provide additional ground-fault protection against a second ground fault until the first one is corrected.

In applying high-resistance grounding, the resistance of the ground circuit should be of a magnitude to pass a ground current at least equal to the charging current of the system. Fig 115 shows a typical high-resistance grounded system with a neutral resistor and a ground alarm relay.

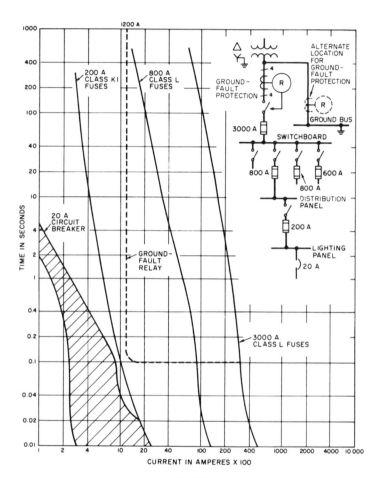

**Fig 106
Fused System Using Ground-Fault
Relays on Main Circuit Only Shows
Coordination with 200 A Fuses and Lack
of Coordination with 800 A Feeder Fuses**

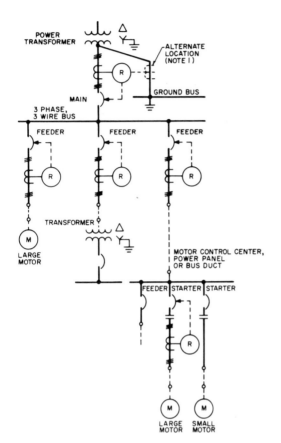

NOTES: (1) The ground-fault sensing device for the main circuit breaker can be in either location shown. When the transformer is remote from the switchgear, the connection to the neutral is not always available.

(2) If a main circuit breaker is not used, then it is desirable to sense ground faults as shown and trip a transformer primary circuit breaker.

Fig 107
Three-Wire Solidly
Grounded System, Single Supply.
Ground Relays Must Have Time Coordination

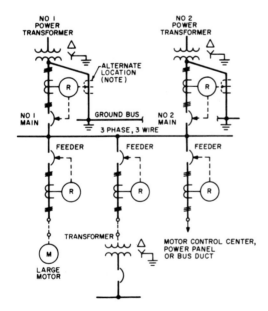

NOTE: The ground-fault sensing device for the main circuit breaker can be in either location shown. When the transformer is remote from the switchgear, the connection to the neutral is not always available.

Fig 108
Three-Wire Solidly
Grounded System, Dual Supply.
Ground Relays Must Have Time Coordination

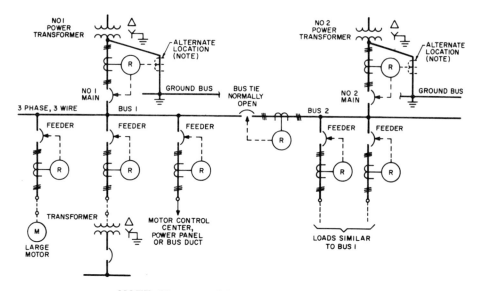

NOTE: The ground-fault sensing device for the main circuit breaker can be in either location shown. When the transformer is remote from the switchgear, the connection to the neutral is not always available.

Fig 109
Three-Wire Solidly
Grounded Secondary Selective System.
Ground Relays Must Have Time Coordination

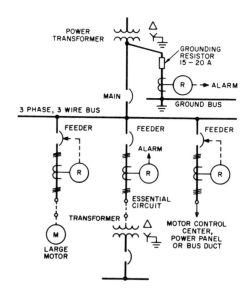

Fig 110
Three-Wire High-Resistance
Grounded System. Ground Relays Must
Have Time Coordination

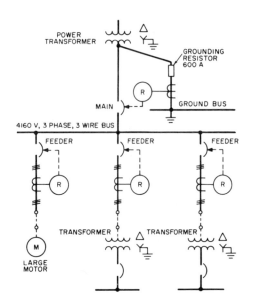

Fig 111
Three-Wire Low-Resistance
Grounded System. Ground Relays Must
Have Time Coordination

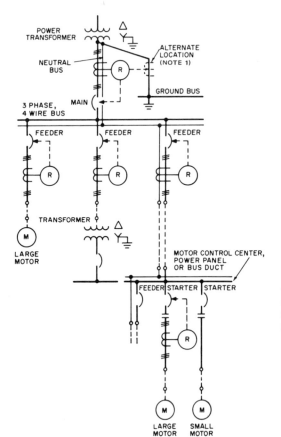

NOTES: (1) The ground-fault sensing device for the main circuit breaker can be in either location shown. When the transformer is remote from the switchgear, the connection to the neutral is not always available.

(2) If a main circuit breaker is not used, then it is desirable to sense ground faults as shown and trip a transformer primary circuit breaker.

Fig 112
Four-Wire Solidly
Grounded System, Single Supply.
Ground Relays Must Have Time Coordination

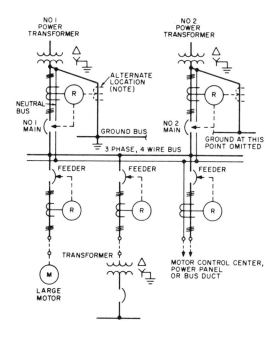

NOTE: The ground-fault sensing device for the main circuit breaker can be in either location shown. When the transformer is remote from the switchgear, the connection to the neutral is not always available.

Fig 113
Four-Wire Solidly
Grounded System, Dual Supply.
Ground Relays Must Have Time Coordination

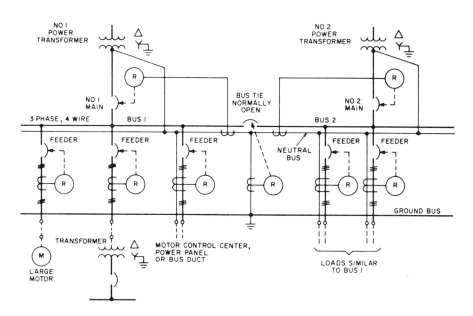

NOTES: (1) Neutral to ground connection must be made as shown. Tap each feeder neutral on proper side of this T point.

(2) Special interlocks must be added to block main ground relays until the tie circuit breaker opens when one transformer only is in service. Consult manufacturer for details.

Fig 114
Four-Wire Solidly Grounded Secondary Selective System. Ground Relays for Feeders, Tie, and Mains Must Have Time Coordination

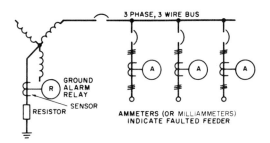

Fig 115
High-Resistance Grounded
System Using Core-Balance Current
Transformers and Fault-Indicating Ammeters

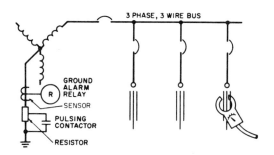

Fig 116
High-Resistance Grounded System with
Pulsing Contactor to Locate Faulted Feeder

Each feeder is equipped with a low-ratio core-balance transformer and ammeter or milliammeter, which will indicate the faulted feeder. No automatic tripping occurs with this scheme.

In lieu of current transformers and ammeters on each feeder, it is possible to arrange a pulsing ground circuit as shown in Fig 116. The pulsing circuit is manually initiated and serves to reduce the resistance to about 50—75 percent of its full value, about once or twice a second. This causes the ground-fault current to vary sufficiently to be detected by a clamp-on ammeter which can be placed in turn around the conductors of each feeder circuit.

If a ground fault is not cleared immediately, a relatively dangerous condition may arise upon the occurrence of a second fault. A second fault has a high probability of occurring because (1) the steady-state voltage on the unfaulted phases has increased, and (2) the initial fault may be intermittent which will cause some transient overvoltages in spite of the resistor grounding. Fig 117 shows how feeder ground relays are applied to trip on the occurrence of the second fault on a different feeder. The feeder ground-relays are set to pick up at a value higher than the maximum initial ground fault current. For example, on a 4.16 kV system a 300 Ω resistor would limit the ground fault current to 8 A. The ground alarm relay would be set to pick up at 5 A or less. The feeder ground relays are set to pick up at a current level higher than that for a single line-to-ground fault, perhaps 10—15 A.

The use of low-voltage grounding resistors coupled with grounding (or standard distribution)

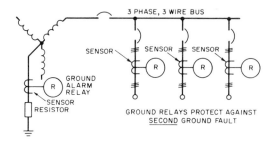

Fig 117
High-Resistance Grounded
System Continues Operation on
First Ground Fault. Ground Relays
Protect Against Second Ground Fault

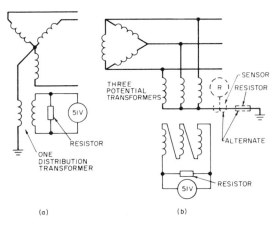

Fig 118
High-Resistance Grounding Using Distribution
or Grounding Transformers
(a) Wye-Connected System
(b) Delta-Connected System

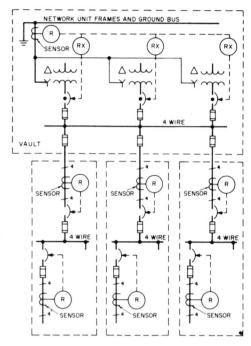

Fig 119
Ground-Fault Protection for
Typical Three-Transformer
Spot Network

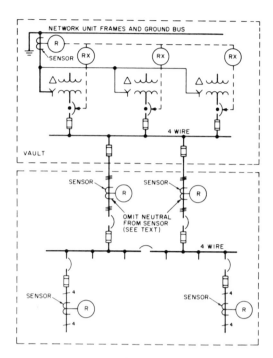

Fig 120
Ground-Fault Protection for
Spot Network when User Switchgear
Is Secondary Selective Type

transformers is preferred by some engineers. Fig 118 shows how these transformers are utilized for (a) three-wire grounded wye systems, and (b) three-wire grounded delta systems.

8.6.2 *Spot-Network Applications.* Spot networks provide continuity of service against the loss of one or more of several utility supply feeders. Each feeder supplies one network transformer at each vault. All transformer secondaries at each vault are parallel at the network service bus from which the user service switchgear is connected.

Fig 119 shows one method of providing ground-fault protection for a three-transformer spot network serving three physically separate service switchboards. Ground protection is provided (1) on each switchboard feeder, (2) in each switchboard main, and (3) in the fault system neutral which will trip out all network protectors in the event of a vault fault. In a typical installation the feeder ground relays are set at about 0.1

s, the main relays at about 0.3 s, and the vault relays at about 0.5 s. The minimum operating current settings for the vault relays are set substantially higher than the setting on the user switchgear so that the vault relays will operate only for a fault in the vault area. Though not shown in the figure, each service switchboard must include a connection from neutral bus to ground bus.

Where it is desired to provide secondary selective flexibility in the user service switchgear, the system sometimes takes the form shown in Fig 120. Here one must be careful in applying the ground sensors (current transformers) in the main circuits. If the ground sensors are installed over all phases and neutral, then a fault on either user bus will always cause tripping of both main circuit breakers. One method of circumventing this is to install the main sensors over the phases only. This will provide the proper selectivity, except that the ground relay minimum operating

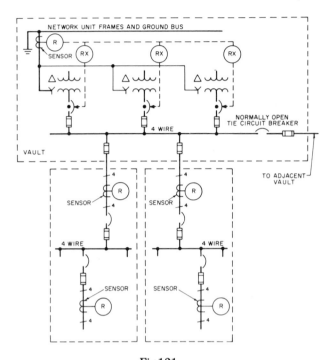

Fig 121
Ground-Fault Protection for
Sectionalized Vault Bus Spot Network

current setting must be set at a value above the normal neutral current. This normal neutral current will consist of the single phase-to-neutral load umbalance plus the third harmonic in the neutral caused by nonincandescent lighting systems. Settings of 1000—2000 A may be required. With this type of system it will be difficult to meet the present new 1200 A minimum setting of the NEC and still accommodate the secondary selective arrangement.

An alternate vault arrangement which has been used is shown in Fig 121. Here the vault is sectionalized in two halves and one of the advantages is that a vault fault does not shut off all electrical service to the user. Ground-fault protection is essentially identical to that shown in Fig 119. An alternate ground-fault protection scheme consisting of heat-sensing systems of the continuous and probe types is described in [5].

8.7 Standards References. The following standards publications were used as references in preparing this chapter.

IEEE Std 142-1972, Grounding of Industrial and Commercial Power Systems (ANSI C114.1-1973)

NFPA No 70, National Electrical Code (1975) (ANSI C1-1975)

8.8 References and Bibliography
8.8.1 *References*

[1] O'CONNOR, J.J. The Threat of Arcing Faults. *Power Magazine*, Jul 1969, p 48.

[2] KAUFMANN, R.H. Application Limitations of Single-Pole Interrupters in Polyphase Industrial and Commercial Building Power Systems. *IEEE Transactions on Applications and Industry*, vol 82, Nov 1963, pp 363-368.

[3] CONRAD, R.R., and DALASTA, D. A New Ground Fault Protective System for Electrical Distribution Circuits. *IEEE Transactions on Industry and General Applications*, vol IGA-3, May/Jun 1967, pp 217-227.

[4] FISHER, L.E. Resistance of Low-Voltage AC Arcs. *IEEE Transactions on Industry and General Applications*, vol IGA-6, Nov/Dec 1970, pp 607-616.

[5] PARVIN, D.G. Thirty-Three Story Office Building Has All-Electric Service. *Transmission and Distribution*, Aug 1969.

8.8.2 *Bibliography*

[6] BAILEY, B.G. Clearing Low-Voltage Ground Faults with Solid-State Trips. *IEEE Transactions on Industry and General Applications*, vol IGA-3, Jan/Feb 1967, pp 60-65.

[7] BAILEY, B.G., and HEILMANN, G.H. Clear Low Voltage Ground Faults with Solid State Trips. *Power Magazine*, Mar 1966, pp 69-71.

[8] BEEMAN, D.L., Ed. *Industrial Power Systems Handbook*. New York: McGraw-Hill, 1955.

[9] BISSON, A.J., and ROCHAU, E.A. Iron Conduit Impedance Effects in Ground Circuit Systems. *AIEE Transactions (Applications and Industry)*, pt II, vol 73, Jul 1954, pp 104-107.

[10] DALASTA, D. Ground Fault Protection for Low and Medium Voltage Distribution Circuits. *Proceedings of the American Power Conference*, vol 32, 1970, pp 885-895.

[11] EDMUNDS, W.H., SCHWEIZER, J.H., and GRAVES, R.C. Protection against Ground Fault Hazards in Industry, Hospital, and Home. *Conference Record, 1968 IEEE Industry and General Applications Group Annual Meeting*, IEEE 68C27-IGA, pp 863-874.

[12] FISHER, L.E. Arcing-Fault Relays for Low Voltage Systems. *IEEE Transactions on Applications and Industry*, vol 82, Nov 1963, pp 317-321.

[13] FISHER, L.E. Proper Grounding Can Improve Reliability in Low-Voltage Systems. *IEEE Transactions on Industry and General Applications*, vol IGA-5, Jul/Aug 1969, pp 374-379.

[14] FISHER, L.E., and SCHWIEGER, W.L. Tripping Network Protectors to Protect Service Entrance Conductors to Industrial and Commercial Buildings. *IEEE Transactions on Industry and General Applications*, vol IGA-5, Sept/Oct 1969, pp 536-539.

[15] FOX, F.K., GROTTS, H.J., and TIPTON, C.H. High-Resistance Grounding of 2400 V Delta Systems with Ground-Fault Alarm and Traceable Signal to Fault. *IEEE Transactions on Industry and General Applications*, vol IGA-1, Sept/Oct 1965, pp 366-372.

[16] GIENGER, J.A., DAVIDSON, O.C., and BRENDEL, R.W. Determination of Ground-Fault Current on Common A-C Grounded-Neutral Systems in Standard Steel or Aluminum Conduit. *AIEE Transactions (Applications and Industry)*, pt II, vol 79, May 1960, pp 84-90.

[17] GROSS, E.T.B. Sensitive Ground Protection for Transmission Lines and Distribution Feeders. *AIEE Transactions*, vol 60, 1941, pp 968-975.

[18] KAUFMANN, R.H. Some Fundamentals of Equipment-Grounding Circuit Design. *AIEE Transactions (Applications and Industry)*, pt II, vol 73, Nov 1954, pp 227-232.

[19] KAUFMANN, R.H. Let's Be More Specific about Grounding Equipment. *Proceedings of the American Power Conference*, vol 24, 1962, pp 913-922.

[20] KAUFMANN, R.H. Ignition and Spread of Arcing Faults. *Conference Record, 1969 IEEE Industrial and Commercial Power Systems and Electric Space Heating and Air Conditioning Joint Technical Conference*, IEEE 69C23-IGA, pp 70-72.

[21] KAUFMANN, R.H., and PAGE, J.C. Arcing-Fault Protection for Low-Voltage Power Distribution Systems — Nature of the Problem. *AIEE Transactions (Power Apparatus and Systems)*, pt III, vol 79, Jun 1960, pp 160-167.

[22] KNOBEL, L.V. A 480 V Ground Fault Protection System. *Electrical Construction and Maintenance*, Mar 1970, pp 110-112.

[23] KREIGER, C.H., LERA, A.P., and CRENSHAW, R.M. Ground Fault Protection for Low Voltage Systems. *Electrical Construction and Maintenance*, Jun 1967, pp 88-92.

[24] MACKENZIE, W.F. Impedance and Induced Voltage Measurements on Iron Conductors. *AIEE Transactions (Communication and Electronics)*, pt I, vol 74, Jan 1955, pp 577-581.

[25] PEACH, N. Protect Low Voltage Systems from Arcing Fault Damage. *Power Magazine*, Apr 1964.

[26] PEACH, N. Get Ground Fault Protection Now. *Power Magazine*, Apr 1968, pp 84-87.

[27] QUINN, G.C. New Devices End an Old Electrical Hazard — Ground Faults. *Modern Manufacturing Magazine*, Apr 1969.

[28] READ, E.C. Ground Fault Protection at Buick. *Industrial Power Systems*, vol 2, Jun 1968.

[29] SHIELDS, F.J. The Problem of Arcing Faults in Low-Voltage Power Distribution Systems. *IEEE Transactions on Industry and General Applications*, vol IGA-3, Jan/Feb 1967, pp 15-25.

[30] SOARES, E.C. Designing Safety into Electrical Distribution Systems. *Actual Specifying Engineer*, Aug 1968, p 38.

[31] VALVODA, F. Protecting against Arcing and Ground Faults. *Actual Specifying Engineer*, May 1967, p 92.

[32] WEDDENDORF, W.A. Evidence of Need for Improved Coordination and Protection of Industrial Power Systems. *IEEE Transactions on Industry and General Applications*, vol IGA-1, Nov/Dec 1965, pp 393-396.

9. Motor Protection

9.1 General Discussion. This chapter is intended to apply specifically to three-phase integral horsepower motors. There are many variables involved in choosing motor protection: motor importance, motor rating (from one to several thousand horsepower), type of motor controller, etc. Therefore it is recommended that protection for each specific motor installation be chosen to meet the requirements of the specific motor and its use. All the items in Sections 9.2 and 9.3 should be referred to as check lists when deciding upon the protection for a given motor installation. After the types of protection have been selected, manufacturers' bulletins should be studied to ensure proper application of the specific protection chosen.

9.2 Items to Consider in Protection of Motors

9.2.1 *Motor Characteristics.* These include type, speed, voltage, horsepower rating, service factor, power factor rating, type of motor enclosure, lubrication arrangement, arrangement of windings and their temperature limits, thermal capabilities of rotor and stator during starting, running and stall conditions, etc.

9.2.2 *Motor Starting Conditions.* Included are full voltage or reduced voltage, voltage drop and degree of inrush during starting, repetitive starts, frequency and total number of starts, and others.

9.2.3 *Ambient Conditions.* Temperature maxima and minima,elevation, adjacent heat sources, ventilation arrangement, exposure to water and chemicals, exposure to rodents and various weather and flood conditions, and others must be considered.

9.2.4 *Driven Equipment.* Characteristics will influence chances of locked rotor, failure to

reach normal speed, excessive heating during acceleration, overloading, stalling, etc.

9.2.5 *Power System.* Type of system grounding, exposure to lightning and switching surges, fault capacity, exposure to automatic reclosing or transfer, possibilities of single-phasing supply (broken conductor, open disconnect or circuit breaker pole, blown fuse), and other loads that can cause voltage unbalance must be considered.

9.2.6 *Motor Importance.* Motor cost, cost of unplanned down time, amount of maintenance and operating supervision to be provided to motor, ease and cost of repair, etc, have to be evaluated.

9.3 Types of Protection

9.3.1 *Undervoltage — Device 27*

(1) *Purpose.* The usual reasons for using undervoltage protection are

(a) To prevent possible safety hazard of motor automatic restarting when voltage returns following an interruption

(b) To avoid excessive inrush to the total motor load on the power system, and the corresponding voltage drop, following a voltage dip, or when voltage returns following an interruption

(2) *Instantaneous or Time Delay.* Undervoltage protection will be either instantaneous (no intentional delay) or of the time-delay type. Time-delay undervoltage protection should be used with motors important to production continuity of service, providing it is satisfactory in all respects, to avoid unnecessary tripping on voltage dips that accompany external short circuits. Examples of nonlatching starters where time-delay undervoltage protection is not satisfactory and

instantaneous undervoltage must be used are the following.

(a) Fused motor starters, having alternating-current voltage held contactors, used on systems of low three-phase fault capacity. With the usual time-delay undervoltage scheme the contactor could drop out on the low voltage accompanying the fault before the fuse blows. The contactor would then reclose into the fault. This problem does not exist if the fault capacity is high enough to blow the fuse before the contactor interrupts the fault current.

(b) Synchronous motors used with starters having alternating-current voltage held contactors. With the usual time-delay undervoltage scheme the contactor could drop out on an externally caused system voltage dip and then reclose reapplying the system voltage to an out-of-phase internal voltage in the motor. The high initial inrush could damage the motor winding, shaft, or foundation. This problem could also occur for large-horsepower high-speed squirrel-cage induction motors. It usually is not a problem with the 1500 hp and smaller induction motors with which voltage held contactor starters are used because the internal voltages of these motors decay quite rapidly.

NOTE: The foregoing two limitations could be overcome by using a capacitor held main contactor, or direct-current battery control on the contactor to prevent its instantaneous dropout. In other words, the time-delay undervoltage feature can be applied directly to the main contactor.

(c) Motors used on systems having fast automatic transfer or reclosing where the motor must be tripped to protect it before the transfer or reclosure takes place.

(d) When the total motor load having time-delay undervoltage protection will result in more inrush and voltage drop after an interruption than the system can satisfactorily cope with. The least important of the motors should have instantaneous undervoltage protection. Time-delay undervoltage protection of selectively chosen delays could be used on the motors whose inrush the system can handle.

(3) *With Latching Contactor or Circuit Breaker.* These motor switching devices inherently remain closed during periods of low or zero alternating-current voltage. The following methods are commonly used to trip (open) them:

(a) Energize shunt trip coil from a direct-current battery.

(b) Energize shunt trip coil from a separately generated reliable source of alternating current. This alternating-current source must be electrically isolated from the motor alternating-current source in order to be reliable.

(c) Energize shunt trip coil from a capacitor charged through a rectifier from the alternating-current system. This is commonly referred to as capacitor trip.

(d) De-energize a solenoid and allow a spring to be released to trip the contactor or circuit breaker. This is commonly referred to as a direct-acting trip scheme.

Items (a)—(c) are usually used in conjunction with voltage-sensing relays [see Section 9.3.1 (6)].

Item (d) could have the solenoid operating directly on the alternating-current system voltage. Alternatively the solenoid could operate on direct current from a battery, in which case a relay would sense loss of alternating-current voltage and de-energize the solenoid. The solenoid could be either instantaneous or time delayed using a dashpot arrangement.

(4) *With Alternating-Current Voltage Held Main Contactor.* Since the main contactor (which switches the motor) will drop out on loss of alternating current, it provides an instantaneous undervoltage function. There are two common approaches to achieve time-delay undervoltage protection:

(a) Permit the main contactor to drop out instantaneously but provide a timing scheme (which will time when alternating-current voltage is low or zero) to reclose the main contactor providing normal alternating-current voltage returns within the preset timing interval. Some of the timing schemes in use are the following:

(i) Capacitor charged through a rectifier from the alternating-current system. The charge keeps an instantaneous dropout auxiliary relay energized for an adjustable interval, which is commonly 2 or 4 s.

(ii) Standard timer which times when de-energized (pneumatic or induction disk, etc).

(iii) Standard timer operated on a direct-current battery.

(b) Use of a capacitor charged through a rectifier from the alternating-current system to hold in the main contactor for a predetermined interval of low or zero alternating-current voltage.

Note that two-wire control is sometimes used with an alternating-current voltage held main contactor. This control utilizes a "maintained closed" start button, or operates from an external contact responsive to some condition such as process pressure, temperature, level, etc. The main contactor drops out with loss of alternating current, but recloses when alternating-current voltage returns. This arrangement does not provide undervoltage protection.

(5) *With Direct-Current Voltage Held Main Contactor.* With this arrangement the contactor remains closed during low or zero alternating-current voltage. Time-delay undervoltage protection is achieved using voltage-sensing relays [see Section 9.3.1 (6)].

(6) *Voltage-Sensing Relays.* The most commonly used type is the single-phase induction disk undervoltage time-delay relay. Since a blown control fuse will cause tripping, it is sometimes desirable to use two or three of these relays connected to different phases and wire them so that all must operate before tripping will occur.

Three-phase undervoltage relays are available. Many operate in response to the area of the voltage triangle formed by the three-phase voltages.

In applications requiring a fixed time delay of a few cycles, an instantaneous undervoltage relay is applied in conjunction with a suitable timer (see Section 9.3.19).

When applying undervoltage protection with time delay, the time delay setting should be chosen so that time-delay undervoltage tripping does not occur before all external fault-detecting relays have an opportunity to clear all faults from the system. This recognizes that the most frequent causes of low voltage are system faults, and when these are cleared most induction motors can continue normal operation. In the case of induction disk undervoltage relays it is recommended that their trip time versus system short-circuit current be plotted to ensure that they do not trip before the system overcurrent relays. This should be done for the most critical coordi-

nation condition which exists when the system short-circuit capacity is minimum.

Typical time delay at zero voltage is 2 to 5 s.

For motors extremely important to continuity of service, such as some auxiliaries in electric generating plants, the undervoltage relays are used to alarm only.

9.3.2 Phase Unbalance — Device 46

(1) *Purpose.* The purpose is to prevent motor overheating damage. Motor overheating occurs when the phase voltages are unbalanced, for two reasons:

(a) Increased phase currents flow in order that the motor can continue to deliver the same horsepower as it did with balanced voltages.

(b) Negative-sequence voltage appears and causes abnormal currents to flow in the rotor. Since the motor negative-sequence impedance is approximately the same as the locked rotor impedance, a small negative-sequence voltage produces a much larger negative sequence current.

(2) *Single Phasing.* Overcurrent protection in each phase is recommended and is required by NFPA No 70, National Electrical Code (1975) (ANSI C1-1975). However, an understanding of the effect of having overload devices in only two phases is useful.

When single phasing occurs at the same voltage level as the motor operates from, one of the phase conductors to the motor carries zero current. If overcurrent devices in only two phases are relied upon for single-phasing protection, and one of them is in the zero current phase, then there is only one overcurrent device available to sense the current. If an overcurrent device is used in each of the three phases, then failure of one to operate still leaves another to sense this single-phase condition.

Single phasing on the supply voltage side of a delta—delta transformer results in zero current in one of the phase conductors to all motors connected to the other side of the transformer.

When a motor is supplied from a delta—wye or wye—delta transformer, single phasing on the supply voltage side of the transformer results in currents to the motor in the ratio of 1:1:2. In two phases the current will be only slightly greater than prior to single phasing, while it will be approximately doubled in the third phase. It is this situation that requires an overload (overcurrent) device in each phase, if the motor does not have suitable phase unbalance protection.

Many motors, especially in the higher horsepower ratings, can be seriously damaged by negative-sequence current heating, even though the stator currents are low enough to go undetected by overload (overcurrent) protection.

Therefore, phase unbalance protection is desirable for all motors where its cost can be justified relative to the cost and importance of the motor.

Phase unbalance protection should be provided in all applications where single phasing is a strong possibility due to the presence of fuses, overhead distribution lines subject to conductor breakage, or disconnect switches which may not close properly on all three phases, etc.

A general recommendation is to apply phase unbalance protection to all motors 1000 hp and above. For motors below 1000 hp, the specific requirements should be investigated.

(3) *Instantaneous or Time Delay*. Unbalanced voltages accompany unbalanced system faults. Therefore it is desirable that phase unbalance protection have sufficient delay to permit the system overcurrent protection to clear external faults without unnecessarily tripping the motor or motors.

Delay is also desirable to avoid the possibility of tripping on motor starting inrush. Therefore unbalance protection having an inherent delay should be chosen, or a suitable additional timer used. If more than 2 or 3 s is used, the motor designer should be consulted.

(4) *Relays*. There are several types of relays available to provide phase unbalance protection, including single phasing. Most of these are described in [1]. Further information about specific relays should be obtained from the various manufacturers. Most of the commonly used relays can be classified as follows:

(a) *Phase Current Balance — Device 46*. These relays detect unbalance in the currents in the three phases. The induction disk type has an inherent time delay. Occasionally a timer may be required to obtain additional delay.

(b) *Negative-Sequence Voltage*. This relay operates instantaneously using a negative-sequence voltage filter. A timer, either internal to the relay or external, is required for time delay.

(c) *Negative-Sequence Current*. An induction disk time-delay relay is available for application

to generators, but is not intended for motors since it has a relatively high pickup. Instantaneous negative-sequence current relays are available with low pickup to provide good motor protection. A timer is required with these to delay tripping.

9.3.3 *Instantaneous Phase Overcurrent — Device 50*

(1) *Purpose*. The purpose is to detect phase short-circuit conditions with no intentional delay. Fast clearing of these faults

(a) Limits damage at the fault

(b) Limits the duration of the voltage dip accompanying the fault

(c) Limits the possibility of fault spreading, fire, or explosion damage

(2) *Instantaneous Overcurrent Relays*. These are normally used with phase current transformers. Relays are required in just two phases if a ground overcurrent relay is also provided (see Sections 9.3.6 and 9.3.7); otherwise one relay per phase is required. However, one relay per phase is often provided, as well as a ground relay. The third phase relay then provides backup protection to the other two phase relays. Requirements of the NEC specify one overcurrent device per phase. These relays are used with the following equipment:

(a) Medium-voltage (2.4—15 kV) circuit breaker type motor starters

(b) Medium-voltage contactor-type starters which do not have power fuses

(c) Low-voltage circuit breaker type motor starters used with motors whose importance or horsepower rating justifies the cost of this protection instead of, or in addition to, direct-acting instantaneous overcurrent protection [see Section 9.3.3 (3)]

These relays are available in several forms:

(a) In individual cases, one relay per case

(b) Grouped, two or three relays per case

(c) As additional element(s) in case with induction overcurrent (device 51) or thermal overcurrent (device 49) element(s).

(3) *Direct-Acting Instantaneous Overcurrent Trip Devices*. The comments in the first paragraph of Section 9.3.3(2) apply here. These trip devices are commonly provided on low-voltage circuit breaker type motor starters.

(4) *Fuses*. These are used to provide fast phase short-circuit protection on medium- and low-

voltage fused-type motor starters. Refer to Section 5.7.4 and Fig 62 for more details on the application of fuses to motor circuits.

(5) *Instantaneous Settings.* Instantaneous overcurrent relays and direct-acting trip devices must have their pickup setting sufficiently high that they do not trip on asymmetrical current occurring:

(a) At initiation of the motor starting inrush

(b) When the motor contributes fault current to an external short-circuit condition

(c) Upon automatic transfer or fast reclosing

For many smaller squirrel-cage induction motors (that are installed on what may be considered a routine basis) it is usual to set the instantaneous pickup at 10 or 11 times the motor full-load current.

For the larger squirrel-cage motors (say, above 200 hp) and synchronous motors it is recommended that the value of maximum symmetrical starting inrush be determined by the motor manufacturer and the instantaneous pickup be set 75 percent above this value. The settings should be even higher for automatic transfer or fast reclosing.

Wound-rotor induction motors usually have reduced inrush due to starting with external rotor resistance. It must be remembered that their contribution to an external fault will exceed their inrush if they are operated with their rotor sliprings short-circuited. To avoid unnecessary tripping, their instantaneous pickup protection must be set on the basis of their contribution to an external fault.

Instantaneous settings are frequently determined by trial and error starting of the specific motor. This approach can result in unnecessary tripping at a later date if the maximum asymmetery possible never occurs during the trial and error starts. A minimum of three trial starts is recommended.

9.3.4 *Time-Delay Phase Overcurrent — Device 51*

(1) *Purpose.* The purpose is to detect

(a) Failure to accelerate to rated speed in the normal starting interval

(b) Motor stalled condition

(c) Low-magnitude phase fault conditions

In many motor protection schemes the overload protection (overcurrent type) is relied upon to provide all three protective functions. Actually, this overload protection is relatively slow, especially the thermal type, since it must not trip on normal motor-accelerating inrush. Schmidt [2] provides data on the magnitude of currents for internal faults in motors.

(2) *Overcurrent Relays and Settings.* The relays normally chosen for this protective function are the induction disk overcurrent type. The long-time relay characteristic is suitable for use with any motor. Sometimes another characteristic is chosen, such as extremely inverse, to get faster operation at high currents or to facilitate overcurrent coordination with the supply feeder relays. However, it must be ensured that this characteristic will not trip on normal accelerating inrush.

These same induction disk relays can also be set to provide overload (overcurrent) protection (see Section 9.3.5), if desired. However, if used for overload protection, they will usually trip much sooner on overload than is necessary to protect the motor. Accordingly they prevent utilizing all the motor inherent thermal overload capability. (This limitation is also true of many thermal overcurrent relays commonly used for overload protection.) The NEC requires one overcurrent device in each phase.

The instantaneous protection (Section 9.3.3) can be provided in the same relay cases with the time-delay elements.

Section 9.3.3(2) applies with regard to the quantity of relays required for phase fault protection. For stall protection, or detection of failure to accelerate normally, only one relay is required, on the assumption of balanced phase currents. For important motors two or three relays are recommended as backup to each other, and also to detect failure to accelerate due to single phasing.

The relay settings are normally chosen as follows:

(a) To also provide overload protection, set pickup at 5—25 percent above the motor continuous service factor rating.

(b) When it is not intended that overload protection be provided, the pickup would be set at 200—350 percent of motor ratings to avoid tripping for overload conditions.

NOTE: In (a) and (b) the time delay would be set to be a small margin longer than that required to prevent tripping on normal acceleration inrush.

(c) In some applications it might be desired to set the pickup slightly above the starting symmetrical inrush. In this case the relays would not "see" motor inrush, and only fault protection is provided. The time delay should then be very short, just sufficient to not trip on inrush asymmetry. The short-time induction disk relay should be chosen in this application.

(3) *Instantaneous Relay and Timer.* These have been used to provide the protection of (c) in combination with (a) or (b) of Section 9.3.4(2). It is available as a standard relay in combination with an induction disk. It is also available as an inverse instantaneous curve in some new solid-state overcurrent relays.

(4) *Direct-Acting and Solid-State Trip Devices.* Trip devices integral with circuit breakers are often used at 600 V and below. Solid-state trip systems generally have a straight-line long-time characteristic with a negative slope 2 on a log—log plot. Refer to Chapter 6.

9.3.5 *Overload (Phase Overcurrent) — Device 49 or 51*

(1) *Purpose.* The purpose is to detect sustained stator current in excess of motor continuous rating and trip prior to motor damage.

On motors having winding temperature devices and close operator supervision, this protection is sometimes arranged just to alarm.

Sometimes two sets of overload protection are provided:

(a) One set to alarm only, at relatively low pickup and fast time setting. This would normally be overcurrent relay(s) as in Section 9.3.5 (3).

(b) Second set to trip at higher pickup or slower time than the overload alarm relays, or both. Use relays as in Section 9.3.5 (3) or (4).

(2) *Quantity of Relays.* In the past it has been common to provide relays in only two phases. However, there is a definite trend to use one relay per phase, or a single relay responsive to individual currents in each of the three phases. This is now required by the NEC.

To the limited extent that overload relays will detect a single-phasing condition, one overload element per phase is desirable in order to respond to single phasing on the supply side of a delta—wye or wye—delta transformer [see Section 9.3.2 (2)].

(3) *Induction Disk Relays and Settings — Device 51.* Long-time overcurrent relays are fre-quently used with circuit breaker type motor starters. For this application they have the following desirable and undesirable features.

Desirable:

(a) Continuously adjustable time delay

(b) Pickup tap settings cover wide range of currents

(c) Quite accurate

(d) Easy and fast to test

(e) Have operation indicator

Undesirable:

(f) Shape of time-delay curve usually results in tripping much faster than necessary, thus preventing all the motor inherent thermal overload capability from being utilized.

(g) Not being thermally operated, they reset quickly after an overload trip, and hence provide no protection against starting again too soon.

(h) Not being thermally operated, they do not "remember" overloads which may come in cycles and progressively overheat the motor.

(i) They are self-resetting and so the hand reset feature is not available without use of a suitable auxiliary relay.

(j) These relays are not significantly affected by change of ambient temperature. This is acceptable and may be considered an advantage for the frequent situation when the motor and relays are in different ambient temperatures. It may be an undesirable characteristic, however, when the motor and relays are in the same ambient and there are significant ambient temperature changes.

(4) *Thermal Overcurrent Relays — Device 49.* These are normally used with contactor-type motor starters and are also often used with circuit breaker type starters. Their desirable and undesirable features are almost the exact opposites of those for induction disk relays [Section 9.3.5 (3)].

Undesirable:

(a) Time delay is usually not adjustable.

(b) Pickup adjustment of many relays is limited to a relatively small range.

(c) They are usually not as accurate as induction disk relays.

(d) They are not as easy and fast to test as induction disk relays.

(e) They usually do not have an operation indicator, and this is undesirable when they are used as a self-resetting protection.

(f) The shape of the time curve of many relays results in tripping much faster than necessary, thus preventing all motor inherent thermal overload capability from being utilized.

(g) After tripping on motor overload, they usually cool down faster than the motor. They permit restarting too soon and then provide a lower degree of overload protection than the restarted motor should have.

Desirable:

(h) Their thermal "memory" provides desirable protection for cyclic overloading and closely repeated motor starts.

(i) A hand reset feature is available in most relays.

(j) Some relays are available either as ambient temperature compensated or as noncompensated. Noncompensated is an advantage when the relay and motor are in the same ambient since the relay time changes with temperature in a similar manner as the motor overload capability changes with temperature.

9.3.6 *Instantaneous Ground Overcurrent — Devices 50G and 50N*

(1) *Purpose.* The purpose is to detect ground-fault conditions with no intentional delay.

(2) *Zero-Sequence Current Transformer or Sensor and Ground Relay — Device 50G.* It is recommended that a zero-sequence (window-type) current transformer that has been designed for this function be used to feed the ground relay (Fig 122).

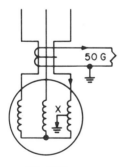

Fig 122
Ground Overcurrent Protection
Using Window-Type Current Transformer

The instantaneous relay is normally set to trip at a primary ground-fault current in the range of 5—20 A.

The following precautions must be observed in applying the relay and zero-sequence current transformer, and in installing the cables through the current transformer:

(a) If the cable passes through the current transformer window and terminates in a pothead on the source side of the current transformer, the pothead must be mounted on a bracket insulated from ground. Then the pothead must be grounded by passing a ground conductor through the current transformer window and connecting it to the pothead.

(b) If metallic covered cable passes through the current transformer window, the metal covering is kept on the source side of the current transformer insulated from ground. The terminator for the metal covering may be grounded by passing a ground conductor through the current transformer window and then connecting it to the terminator.

(c) Cable shield(s) should be grounded by passing a ground conductor through the current transformer window and then connecting it to the shield(s).

(d) It is important to test the overall current transformer and ground relay scheme by passing current in a test conductor through the current transformer window. Since normally there is no current in the relay, an open circuit in the current transformer secondary or wiring to the relay can only be discovered by this overall test.

(3) *Residually Connected Current Transformers and Ground Relay—Device 50N.* Applications have been made using the residual connection from three current transformers (one per phase) to feed the relay. This arrangement is not ideal since high phase currents (due to motor starting inrush or phase faults) may cause unequal saturation of the current transformers and produce a false residual current resulting in undesired tripping of the ground relay.

Sometimes decreasing the relay pickup setting will overcome the problem because this has the effect of increasing the relay burden. In cases where this is effective, decreasing the relay pickup setting actually increases the relay pickup in terms of primary current since the higher relay burden requires more current transformer exciting current than is gained with the lower relay setting.

Some improvement is obtained by inserting an impedance in the residual connection in series with the relay. Usually, the best solution is to use device 51N (Section 9.3.7).

(4) *Combination-Type Starters.* When an instantaneous ground relay is applied to this type of starter on a solidly grounded system, it must be remembered that the contactor might not have the capability to clear the maximum ground-fault capacity. Therefore it is necessary to ensure that the fuses or circuit breaker instantaneous trip devices will clear a ground fault before the contactor is damaged in trying to clear it. In some cases it may be necessary to delay tripping from the ground relay or use a time-delay type ground relay (Section 9.3.7).

(5) *When Surge Arresters are Installed at Motor Terminals.* There is a possibility that a surge discharge through an arrester would cause the ground relay to trip unnecessarily. To avoid this possibility it has been usual to recommend that an overcurrent time-delay relay be used in this situation. If the instantaneous relay has sufficient inertia, however, it may override a surge discharge without tripping.

9.3.7 *Time-Delay Ground Overcurrent — Devices 51G and 51N*

(1) *Purpose.* The purpose is to detect ground fault conditions. Early applications of ground protection used current transformers and relays. However, both instantaneous and time-delay ground-fault protection is now available with solid-state tripping systems on low-voltage (up to 600 V) circuit breakers (see Chapter 6).

(2) *Zero-Sequence Current Transformer and Time-Delay Ground Relay — Device 51G.* When the zero-sequence current transformer is used for motor ground protection, it is usual to use an instantaneous overcurrent ground relay. When a time-delay relay is used [Section 9.3.6(5)] it is usually a short time, or an extremely inverse, induction disk relay set at 0.5 A tap and 1.0 time dial. The comments in Section 9.3.6 (2) apply here.

(3) *Residually Connected Current Transformers and Ground Relay — Device 51N.* The relay is usually a short time, or an extremely inverse, induction disk type set at 0.5 A tap and 1.0 time

dial. To get lower pickup, with high ratio current transformers, a 0.2 A relay is sometimes used.

If one of the current transformer secondary phase conductors becomes open circuited, the other two current transformers feed phase current through the residual ground relay causing it to trip.

(4) *Choice of Resistor for System Grounding.* The object of resistance grounding is normally to limit the motor damage caused by a ground fault. (In mine distribution systems the object is to limit equipment frame to earth voltages for safety reasons.) However, the ground-fault current should not be limited to the extent that very much of the neutral end of motor wye windings goes unprotected. In the past, protection to within 5—10 percent of the neutral has often been considered adequate. Fawcett [3] recommends that the ground overcurrent relay should have at least 1.5 multiples of pickup for a ground fault one turn away from the neutral.

Noting the foregoing, it is recommended that the ground resistor rating and the ground protection be chosen together after having determined the winding arrangements of the various motors to be served. On this basis, the ground resistor chosen will normally limit the ground-fault current within the range of 100—2000 A. A 10 s time rating is usually chosen for the resistor.

Note also that to avoid excessive transient overvoltages the resistor should be chosen so that the following impedance ratio is achieved:

$$\frac{R_0 \text{ (zero-sequence resistance)}}{X_0 \text{ (zero-sequence reactance)}}$$

must be equal to or greater than 2.

9.3.8 *Phase Current Differential — Device 87*

(1) *Purpose.* The purpose is to quickly detect fault conditions.

(2) *Conventional Phase Differential.* This scheme uses six identical current transformers (one pair for each phase) and three relays (one per phase). Since the current transformers carry load current, they must have primary current ratings chosen accordingly. (See Fig 123.)

The currents from each pair of current transformers are subtracted and their difference is fed to the relay of the associated phase. For normal (nonfault) conditions the two currents in each pair are equal and their difference is, therefore,

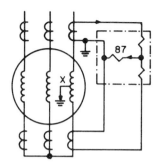

Fig 123
Conventional Phase Differential
Protection Using Three Percentage
Differential Relays (One Shown)

zero. For a fault located between the two current transformers of any pair, the currents from the two current transformers will differ. This difference will operate the relay of the associated phase.

While sometimes applied to delta-connected motors, this scheme is usually used with wye-connected motors. (Wye-connected motors are much' more common than delta-connected ones in the larger horsepower ratings.) With the wye-connected motor three of the current transformers are normally located at the starter (or motor switchgear) and the other three in the three phases at the motor winding neutral. Occasionally the three neutral phases are cabled back to the starter; for example, if there is a neutral starting reactor and it is located at the starter. In this case the neutral current transformers would also be at the starter.

There are three types of relays generally applicable for conventional phase differential protection:

(a) High-speed (instantaneous) differential. While this protection is the most expensive, it is not much more expensive when the total cost of relays, wiring, current transformers, and current transformer mounting space and installation is considered. Therefore, this is the type now recommended. There are no settings to be chosen for these relays.

(b) Slow-speed induction disk percent differential. In the past this type has usually been used. There are no settings to be chosen for these relays.

(c) Standard induction disk overcurrent relays differentially connected. These relays have not been used very frequently. However, they are less expensive than those in (b) and quite satisfactory provided they are used with identical current transformers [which is also the normal recommendation for those in (a) and (b)]. They would normally be set between 0.5 and 2.5 A tap and 1.0 time dial. However, the high-speed type relays are the best choice.

(3) *Self-Balancing Differential Using Zero-Sequence Current Transformers.* Three window-type current transformers are used. These are normally installed at the motor. One current transformer per phase is used with the motor line and neutral leads of one phase passed through it such that the two currents normally cancel each other. A winding phase-to-phase or ground fault will result in an output from the current transformer of the associated phase and operate the associated relay (see Fig 124).

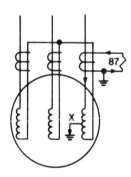

Fig 124
Self-Balancing Differential
Protection (One Relay Shown)

The current transformers and relays would normally be the same as those used for zero-sequence instantaneous ground overcurrent protection [Section 9.3.6(2)] with the relay set between 0.25 and 1.0 A pickup. Therefore this differential scheme will usually have a lower primary pickup in amperes than the conventional differential scheme (since the current transformer ratio is usually greater with the conventional scheme). This differential scheme has a slight advantage over that of Section 9.3.6(2) in detecting ground faults. For motors installed on grounded systems this is quite significant, since most faults will be ground faults. If the fault does not involve ground, then the stator iron is not being damaged. The usual objective of motor-fault protection is to remove the fault before the stator iron is significantly damaged.

With the current transformers located at the motor, this scheme does not detect a fault in the cables to the motor. A fault in these cables would normally be detected by the overcurrent protection of Sections 9.3.3 through 9.3.7. In the case of large motors it is often a problem to coordinate the supply phase overcurrent protection. The presence of motor differential protection is sometimes considered to make the above coordination less essential. In this regard the conventional differential is better than the zero-sequence differential since the motor cables are also included in the differential protection zone, and hence coordination between the motor differential and supply phase overcurrent relays is complete.

As with zero-sequence ground overcurrent protection, it is important during initial startup to test the overall current transformer and relay combinations by passing current in a test conductor through the window of each current transformer. Since normally the relays do not carry current, an open circuit in a current transformer secondary or wiring to a relay can only be discovered by this overall testing.

(4) *Contactor-Type Starters with Power Fuses.* The comments of Section 9.3.6(4) also apply when using differential protection with these starters.

(5) *When to Apply Differential Protection.* The following general recommendations are made:

(a) With all motors 1000—2000 hp and above used on ungrounded systems.

(b) With all motors 1000—2000 hp and above used on grounded systems where the ground protection applied is not considered sufficient without differential protection to protect against phase-to-phase faults.

(c) For large horsepower ratings (2500—5000 hp) the cost of differential protection compared to the cost of the motor would generally justify the use of this relay. However, differential protection is frequently justified for much smaller motors, especially at voltages above 2400 V.

9.3.9 *Split Winding Current Unbalance — Device 61*

(1) *Purpose.* The purpose is to quickly detect low-magnitude fault conditions. This protection also serves as backup to instantaneous phase overcurrent (Section 9.3.3) and ground overcurrent protection (Sections 9.3.6 and 9.3.7).

This protection is normally only applied to motors having two (or three) winding paths in parallel per phase (see Fig 125).

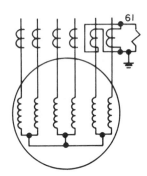

Fig 125
Split-Phase Motor Overcurrent Protection
Can Be Used with Two Paths per Phase
(One Relay Shown)

(2) *Arrangement of Current Transformers and Relays.* The usual application is with a motor having two winding paths in parallel per phase. The six line leads (two per phase) of the motor must be brought out and one current transformer is connected in each of the six leads. The primary current rating of the current transformers must be chosen to carry full-load current.

The current transformers may be installed at the motor. It is often convenient, however, to use six cables to connect the motor to its starter (or switchgear), and in this case the current transformers can be located in the starter.

The currents from each pair of current transformers, associated with the same phase, are subtracted and their difference is fed to a short-time induction disk overcurrent relay. Three of these relays are required (one per phase), and each is set at 1.0 time dial and between 0.5 and 2.5 A. The relay must be set above the maximum current unbalance that can occur between the two parallel windings for any motor loading condition.

(3) *Evaluation of this Protection*

(a) Total cost would be somewhat less than conventional phase differential [Section 9.3.8 (2)] and more than self-balancing differential [Section 9.3.8 (3)].

(b) The primary pickup current for this protection would be about half that of conventional phase differential (since both schemes require the current transformer primaries to be rated to carry normal load currents). Self-balancing differential would usually have a lower primary pickup in amperes.

(c) This protection has a slight time delay compared to the phase differential schemes.

(d) When the current transformers are located in the motor starter, split winding protection has the same advantage over self-balancing differential as does conventional phase differential, namely, it detects a fault in the motor cables and may facilitate coordination with the supply feeder overcurrent relays. [This is discussed in Section 9.3.8(3).]

(e) If one of the current transformer secondary conductors becomes open circuited, the current from the other current transformer of the same pair will cause the relay to trip. This is undesirable from a continuity of service point of view.

(f) The salient feature which this protection provides, and no other motor protection has, is the ability to sense short-circuited winding turns. The number of turns which must be short-circuited in order that detection will occur depends upon the motor winding arrangement as well as the relay pickup and current transformer ratio. An analysis of the specific motor winding would be required to determine how worthwhile this feature is. It may be that the short-circuited turns would cause a ground fault and be detected by the self-balancing differential scheme before this split winding protection would sense the short-circuited turns condition.

(g) This protection could be applied to a motor with four winding paths in parallel per phase by grouping them as two pairs and then treating them as if there were only two paths in parallel (that is, six current transformers and three relays are used).

(h) A split differential scheme is often effectively used where one current transformer is in one of the parallel paths and the other current transformer sees the total phase current.

(4) *When to Apply Split Winding Protection.* This protection has not been used in the past with very many motors. It is probably desirable on important motors that have two or four winding paths in parallel per phase and are rated above 5000 to 10 000 hp.

9.3.10 *Stator Winding Overtemperature — Device 49 or 49S*

(1) *Purpose.* The purpose is to detect excessive stator winding temperature prior to the occurrence of motor damage. This protection is often arranged just to alarm on motors operated with competent supervision. Sometimes two temperature settings are used, the lower setting for alarm, the higher setting to trip.

(2) *Resistance Temperature Detectors.* Six of these detectors are commonly provided (when specified) in motors rated 1500 hp and above. They are installed in the winding slots when the motor is being wound. The six are spaced around the circumference of the motor core to monitor all phases.

Commonly used types are rated either 10 or 120 Ω, at a specified temperature. Their resistance increases with temperature and wheatstone bridge devices are used to provide temperature indication or contact operation, or both.

For safety reasons the detectors should be grounded. This places a ground on the wheatstone bridge control and, therefore, makes it undesirable to directly operate the wheatstone bridge control from a switchgear direct-current battery (since these direct-current control schemes should normally operate ungrounded in order to achieve maximum reliability). However, use of alternating-current control has the following limitations:

(a) Loss of alternating current due to a fuse blowing, etc, removes the protection, unless (b) applies.

(b) If the null point of the wheatstone bridge, for loss of alternating-current power, is near the trip contact setting, then loss of alternating current or a voltage dip may cause a false trip of the motor.

The foregoing must be evaluated in applying this protection.

An open circuit in the detector leads appears as an infinite resistance and will cause a false trip since the output is the same as a very high-temperature condition.

The following arrangements are frequently used:

(a) Determine which detector normally runs hottest and permanently connect a trip relay to it. Use one temperature indicator and a manual switch to monitor the other five detectors.

(b) Use manual switch and combination indicator and alarm relay. *Precaution:* An open circuit in the switch contact will cause a false trip.

(c) Use manual switch and indicator only.

(d) Use one, two, or three (one per phase) alarm relays and one, two, or three (one per phase) trip relays set at a higher temperature.

(3) *Thermocouples.* Thermocouples are used to indicate temperatures and for alarm and trip functions, in a similar manner as resistance temperature detectors.

An open circuit in the thermocouple leads will not cause a trip since the output is the same as a low-temperature condition.

The output from thermocouples is compatible with central control room temperature monitoring and data logging schemes.

(4) *Thermistors.* These are used to operate relays for alarm or trip functions, or both. They are not used to provide temperature indication. However, they are available combined with thermocouples: the thermocouples to provide indication and the thermistor to operate a relay. Relays to operate from thermistors are relatively inexpensive.

There are two types of thermistors:

(a) *Positive-Temperature-Coefficient Type.* With this type its resistance increases with temperature. An open circuit in the thermistor would appear as a high-temperature condition and would operate the relay. This is the failsafe type of arrangement.

(b) *Negative-Temperature-Coefficient Type.* Resistance decreases as temperature increases. An open circuit in the thermistor appears as a low-temperature condition and does not cause relay operation.

(5) *Thermostats and Temperature Bulbs.* These devices are used on some motors. They have not been commonly used on the larger motors and will not be discussed here.

(6) *When to Apply Stator Winding Temperature Protection.* It is commonly applied to all motors rated 1500 hp and above. In the following situations it is particularly desirable:

(a) Motors in high ambient temperatures or at high altitudes

(b) Motors whose ventilation systems tend to become dirty, losing cooling effectiveness

(c) Motors subject to periodic overloading due to load characteristics of the drive or process

(d) Motors likely to be subjected to steady-state overloading in order to increase production

(e) Motors for which continuity of service is very important

9.3.11 *Rotor Overtemperature — Device 49 or 49R*

(1) *Synchronous Motors.* Rotor winding overtemperature protection is available for brush-type synchronous motors, but is not normally used. One well-known approach is to use a Kelvin bridge type strip chart recorder with contacts adjustable to the temperature settings desired. The Kelvin bridge measures the field resistance in order to determine the field winding temperature, using field voltage and field current (from a shunt) as inputs.

(2) *Wound Rotor Induction-Motor Starting Resistors.* Some form of temperature protection should be applied for these resistors on drives having severe starting requirements such as long acceleration intervals or frequent starting. Resistance temperature detectors as well as other types of temperature sensors have been used in proximity to the resistors.

9.3.12 *Synchronous Motor Protection*

(1) *Damper Winding Protection — Device 26.* When a synchronous motor is starting, high currents are induced in its rotor damper winding. If the motor takes longer to accelerate than it has been designed to be suitable for, the damper winding may overheat and be damaged.

Several different relay type and static type protective schemes are available. None of these schemes directly senses damper winding temperature. Instead, they try to simulate what it must be by evaluating two or more of the following quantities:

(a) Magnitude of induced field current which flows through the field discharge resistor. This is a measure of the relative magnitude of induced damper winding current.

(b) Frequency of induced field current which flows through the discharge resistor. This is a measure of rotor speed and provides an indication, therefore, of the increase in damper winding thermal capability resulting from the ventilation effect and the decrease of induced current.

(c) Time interval after starting.

(2) *Field Current Failure Protection — Device 40.* Field current may drop to zero or a low value when a synchronous motor is operating for several reasons.

(a) Tripping of the remote exciter, either motor-generator set type or static type. (Note that control for these should be arranged so that it will not drop out on an alternating-current voltage dip.)

(b) Burnout of the field contactor coil. (Note that the control should be arranged such that the field contactor will not drop out on an alternating-current voltage dip.)

(c) Accidental tripping of the field circuit breaker. (Note that field circuit overcurrent protection is usually omitted from field circuit breakers and contactors in order to avoid unnecessary tripping. The field circuit is usually ungrounded and should have ground detection lights or relay applied to it to detect the first ground fault before a short circuit occurs [see Section 9.3.15(1)].

(d) High resistance contact or open circuit between slip ring and brushes due to excessive wear or misalignment.

Reduced field current conditions should be detected because

(a) Heavily loaded motors will pull out of step and stall.

(b) Lightly loaded motors are not capable of accepting load when required to.

(c) Intermediate loaded motors, which do not pull out of step, are likely to do so on an alternating-current voltage dip that they otherwise might ride through.

(d) The excitation drawn from the power system by large motors may cause a serious voltage drop and endanger continuity of service of other motors.

A common approach is to use an instantaneous direct-current undercurrent relay to monitor field current. This application should be investigated to ensure that there are no transient conditions which would reduce the field current and cause unnecessary tripping of the instantaneous relay. A timer could be used to obtain a delay of

one or more seconds, or the relay could be connected to alarm only where competent supervising personnel are available.

Field current failure protection is also obtained by the generator-type loss of excitation relay which operates from potential transformers and current transformers monitoring motor stator voltages and current. This has been done on some large motors (4000 hp and above). This relay may also provide pullout protection [Section 9.3.12(4)].

(3) *Excitation Voltage Availability — Device 53.* A simple voltage relay is used as a permissive start to ensure that voltage is available from the remote exciter. This avoids starting and then having to trip because excitation was not available. Loss of excitation voltage is not normally used as a trip; the field current failure protection is used for this function.

(4) *Pullout Protection — Device 78 (also 55 and 95).* Pulling out of step is usually detected by one of the following relay schemes:

(a) An instantaneous relay connected in the secondary of a transfomer whose primary carries the direct field current. The normal direct field current is not transformed. When the motor pulls out of step, alternating currents are induced in the field circuit and these are transformed and operate the pullout relay. This relay, while inexpensive, is sometimes subject to false tripping on alternating-current transients accompanying external system fault conditions and also alternating-current transients caused by pulsations in reciprocating compressor drive applications. Device 95 has sometimes been used to designate this relay.

(b) A power factor type relay responding to motor stator voltage and current obtained from potential and current transformers. This relay is often designated as device 55.

(c) The generator-type loss of excitation relay has been used [see Section 9.3.12(2)].

(5) *Incomplete Starting Sequence — Device 48.* This protection is normally a timer which blocks tripping of the field current failure protection [Section 9.3.12(2)] and the pullout protection [Section 9.3.12(4)] during the normal starting interval. The timer is started by an auxiliary contact on the motor starter and times for a preset interval that has been determined during test starting to be slightly greater than the normal

interval from start to reaching full field current. The timer puts the field current failure and pull-out protection in service at the end of its timing interval.

This timer is often the de-energize-to-time type so that it is failsafe with regard to applying the field current failure and pullout protection.

(6) *Operation Indicator for Foregoing Protections.* Many types of the foregoing protective devices do not have operation indicators. It is suggested that separate operation indicators be used with these protective devices.

9.3.13 *Induction Motor Incomplete Starting Sequence Protection — Device 48.* It is recommended that wound-rotor induction motors and reduced voltage start motors have a timer applied to protect against failure to reach normal running conditions within the normal starting time. The timer would be started by an auxiliary contact on the motor starter and would time for a preset interval that has been determined during test starting to be slightly greater than the normal starting interval. The timer trip contact would be blocked by an auxiliary contact of the final device that operates to complete the starting sequence. This final device would be the final secondary contactor in the case of a wound-rotor motor, or it would be the device which applies full voltage to the motor stator.

This timer is often the de-energized-to-time type so that it is failsafe from a protection point of view.

Incomplete sequence protection should also be applied to part winding and wye—delta motor starting control, as well as pony motor and other sequential start arrangements.

9.3.14 *Protection against too Frequent Starting.* The following protections are available:

(a) A timer, started by an auxiliary contact on the motor starter, with contact arranged to block a second start until the preset timing interval has elapsed.

(b) Stator thermal overcurrent relays [Section 9.3.5(4)] provide some protection. The degree of protection depends upon

(a) The normal duration and magnitude of motor inrush

(b) The relay operating time at motor inrush and the cool-down time of the relay

(c) The thermal type damper winding protection on synchronous motors [Section 9.3.12(1)]

(d) Rotor overtemperature protection (Section 9.3.11)

Note that large motors are often provided with nameplates giving their permissible frequency of starting.

9.3.15 *Rotor Winding Protection*

(1) *Synchronous Motors.* The field and field supply should not be grounded. While the first ground does not cause a fault, a second ground probably will. Therefore it is important to detect the first ground. The following methods are used.

(a) Connect two lights in series between field positive and negative with the midpoint between the lamps connected to ground. A ground condition will show by unequal brilliancy of the two lamps.

(b) Connect two resistors in series between field positive and negative with the midpoint between the resistors connected through a suitable intantaneous relay to ground. The maximum resistance to ground which can be detected depends upon the relay sensitivity and the resistance in the two resistors.

This scheme will not detect a ground fault at midpoint in the field winding. If a varistor is used instead of one of the resistors, then the point in the field winding at which a ground fault cannot be detected changes with the magnitude of the excitation voltage. This approach is used to overcome the limitation of not being able to detect a field midpoint ground fault.

(c) Some schemes apply a low alternating current voltage between the field circuit and ground and monitor the alternating-current flow to determine when a field circuit ground fault occurs. Before using one of these schemes, it should be determined that a damaging alternating current will not flow through the field capacitance to the rotor iron and then through the bearings to ground, thus causing damage to the bearings.

(d) If a portion of the field becomes faulted, damaging vibration may result. Vibration detectors should be considered [Section 9.3.21(3)].

(2) *Wound Rotor Induction Motors.* Item (1) of this section applies here, except that the rotor winding is three-phase alternating current instead of a direct-current field (see Fig 126). Yuen et al. [4] give experience confirming the desirability and effectiveness of this protection.

Wound-rotor motor damage can occur due to

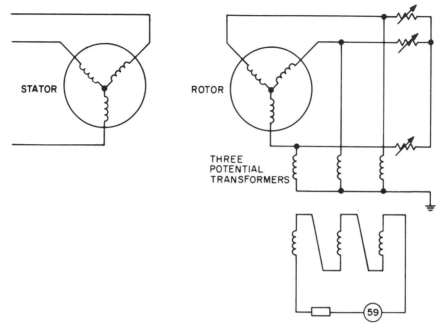

Fig 126
Rotor Ground Protection of
Wound-Rotor Motor

high resonant torques resulting from operation with unbalanced impedances in the external rotor circuit on speed-controlled motors. Protection to detect this is available although it has seldom been used.

9.3.16 *Lightning and Surge Protection*

(1) *Types of Protection.* Surge arresters are often used, one per phase connected between phase and ground, to limit the voltage to ground impressed upon the motor stator winding due to lightning surges and switching surges.

Surge capacitors are used, connected between each phase and ground, to decrease the slope of the wavefront of lightning surge voltages and switching surge voltages. As the surge voltage wavefront travels through the motor winding, the surge voltage between adjacent turns and adjacent coils of the same phase will be lower for a wavefront having a decreased slope. (A less steep wavefront is another way of designating a wavefront having a decreased slope.)

(2) *Locations of Surge Protection.* The surge arresters and surge capacitors should be con-nected within 3 circuit feet of the terminals of each motor. The supply circuit should connect to the surge equipment first and then go to the motor terminals. The two important points are

(a) The surge protection should be as close to the motor terminals (in circuit feet) as feasible

(b) The supply circuit should connect directly to the surge equipment first and then go to the motor

It is becoming more common to specify the surge protection to be supplied in a terminal box on the motor or in a terminal box adjacent to the motor.

If the motors are within 100 ft of their starters or the supply bus, it is sometimes compromised for cost reasons to locate the surge protection in the starters or supply bus switchgear. In the latter case, one set of surge protection is used for all the motors within 100 ft of the bus. Alternatively, this approach may be used for the smaller motors, and individual surge protection installed at each larger motor.

When surge protection is supplied in a motor

terminal box, it is necessary to disconnect it before high-voltage testing the motor. This is a recognized inconvenience of this arrangement.

(3) *When to Apply Surge Protection.* The following general guides are given.

(a) Apply to each medium-voltage motor rated above about 500 hp.

(b) Apply to each motor rated above 200 hp that is connected to open overhead lines at the same voltage level as the motor.

(c) When there is a transformer connecting the motor(s) to open overhead lines, surge protection is still required sometimes to protect against lightning surges. Techniques are available to analyze this situation. If in doubt it is best to provide surge protection. Refer to (2) for surge protection on the supply bus for motors located within 100 ft of the bus.

9.3.17 *Protection against Overexcitation from Shunt Capacitance*

(1) *Nature of Problem.* When the supply voltage is switched off, an induction motor initially continues to rotate and retain its internal voltage. If a capacitor bank is left connected to the motor, or if a long distribution line having significant shunt capacitance is left connected to the motor, the possibility of overexcitation exists. Overexcitation results when the voltage versus current curves of the shunt capacitance and the motor no-load excitation characteristic intersect at a voltage above the rated motor voltage.

The maximum voltage that can occur is the maximum voltage on the motor no-load excitation characteristic (sometimes called magnetization or saturation characteristic). This voltage, which decays with motor speed, can be damaging to a motor (see Fig 127 as an example).

Damaging inrush can occur if automatic reclosing or transfer takes place on a motor which has a significant internal voltage due to overexcitation.

(2) *Protection.* It is assumed here that sufficient knowledge of the system and motor exists to determine whether protection against overexcitation is required. The simplest protection would be instantaneous overvoltage relays.

An alternative is to use a high-speed underfrequency relay. However, this may not be fast enough on drives having high inertia or light loading.

The underfrequency relay is not suitable for drives whose frequency may not decrease follow-

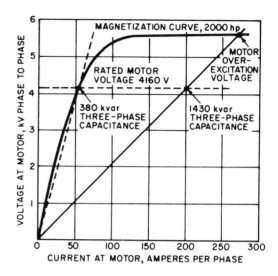

Fig 127
Excess Shunt Capacitance
from Utility Line Is Likely
to Overexcite a Large High-Speed
Motor. Voltage May Shoot up to 170 Percent

ing loss of the supply circuit breaker. Examples of these are

(a) Mine hoist with overhauling load characteristic at time of loss of supply circuit breaker

(b) Motor operating as induction generator on shaft with process gas expander

With these applications a loss of power relay could be used [see Section 9.3.18(2), item (f)].

9.3.18 *Protection against Automatic Reclosing or Automatic Transfer*

(1) *Nature of Problem.* When the supply voltage is switched off, motors initially continue to rotate and retain an internal voltage. This voltage decays with motor speed and internal flux. If system voltage is restored out of phase with a significant motor internal voltage, high inrush can occur and damage the motor windings or produce torques damaging to the shaft, foundation or drive coupling, or gears.

IEEE Std 288-1969, Induction Motor Protection (ANSI C37.92-1972), discusses considerations as to the probability of damage occurring for various motor and system parameters.

(2) *Protection.* The following protection is used.

(a) Delay restoration of system voltage, using timer, for a preset interval known from actual tests to be sufficient for adequate decay of the motor internal voltage.

(b) Delay restoration of system voltage until the internal voltage fed back from the motor(s) has dropped to a low enough value. This value is commonly considered to be 25 percent of rated voltage. Note that the frequency also decreases as the voltage decays (due to the motor(s) slowing down). The undervoltage relay and its setting should be chosen accordingly. Sometimes a full-wave rectifier and direct-current relay are chosen so that the relay dropout will be independent of frequency. If an alternating-current frequency-sensitive relay is used, it should be set, based on motor and system tests, to actually drop out at 25 percent voltage (and the frequency that will exist when 25 percent voltage is reached).

(c) Use of a high-speed underfrequency relay to detect the supply outage and trip the motor before supply voltage is restored. The limitations of this relay given in Section 9.3.17(2) also apply here. A further limitation exists if the motor operates at the same voltage level as the supply lines on which faults may occur followed by an automatic reclosing or transfer operation. The problem is that the underfrequency relay requires some voltage in order to have operating torque. If there is no impedance (such as a transformer) between the motor and the system fault location, then there may not be sufficient voltage to permit the underfrequency relay to operate.

(d) Single-phase or three-phase undervoltage relays are used as follows:

(i) One relay with a sufficiently fast time setting can be connected to the same potential transformer as the underfrequency relay [see (c)] and take care of the fault condition which results in insufficient voltage to operate the underfrequency relay.

(ii) One, two, or three relays (each relay connected to a different phase) can be used to detect the supply outage and trip the motor(s) when sufficient time delay exists before the supply will be restored.

(e) Loss of power relay. This relay must be sufficiently fast and sensitive. A high-speed three-phase relay has been used frequently. Being three-phase, it is more difficult to apply and connect properly than the other schemes listed.

Since this is a loss of power relay, it must be blocked at start-up until sufficient load is obtained on the circuit or motor with which it is applied.

(f) Reference is sometimes made to using a reverse power relay to detect a separation between motors and their source. While this approach is suitable in some circumstances, its limitations should be recognized:

(i) During the time the fault is on and the source still connected to the motors, net power flow will continue into the motors for most faults. This is not the case for three-phase bolted faults; however, for these faults there is no impedance into which reverse power can flow.

(ii) In view of the foregoing, tripping by reverse power can usually only be relied upon if there is a definite load remaining to absorb power from high-inertia motor drives after the source fault detecting relays isolate the source from the motors.

(iii) Reverse power relays responsive to reactive power (vars) instead of real power (watts) usually do not provide a suitable means of isolating motors prior to automatic reclosing or automatic transfer operations.

Therefore, the loss of power relay application is generally more suitable than the reverse power relay application.

9.3.19 *Protection against Excessive Shaft Torques Developed during Phase-to-Phase or Three-Phase Short Circuits Near Synchronous Motors.* A phase-to-phase or three-phase short circuit at or near the motor terminals produces very high shaft torques that may be damaging to the motor or driven machine. Computer programs have been developed for calculation of these torques. Refer to IEEE Std 329-1971, Synchronous Motor Protection (ANSI C37.94-1972), for information on this potential problem.

To minimize exposure to damaging torques, a three-phase high-speed voltage relay can be applied to detect severe phase-to-phase or three-phase short-circuit conditions for which the motor(s) should be tripped. This relay is often the type whose torque is proportional to the area of the three-phase voltage triangle. When a severe reduction in phase-to-phase or three-phase voltage occurs, a spring in the relay will overcome the torque produced by the voltage and cause tripping. An additional delay before tripping of 1 to 8 cycles may be satisfactory from a protection point of view, and desirable to avoid unnecessary shutdowns. This can be achieved using a suitable form of timer. Selection of protection and settings for this application should be done in conjunction with the motor and driven machine suppliers as well as the protection supplier.

9.3.20 *Protection against Failure to Rotate or Reverse Rotation*

(1) *Failure to Rotate.* This condition will occur if the supply is single phased, or if the motor or driven machine is jammed in some way. The following protection is available:

(a) Section 9.3.2(4) discusses types or relays to detect single phasing.

(b) The direct means to detect failure to rotate is to use a shaft speed sensor and timer to check whether a preset speed has been reached by the end of a short preset time interval after energizing the motor. This protection is desirable for induction and brushless synchronous motors that have a permissible locked rotor time less than normal acceleration time.

(c) For induction and brushless synchronous motors having a permissible locked rotor time greater than normal acceleration time, it is normal to rely upon the time-delay phase overcurrent relays (Section 9.3.4 or 9.3.5).

(d) For brush-type synchronous motors having a permissible locked rotor time less than normal acceleration time, it is possible to use a frequency-sensitive relay connected to the field discharge resistor and a timer to achieve this protection. This is because the frequency of the induced field current flowing through the discharge resistor indicates the motor speed. An induction disk frequency-sensitive adjustable time-delay voltage relay is available to provide this protection.

(e) For brush-type synchronous motors having a permissible locked rotor time greater than normal acceleration time, it is normal to rely upon the damper winding protection [Section 9.3.12(1)] and incomplete starting sequence protection [Section 9.3.12(5)].

(2) *Reverse Rotation.* A directional speed switch mounted on the shaft and a timer can be used to detect starting with reverse rotation. Some motor drives are equipped with a ratchet arrangement to prevent reverse rotation.

A reversal in the phase rotation can be detected by a reverse phase voltage relay [Section 9.3.2(4)] if the reversal occurs in the system on the supply side of the relay. This relay cannot detect a reversal which occurs between the motor and the point at which the relay is connected to the system.

9.3.21 *Mechanical and Other Protection*

(1) *Bearing and Lubricating Systems.* Various types of temperature sensors are used on sleeve bearings to detect overheating: resistance temperature detectors, thermocouples, thermistors, thermostats, temperature bulbs. Excessive bearing temperature may not be detected soon enough to prevent bearing damage. However, if the motor is tripped before complete bearing failure occurs, more serious mechanical damage to the rotor, and hence to the stator, may be prevented. Accordingly, for maximum effectiveness it is recommended that:

(a) A fast-responding type of temperature sensor be used

(b) The temperature sensor be located in the bearing metal where it is close to the source of overheating

(c) The temperature sensor be used for tripping instead of alarm; there may be situations where both alarm and trip sensors can be used, the former having a lower temperature setting

Alarm and trip devices should be provided with bearing lubricating systems to monitor

(a) Lubricating oil temperature, preferably from each bearing, and also to the bearings

(b) Cooling water temperature, both temperature in and out

(c) Lubricating oil flow and cooling water flow

(d) In lieu of the flow monitoring recommended in (c), a suitable arrangement of pressure switches is often used. However, flow monitoring is strongly recommended for important or high-speed machines.

It is generally considered that temperature sensors cannot detect impending failure of ball or roller bearings soon enough to be effective. Vibration detectors should be considered [Section 9.3.21(3)].

Protection to detect currents that may cause bearing damage should be considered for motors having insulated bearing pedestals.

(2) *Ventilation and Cooling Systems.* Alarm and trip devices should be considered

(a) To detect high-pressure drop across filters in motor ventilation systems

(b) To detect loss of air flow from external blower(s) in motor ventilation systems

(c) In lieu of the air flow monitoring in (b), a suitable arrangement of pressure switches is often used; flow monitoring is preferable, however

(d) With water-cooled motors, water temperature, flow, or pressure monitoring

(e) With inert gas-cooled motors, suitable pressure and temperature sensors

(3) *Vibration Detectors.* Experience indicates, especially on the higher speed drives, that serious damage can be prevented using vibration detectors for tripping, or alarm and tripping.

(4) *Liquid Detectors.* On large machines liquid detectors are sometimes provided to detect liquid (usually water) inside the stator frame. This can occur because of one of the following.

(a) Excessive condensation. Use of motor space heaters, or low-voltage winding heating with automatic control should normally avoid this problem.

(b) Exposure to hosing down cleaning operations, flooding, or outdoor weather conditions. A suitable choice of motor enclosure should normally avoid this problem.

(c) Leak from motor cooler. Double-tube coolers are sometimes used to help avoid this problem.

(5) *Fire Detection and Protection.* The following should be considered.

(a) Installation of suitable fire detectors to alert operators to use suitable hand-type fire extinguishers.

(b) Installation of suitable fire detectors and automatic system to apply carbon dioxide into motor.

(c) Some old large motors have internal piping to apply water for fire extinguishing. Possible false release of the water is a serious disadvantage.

(d) Use of synthetic lubricating oil which does not burn is worth considering for drives having large lubricating systems and reservoirs and also for use in hazardous atmospheres.

9.4 Standards References. The following standards publications were used as references in preparing this chapter.

IEEE Std 288-1969, Induction Motor Protection (ANSI C37.92-1972)

IEEE Std 329-1971, Synchronous Motor Protection (ANSI C37.94-1972)

NFPA No 70, National Electrical Code (1975) (ANSI C1-1975)

9.5 References

[1] GLEASON, L.L., and ELMORE, W.A. Protection of Three-Phase Motors against Single-Phase Operation. *AIEE Transactions (Power Apparatus and Systems),* pt III, vol 77, Dec 1958, pp 1112-1120.

[2] SCHMIDT, R.A. Calculation of Fault Currents for Internal Faults in AC Motors. *AIEE Transactions (Power Apparatus and Systems),* pt III, vol 75, Oct 1956, pp 818-824.

[3] FAWCETT, D.V. Protection of Large Three-Phase Motors. *IEEE Transactions on Industry and General Applications,* vol IGA-3, Jan/Feb 1967, pp 52-55.

[4] YUEN, M.H., RITTENHOUSE, J.D., and FOX, F.K. Large Wound-Rotor Motor with Liquid Rheostat for Refinery Compressor Drive. *IEEE Transactions on Industry and General Applications,* vol IGA-1, Mar/Apr 1965, pp 140-149.

10. Transformer Protection

10.1 General Discussion. Increased use of electric power in industrial plants has been requiring larger and more expensive primary and secondary substation transformers. This chapter is directed toward the proper protection of these transformers.

Primary substation transformers normally range in size between 1000 and 12 000 kVA with a secondary voltage between 2400 and 13 800 V. Secondary substation transformers normally range in size between 300 and 2500 kVA with secondary voltages of 208, 240, or 480 V. Larger and smaller transformers may also be protected by the devices discussed in this chapter.

10.2 Need for Protection. Proper protection is important on transformers of all sizes, even though they may be some of the simplest and most reliable components of the plant's electric system.

A previous study [1] indicated that transformers above 500 kVA had a failure rate which was lower than that of most other system components. In that study, transformers averaged only 76 failures per 10 000 transformer-years. This might incorrectly be taken to imply that little or no transformer protection is required.

The need for transformer protection is strongly indicated when the average forced hours of downtime per transformer-year is considered. In this category, transformers ranked just below the utility power supply in most cases, as shown in Table 29. The large 253 h average out-of-service time per transformer failure is a challenge to the system engineer to properly protect the transformer and minimize any damage that could occur.

Protection is achieved by the proper combination of system design and protective devices needed to economically meet the requirements of the application. Protection should be designed to include

(1) Protection of the transformer from harmful conditions occurring on the system to which the transformer is connected

(2) Protection of the electric system from the effects of transformer failure

(3) Detection and indication of conditions occurring within the transformer which might cause damage or failure

The system and devices should be selected to provide protection in each of the above areas. Some devices will include more than one of these areas in their protective operation.

In addition to electrical protection for each installation, physical and environmental protection must be included. Special consideration must be given to such things as damage from vehicles operating in the area, harmful fumes or vapors, and other unusual service conditions.

10.3 Liquid Preservation Systems. Liquid preservation systems are used to preserve the amount of the insulating liquid and prevent its contamination from airborne vapors, especially water vapor. The importance of maintaining the purity

Table 29
Reliability of Major System Components [1]

System Component	Average Failures per Year	Average Hours per Failure	Average Forced Hours Downtime per Year
Electric utility power supply	0.963	2	1.93
Power cables (per 1000 circuit ft)			
underground, nonleaded conduit	0.0397	36	1.43
above ground, nonleaded conduit	0.0252	44	1.11
Drawout metal-clad circuit breakers			
600 V and below	0.0163	25	0.41
above 600 V	0.0104	79	0.82
Transformers (primary below 15 kV)	0.0076	253	1.92

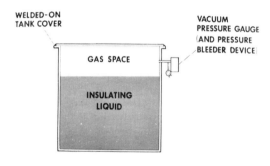

Fig 128
Sealed-Tank Construction Is the
Most Common Type of Oil Preservation System
Used for Industrial Substation Transformers

of insulating oil becomes increasingly critical at higher voltages, especially above 69 kV, because of increased electrical stress on the insulating oil. Different types of systems are used.

10.3.1 *Sealed Tank.* The sealed-tank design is the one most commonly used and is standard on most substation transformers. Fig 128 shows this design, where, as the name implies, the transformer tank is sealed, isolating it from the atmosphere.

A gas space equal to about one sixth the liquid volume is maintained above the liquid to allow for thermal expansion. This space is purged of air and is usually nitrogen filled.

A pressure-vacuum gauge and bleeder device are furnished on the tank to allow monitoring the

internal pressure or vacuum and to allow relieving any excessive pressure buildup.

10.3.2 *Positive-Pressure Inert Gas.* This design shown in Fig 129 is similar to the sealed-tank design with the addition of a gas (usually nitrogen) pressurizing assembly. This assembly provides a slight positive pressure in the gas supply line to prevent air from entering the transformer during operating mode or temperature changes. Transformers of 69 kV primary winding and above and of 7500 kVA and above can be equipped with this device.

10.3.3 *Gas-Oil Seal.* This design incorporates a captive gas space which isolates a second auxiliary oil tank from the main transformer oil as shown in Fig 130. The auxiliary oil tank is open to the atmosphere and provides room for thermal expansion of the main transformer oil volume.

The main tank oil expands or contracts due to changes in its temperature causing the level of the oil in the auxiliary tank to raise or lower as the captive volume of gas is forced out of or allowed to reenter the main tank. The pressure of the auxiliary tank oil on the contained gas maintains a positive pressure in the gas space preventing atmospheric vapors from entering the main tank.

This design may be used where reduced overall height of a transformer is required.

10.3.4 *Conservator Tank.* The conservator-tank design shown in Fig 131 does not have a gas space above the oil in the main tank. It includes a second oil tank above the main tank cover with a gas space adequate to absorb the thermal expan-

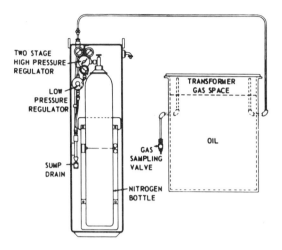

Fig 129
Positive-Pressure Inert-Gas Assembly
Is Often Used on Sealed-Tank Transformers
above 7500 kVA and 69 kV Primary Voltage

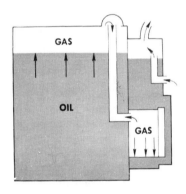

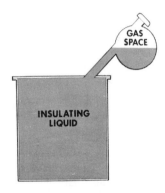

Fig 130
Gas-Oil Seal Design Isolates Main Tank
Oil from Atmosphere. This Design is not
Commonly Used in the United States Today

Fig 131
Conservator-Tank Construction Must Be
Used with "Bucholtz" Type Sudden Pressure
Relay or with Gas Accumulator Relay

Fig 132
Liquid-Level Indicator Depicts
Level of Liquid with Respect to a
Predetermined Level, Usually 25°C

Fig 134
Gas-Pressure Gauge Indicates
Internal Gas Pressure Relative
to Atmospheric Pressure. Bleeder
Allows Pressure to Be Equalized Manually

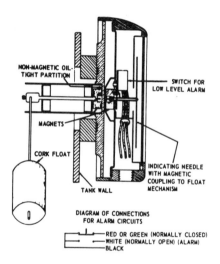

Fig 133
Liquid-Level Indicating
Needle Is Driven by a Magnetic
Coupling to the Float Mechanism

sion of the main tank oil volume. The second tank is connected to the main tank by an oil-filled tube or pipe.

Since a gas space is not required in the main tank, this can result in a lower height unit. This type of design must be used when the "Bucholtz" type sudden-pressure relay and the gas accumulator relay are used. These relays are described later in the chapter.

10.3.5 *Liquid-Level Gauge.* The liquid-level gauge, shown in Figs 132 and 133, is used to measure the level of insulating liquid within the tank with respect to a predetermined level, usually indicated as the 25°C (77°F) level. An excessively low level could indicate the loss of insulating liquid which would lead to internal flashovers if not corrected. Periodic observation is normally employed to check that the liquid level is within acceptable limits. Alarm contacts for low liquid level are normally available as a standard option when specified.

10.3.6 *Pressure-Vacuum Gauge.* The pressure-vacuum gauge in Fig 134 indicates the difference between the transformer internal gas pressure and the atmospheric pressure. It is used on transformers with sealed-tank oil preservation systems. Both the pressure-vacuum gauge and the sealed-tank oil preservation system are standard on most small and medium power transformers. This is likely the simplest and most maintenance-free of all oil preservation systems.

The pressure in the air space is normally related to the thermal expansion of the insulating liquid and will vary with load and ambient temperature changes. Large positive or negative pressures could indicate an abnormal condition, particularly if the transformer had been observed to remain within normal pressure limits for some time.

10.4 Transformer Failures. Failure of a transformer can be caused by any of a number of internal or external conditions which make the unit incapable of performing its proper function electrically or mechanically. Transformer failures may be grouped as follows:

(1) Winding failures are the most frequent cause of transformer failure. Reasons for this type of failure include insulation deterioration or defects, overheating, mechanical stress, vibration, and voltage surges.

(2) Terminal boards and no-load tap changers. Failures are attributed to improper assembly, damage during transportation, excessive vibration, or inadequate design.

(3) Bushing failures can be caused by vandalism, contamination, aging, cracking, or animals.

(4) Load tap changer failures can be caused by mechanism malfunction, contact problems, insulating liquid contamination, vibration, improper assembly or excessive stresses within the unit. Load tap changing units are normally applied on utility systems rather than on industrial systems.

(5) Miscellaneous failures would include core insulation breakdown, bushing current transformer failure, liquid leakage due to poor welds or tank damage, shipping damage, and foreign materials left within the tank.

Failure of other equipment within the transformer protective device zone could cause the loss of the transformer to the system. This would include any equipment between the next upstream protective device and the next downstream device. Included may be such components as cables, bus ducts, switches, instrument transformers, lightning arrestors, and neutral grounding devices.

10.5 Protection. Transformer failures, other than from physical or environmental reasons, are caused by three distinct undesirable conditions, overloads, short circuits, and overvoltages. Protection against these conditions is achieved by the proper combination of sensing device to detect the condition and equipment to reduce the condition to a desirable level or to disconnect the transformer from the system or from the undesirable condition.

Sensing devices must be able to distinguish between the various normal operating conditions and abnormal conditions. For example, primary overload protection must not operate on high magnetizing current inrush, but it should operate on a low, long-time overload. Numerous special sensors are available to aid in correctly identifying the presence of an internal fault condition.

Disconnecting equipment, usually circuit breakers or fused switches, must be capable of carrying all normal load currents and they also must be able to interrupt the maximum possible fault current at the point of application. In both cases, allowance for possible changes in the mode of operation or increases in system fault capacity should be made.

Mechanical protective devices usually include liquid-level gauge, pressure-vacuum gauge, and some form of liquid preservation system.

10.6 Overload Thermal Protection. An overload will cause a rise in the temperature of the various transformer components. If the final temperature is above the design temperature limit, deterioration of the insulation system will occur, causing a reduction in the useful life of the transformer. The insulation may be weakened such that a moderate overvoltage may cause insulation breakdown before expiration of expected service life.

Protection against overloads consists of both load limitation and overload detection. Transformer loads may be limited by designing a system where the transformer capacity is greater than the total connected load assuming a diversity in load usage. This is an expensive method to provide overload protection. Load growth and change in operating procedures would quite often eliminate the extra capacity needed for this protection.

Load limitation by disconnecting part of the load can be done automatically or manually. Automatic operation, because of its cost, is restricted to larger units. However, manual operation is often preferred as it gives greater flexibility in selecting the expendable loads.

The major load limitation that can be properly applied to a transformer is one that responds to transformer temperature. By monitoring the temperature of the transformer, overload conditions can be detected. A number of monitoring

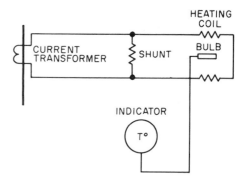

Fig 135
Liquid Temperature Indicator Is
the Most Common Transformer
Temperature Sensing Device

Fig 136
Thermal (or Winding Temperature)
Relay Uses a Heating Element to
Duplicate Effects of Current in Transformer

devices which mount on the transformer are available as standard or optional accessories.

These devices are normally used for alarm or to initiate secondary protective device operation. They include the following.

10.6.1 *Liquid Temperature Indicator.* The liquid temperature indicator shown in Fig 135 measures the temperature of the insulating liquid at the top of the transformer. Since the hottest liquid is less dense and rises to the top of the tank, the temperature of the liquid at the top reflects the temperature of the transformer windings and is related to the loading of the transformer. A high reading could indicate an overload condition.

The liquid temperature indicator is normally furnished as a standard accessory on power transformers. It is equipped with a temperature indicating pointer and a drag pointer which shows the highest temperature reached since it was last reset.

This device can be equipped with one to three contacts which operate at preset temperatures. The single contact can be used for alarm purposes. When forced air cooling is employed, the first contact (normally 60°C) initiates the first stage of fans. The second contact (90°C) either initiates a second stage of fans, if furnished, or an alarm. The third contact (115°C), if furnished, is used for the final alarm or to initiate load reduction on the transformer. The indicated temperatures would change for different temperature insulation system designs.

10.6.2 *Thermal Relays.* Thermal relays, diagramatically shown in Fig 136, are used to give a more direct indication of winding temperatures than the liquid temperature relay. A current transformer, mounted on a transformer bushing, supplies current to the thermometer bulb heater coil which contributes the proper heat to closely simulate the transformer hot-spot temperature.

The indicator is a bourdon gauge connected through a capillary tube to the thermometer bulb. The fluid in the bulb will expand or contract proportionally to the temperature changes and is transmitted through the tube to the gauge. Coupled to the shaft of the gauge indicator are three cams which operate individual switches at preset levels of indicated transformer temperature.

Thermal relays are used more often on transformers 10 000 kVA and above than on smaller transformers. They can be used on all sizes of substation transformers.

10.6.3 *Hot-Spot Temperature Equipment.* Hot-spot temperature equipment in Fig 137 is similar to the thermal relay equipment on a transformer, since it duplicates the hottest spot temperature of the transformer. While the thermal relay does it with fluid expansion and a bourdon gauge, the hot-spot temperature equipment does it electrically using a Wheatstone bridge method. Since this can be used with more than one detector coil location, temperatures of several locations within the transformer can be checked.

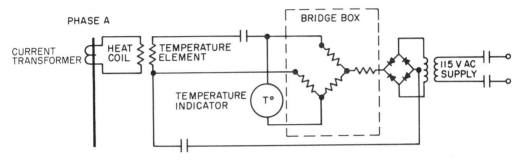

**Fig 137
Hot-Spot Temperature Indicator
Utilizes Wheatstone Bridge Method
to Determine Transformer Temperature**

**Fig 138
Forced-Air Fans Are Normally
Controlled Automatically from a Top Liquid
Temperature or Winding Temperature Relay**

The hot-spot temperature equipment, however, is not as flexible as the thermal relay because it merely indicates temperature and cannot perform such functions as starting fans, activating alarms, or tripping circuit breakers. This equipment should be used only in such cases where the transformer operation is closely monitored.

10.6.4 *Forced-Air Cooling.* Another means of protecting against overloads is to increase the transformer capacity by auxiliary cooling as shown in Fig 138. Forced-air cooling equipment is used to increase the capacity of a transformer by 15 to 33 percent of base rating, depending upon transformer size and design. Dual cooling by a second stage of forced-air fans or a forced-oil system will give a second increase in capacity applicable to transformers 12 000 kVA and above.

Auxiliary cooling of the insulating liquid helps keep the temperature of the windings and other components below the design temperature limits. Usually the cooling equipment is automatically initiated by the top liquid temperature indicator or the thermal relay, after a predetermined temperature is reached.

10.7 Short-Circuit-Current Protection.

In addition to thermal damage from prolonged overloads, transformers are adversely affected by internal or external short-circuit conditions. Internal short-circuit damage is caused by electromagnetic forces, temperature rise, and arc-energy release.

External short circuits could subject the transformer to short-circuit-current magnitudes limited only by the sum of transformer and supply-system impedance. Transformers with unusually low impedance may experience mechanical damage. Prolonged flow of a short-circuit current of lesser magnitude can inflict thermal damage.

Protection of the transformer against both internal and external faults and overloads should be as rapid as possible to reduce damage to a minimum. This may be restricted by selective coordination system design and operating procedure limitations.

There are several sensing devices available which provide varying degrees of short-circuit protection. The devices sense two different aspects of a short circuit. The first group of devices

Fig 139
Pressure-Relief Device Limits
Internal Pressure to Prevent Tank
Rupture under Internal Fault Conditions

sense the formation of gases consequent to a fault and are used to detect internal faults. The second group senses the magnitude of short-circuit current directly. The overcurrent relays can also provide overload protection.

The gas-sensing devices include a pressure-relief device, pressure relay, gas-detector relay, and combustible-gas relay. The current-sensing devices include fuses, overcurrent relays, and differential relays.

10.7.1 *Gas-Sensing Device.* Low-magnitude faults in the transformer normally cause gases to be formed by the decomposition of insulation exposed to high temperature at the fault. Detection of the presence of these gases can allow the transformer to be taken out of service before extensive damage occurs. In some cases, gas may be detected a long time before the unit fails.

High-magnitude fault currents will usually be first sensed by other detectors, but the gas-sensing device will respond with modest time delay.

10.7.2 *Pressure-Relief Device.* A pressure-relief device is a standard accessory on all liquid insulated substation transformers, except on secondary substation oil-insulated units, where it is optional. This device shown in Fig 139 can relieve both minor and serious internal pressures. When the internal pressure exceeds the tripping pressure ($10 \text{ lb}_f/\text{in}^2 \pm 1 \text{ lb}_f/\text{in}^2$ gauge), the device snaps open allowing the excess gas or fluid to be released. Upon operation, a pin (standard), alarm contact (optional), or semaphore signal (optional) is actuated to indicate operation. The device has normally automatic resetting and requires little or no maintenance or adjustment.

The major function of the pressure-relief device is to prevent rupture or damage to the transformer tank due to internal fault conditions.

10.7.3 *Pressure Relay.* The pressure relay is normally used to initiate isolation of the transformer from the electric system and to limit damage to the unit when there is an abrupt rise in the transformer internal pressure. The abrupt pressure rise is due to the vaporization of the insulating liquid by an internal fault. The bubble of gas formed in the insulating liquid creates a pressure wave which activates the relay promptly.

One type of relay, shown in Fig 140, uses the insulating liquid to transmit the pressure wave to the relay bellows. Inside the bellows a special oil transmits the pressure wave to a piston which will actuate a set of switch contacts. This type of relay is mounted on the transformer tank below oil level.

Another type of relay shown in Fig 141 uses the inert gas above the insulating liquid to transmit the pressure wave to the relay bellows. Expansion of the bellows actuates a set of switch contacts. This type of relay is mounted on the transformer tank above oil level.

Both types of relays have a pressure-equalizing opening to prevent operation of the relay on gradual rises in internal pressure due to changes in loading or ambient conditions.

Both types of pressure relays are very sensitive to the rate of rise in the internal pressure. The time for the relay switch to operate is in the magnitude of 4 cycles for high rates of pressure rise [25 (lb_f/in^2)/s of oil pressure rise; 5 (lb_f/in^2)/s of air pressure rise]. These relays are designed to be insensitive to mechanical shock and vibration, through faults, and magnetizing current inrush.

The use of pressure relays normally increases as the size and value of the transformer increase. Most transformers 5000 kVA and above are equipped with this type of device. This relay provides valuable protection at low cost.

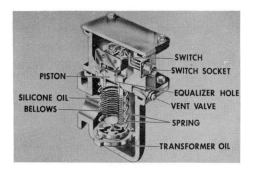

Fig 140
Pressure Relay Type Mounted on
Transformer Tank below Normal Oil Level

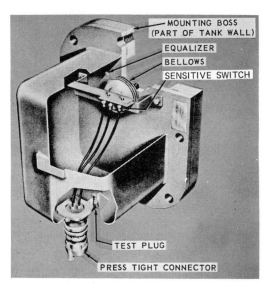

Fig 141
Pressure Relay Type Mounted on
Transformer Tank above Normal Oil Level

10.7.4 *Gas-Detection Relay.* The gas-detection relay shown in Fig 142 is a special device used to detect and indicate an accumulation of gas from a transformer. Incipient winding faults or hot spots in the core normally generate small amounts of gas which are channeled to the top of the special domed cover. From there the bubbles enter the accumulation chamber of the relay through a pipe. An accumulation of gas of 100 cm^3, for example, will lower a float and operate an alarm switch. The gas can then be withdrawn for analysis and recording. Note that the conservator-type tank design must be used.

10.7.5 *Combustible-Gas Relay.* The combustible-gas relay as shown in Fig 143 is a special device used to detect and indicate the presence of combustible gas coming from the transformer. The combustible gas is formed by the decomposition of insulating materials within the transformer by a low-level fault. These faults are normally not detected until they develop into larger and more damaging ones.

The combustible-gas relay can be used on transformers with positive-pressure inert gas-oil preservation systems. The relay periodically takes a sample of the gas in the transformer and tests it with a heated wire. If combustible gases are in the sample, they will ignite, further heating the wire which in turn changes its resistance. The change is detected by a bridge network and activates a signal relay. This relay is expensive and is not normally applied on substation transformers.

Portable gas-analysis equipment can be used to test the composition of gases in the transformers. By analyzing the percentage of unusual or decomposed gases in the transformer, it can be determined if the transformer has a low-level fault, and if so what type of fault has occurred. This type of device is normally used on utility systems having large numbers of large-capacity transformers.

10.7.6 *Current-Sensing Devices.* Fuses, overcurrent relays, and differential relays must be designed into a system to properly provide protection to the transformer. This system of protection must meet the requirements of applicable standards and codes as well as needs of the power system. For the pertinent provisions of NFPA No 70, National Electrical Code (1975), (ANSI C1-1975), refer to Chapter 7, Section 7.3.5.

The NEC requirements are upper limits that must be met when selecting overcurrent protective devices. These requirements are not guidelines to the design of a system with maximum protection for transformers. Setting a transformer primary or secondary overcurrent protective device at 2.5 times rated current could allow that transformer to be damaged without the device operating.

The best protection for the transformer would be its own primary and secondary circuit breakers set to operate at minimum values. Common practice is for the secondary circuit breaker to protect the transformer for loading in excess of 125 percent of maximum rating. Using a primary circuit breaker for each transformer is expensive, especially for small-capacity and small-value transformers.

In many industrial systems the economical compromise of installing one circuit breaker to feed two to six relatively small transformers is used. Each transformer has its own secondary circuit breaker. Overcurrent protection must satisfy the requirements prescribed by the NEC. The major disadvantage of this system is that all transformers will be de-energized by the opening of the primary circuit breaker.

By using fused switches on the primary where possible, additional short-circuit protection can be provided for the transformer and additional selectivity provided for the system. It does not provide additional low-level overcurrent protection nor does it provide protection against single-phase operation. Coordination becomes more difficult with the limited range and characteristics of fuses available. Fuses must be sized large enough to pass transformer magnetizing current inrush and so are less sensitive to low-magnitude fault conditions.

Any protective device on the primary of the transformer must be set high enough to allow the magnetizing current inrush to pass without operation of that device. Magnetizing current inrush time and magnitude vary between different designs of transformers. Inrush currents of eight or twelve times normal full-load current for 0.1 s are frequently used for coordination purposes.

Transformers must be protected against fault currents which exceed the through-fault current withstand rating.

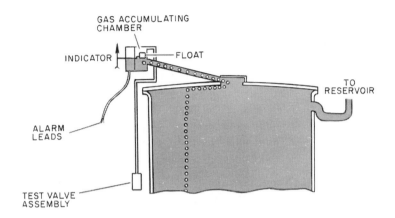

Fig 142
Gas-Detector Relay Accumulates
Gases from Top Air Space of Transformer

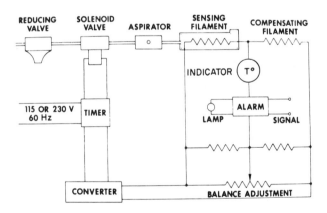

Fig 143
Combustible Gas Relay Periodically Samples Gas in
Transformer to Detect Any Minor Internal Fault
Before it Can Develop into a Serious Fault

Table 30
Transformer Through-Fault-Current Withstand

Transformer Impedance (percent)	Withstand Multiple of Base Current	Time Period (seconds)
4.0 and lower	25.0	2
5.0	20.0	3
5.5	18.2	3.5
5.75	17.4	3.75
6.0	16.6	4
7.0 or above	14.4 or less	5

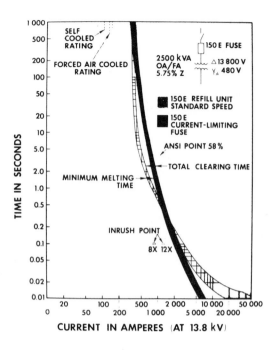

Fig 144
Comparison of Expulsion
(or Power) Fuse and Equivalent
Current-Limiting Fuse Emphasizing
Difference in Characteristics of Two Fuse Types

Transformers are designed to withstand the effects of through-fault currents as defined by IEEE Std 462-1973, General Requirements for Distribution, Power, and Regulating Transformers (ANSI C57.12.00-1973). Thermally, the transformer must withstand the temperature rise that occurs when the fault current flows for a designated time period. The time period is related to the transformer impedance.

Mechanically the transformer must withstand the electromagnetic forces caused by the through-fault current to the extent designated by IEEE Std 462-1973. The through-fault current magnitude is limited by the transformer impedance and influenced by the instant on the excitation wave when the fault current is initiated.

IEEE Std 462-1973 requires that the transformer withstand capability be designed based on the assumption that it is connected to an infinite bus and that a fault occurs at the instant necessary to provide maximum offset.

The system designer and coordinator must protect the transformer from being subjected to through-fault currents on any winding of greater magnitude or longer time than indicated in Table 30, based on IEEE Std 462-1973, Section 10.1.

For delta primary, wye secondary connected transformers the withstand magnitude as seen by the primary device must be reduced to 58 percent of the value given in Table 30.

10.7.7 *Fuses.* Fuses are relatively simple and inexpensive one-time devices that can provide short-circuit protection for a transformer primary. Fuses are normally applied in conjunction with a magnetizing current break switch which is interlocked with the secondary circuit breaker to prevent the switch from operating under load conditions.

Fuse selection considerations include having an interrupting capacity higher than the system fault capacity at the point of application, having a continuous current above the maximum continuous load under various operating modes, and having time—current characteristics which will pass the magnetizing current inrush without fuse damage and will interrupt before the transformer withstand point is reached.

A number of features of relay-protected systems are lacking when fused switches are used. Low-level overcurrent protection is not available. Single-phase operation can occur when only one fuse blows. Selective coordination is difficult due to limited characteristics of available fuses. For large-size transformers and some connections fuses cannot be properly applied. Fig 144 shows a comparison between two types of fuses normally used for transformer protection. These fuses are further described in Chapter 5.

10.7.8 *Overcurrent Relays.* Overcurrent relays described in Chapter 3 are a means of providing overcurrent and short-circuit protection for transformers. They are applied in conjunction with current transformers and a circuit breaker, sized for the maximum continuous and interrupting duty requirements of the application. A typical application is shown in Fig 145.

Overcurrent relays are selected to provide a range of overcurrent settings above the permitted overloads and instantaneous settings when possible within the transformer through-fault current withstand rating. The characteristics should be selected to coordinate with upstream and downstream protective devices.

Ground faults occurring in the substation transformer secondary or between the transformer secondary and main secondary circuit breaker cannot be isolated by the main secondary circuit breaker, which is located on the load side of the ground fault. These ground faults, when limited by a neutral grounding resistor, are not seen by the transformer primary fuses and can be isolated only by a primary circuit breaker tripped by a ground relay in the grounding-resistor circuit.

10.7.9 *Differential Relays.* Differential relays operate on current unbalance between the primary and secondary windings of a transformer. When this unbalance occurs, the relays initiate disconnection of the transformer from the system. Since the relays operate on the difference in currents, which is normally small, they can be sensitive to fault conditions within the zone of protection. Refer to Chapter 3.

The differential relay zone of protection extends from the upstream current transformer location to the downstream current transformer location. Faults within will be detected by the relays when properly applied. A typical application is shown in Fig 146.

Several problems are involved in the application of differential relays.

(1) The system must be designed so that the relays can operate a transformer primary circuit breaker. If a remote circuit breaker can be operated, a remote trip system must be used, by using either a pilot wire or a high-speed grounding switch. Normally the utility controls the remote circuit breaker and may not allow it to be tripped. Operation of a user-owned local circuit breaker presents little problem.

(2) Current transformers associated with each winding have different ratios, ratings, and characteristics when subjected to heavy loads and short circuits. Auxiliary current transformers or relays with tapped operating and restraint coils may be used.

(3) Transformer taps can be operated changing the effective turns ratio. By selecting the ratio and taps for midrange, the maximum unbalance will be equivalent to half the transformer tap range.

(4) Magnetizing current inrush appears as an internal fault to the differential relays. The relays must be desensitized to the current inrush, but they should be sensitive to short circuits within the zone during the same period. This can be accomplished in the more expensive relays with harmonic restraint. The magnetizing current inrush has a large harmonic component which is not present in short-circuit currents. This is used by the harmonic-restraint relays to distinguish between faults and inrush. For some less critical applications, voltage relays are used to desensitize or control the tripping of the differential relays.

(5) Transformer connections often introduce a phase shift between high- and low-voltage currents. This is compensated for by proper current transformer connections. For a delta primary, wye secondary connected transformer, current transformers are normally wye connected in the primary and delta connected in the secondary.

(6) Heavy currents for faults outside the zone of protection can cause an unbalance between the current transformers. Percentage differential relays shown in Fig 147, which operate when the difference is greater than a definite percentage of the phase current, are designed to overcome this problem. Percentage differential relays also help in solving the tap-changing problem and the current transformer ratio balance problem. Percentage slopes available are 15 percent for standard transformers, 25 percent for load tap-changing transformers, and 40 percent for special applications.

Differential relays are often recommended on all transformers 1000 kVA and above and with 2400 V and above secondary voltage. Harmonic-

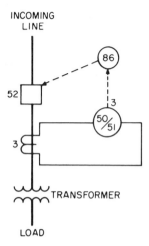

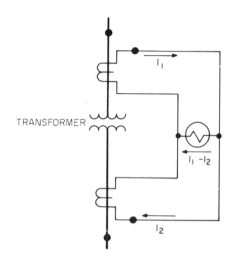

Fig 145
Overcurrent Relays Are Frequently
Used to Provide Transformer Protection
in Combination with Primary Circuit Breaker

Fig 146
Standard Overcurrent Relays Carefully
Applied Can Be Used for Differential
Protection of Transformers. However,
Percentage Differential Relays Are Preferred

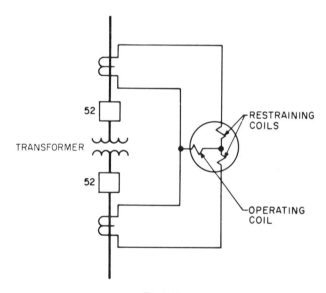

Fig 147
Percentage Differential Relays Provide Increased Sensitivity
while Minimizing False Operation as a Result of Current
Transformer Mismatch Errors for Heavy Through Faults

restraint percentage differential relays are recommended for transformers 2000 kVA and above.

10.8 Protection against Overvoltages. Overvoltages caused by lightning, switching surges, and faults must be considered in designing transformer protection. Liquid-insulated transformers have relatively high basic insulation level ratings when compared with standard ventilated dry and sealed dry transformers. Special sealed dry and ventilated dry transformers with basic insulation level ratings approaching that of liquid transformers are available.

Ordinarily, if the liquid-insulated transformers are supplied by enclosed conductors from the secondaries of transformers with adequate primary surge protection, no additional protection is required. However, if the transformer primary or secondary is connected to conductors which are exposed to lightning, the installation of lightning arresters is necessary. For best protection, the lightning arresters should be mounted as close as possible to the transformer terminals, preferably within 3 ft.

The degree of protection is determined by the amount of exposure, the size and importance of the transformer to the system, and the type and cost of the arresters. In descending order of cost and degree of protection, the arresters available are station type, intermediate type, and distribution type. Ventilated dry and sealed dry type transformers are normally used indoors, connected to shielded systems and furnished with surge protection. These transformers are sensitive to transmitted and reflected surges caused by lightning and system disturbances. Special distribution-type arresters have been developed for the protection of dry type transformers and rotating machinery.

10.9 Conclusion. Protection of today's larger and more expensive transformers can be achieved by the proper selection and application of protective devices. Published application guides covering transformers are few in number; see, for example, IEEE Std 273-1967, Protective Relay Applications to Power Transformers (ANSI

C37.91-1972). The system design engineer must rely heavily on his sound engineering judgment to achieve an adequate protection system.

10.10 Standards References. The following standards publications were used as references in preparing this chapter.

IEEE Std 273-1967, Protective Relay Applications to Power Transformers (ANSI C37.91-1972)

IEEE Std 462-1973, General Requirements for Distribution, Power, and Regulating Transformers (ANSI C57.12.00-1973)

NFPA No 70, National Electrical Code (1975) (ANSI C1-1975)

10.11 References and Bibliography

10.11.1 *References*

[1]   DICKINSON, W.H. Report on Reliability of Electric Equipment in Industrial Plants. *AIEE Transactions (Applications and Industry)*, pt II, vol 81, Jul 1962, pp 132-151.

10.11.2 *Bibliography*

[2]   AIEE COMMITTEE REPORT. Bibliography of Industrial System Coordination and Protection Literature. *IEEE Transactions on Applications and Industry*, vol 82, Mar 1963, pp 1-2.

[3]   BEEMAN, D.L., Ed. *Industrial Power Systems Handbook.* New York: McGraw-Hill, 1955.

[4]   BOYARIS, E., and GUYOT, W.S. Experience with Fault Pressure Relaying and Combustible Gas Detection in Power Transformers. *Proceedings of the American Power Conference*, vol 33, Apr 1971, pp 1116-1126.

[5]   BURGIN, E.R. A Comparison of Protective Methods and Devices for Industrial Power Transformers. *Proceedings of the American Power Conference*, vol 26, Apr 1964, pp 931-938.

[6]   MASON, C.R. *The Art and Science of Protective Relaying.* New York: Wiley, 1956.

[7]   *The Art of Protective Relaying.* Philadelphia, PA: General Electric Company, Bulletin 1768.

[8]   *Applied Protective Relaying.* Newark, NJ: Westinghouse Electric Corporation.

11. Conductor Protection

11.1 General Discussion. This chapter deals with power cable protection as well as with busway protection. The primary considerations are presented along with some methods of application.

The proper selection and rating or derating of power cables is as much a part of cable protection as the application of the short-circuit and overcurrent protection devices. The whole scheme of protection is based on a cable rating that is matched to the environment and operating conditions. Methods of asigning these ratings are discussed.

Power cables require short-circuit-current, overload, and physical protection in order to meet the requirements of NFPA No 70, National Electrical Code (1975), (ANSI C1-1975), (NEC). A brief description of the phenomena of short-circuit current, overload current, and their temperature rises is presented, followed by a discussion of the time—current characteristics of both cables and protective devices. In addition, typical cases of cable systems and sample selection and coordination of protective devices are offered.

Because of their rigid construction, busways provide their own mechanical protection. However, they do require short-circuit-current and overload protection. A brief discussion of the types of faults on the busways is presented, followed by a discussion of various methods of fault protection.

The general intent of this chapter is to provide a basis for design, pointing out the problems involved and providing guidance in the application of cable and busway protection. Each specific case and type of cable or busway requires

attention. In most cases, the attention is routine, but the out-of-the-ordinary cable and busway schemes require careful consideration.

A. Cable Protection

11.2 Cable Protection. Cables have always been the lifeline of any industrial and commercial facility or process. As today's plants are built and expanded, equipment and production units are increasing in size for greater efficiency and production. Cables transmit the ever increasing power to these machines or complexes, and they also transmit the information and control signals which are essential to the closely controlled uniform products expected of today's technology. Thus cables are generally classified as either power or control types. Power cables are divided into two voltage classes; 600 V and below, and above 600 V. Control cables include those used in the control of equipment and also for voice communication, metering, and data transmission.

The amount of damage caused by the faulting of power cable has been illustrated many times. As power and voltage levels increase, the potential hazards also increase. High temperature is probably the most frequent cause of decreased cable life and failure. Power cables, internally heated as a result of their resistance to the current being carried, can undergo insulation failure if the temperature buildup becomes excessive. Proper selection and rating ensures that the cable is large enough for the expected current. In addition to insulation breakdown, protection is also required against unexpected overload and short-circuit current. Overcurrent can occur due to an increase in the number of connected loads or due to overloading of existing equipment.

Short circuits cause cable damage because of excessive temperature. However, in this case large quantities of energy may be liberated at one place, and cables may be heated for their entire lengths at a rate which for practical considerations precludes dissipation of the heat.

Physical conditions can also cause cable damage and failure. Failure due to excessive heat may be caused by high ambient temperature conditions or fire. Mechanical damage may result in short circuits or reduced cable life and may be caused by persons, equipment, animals, insects, or fungus.

Cable protection is required to protect personnel and equipment and to ensure continuous service. From the standpoint of equipment and process the type of protection selected is generally determined by economics and the engineering requirements. Personnel protection also recieves careful engineering attention as well as special consideration to assure compliance with the various codes that may be applicable to a particular installation.

Protection against overcurrent is generally achieved by means of a device sensitive to current and the length of time during which it flows. Short-circuit protective devices are sensitive to much greater currents and shorter times. Protection against environmental conditions takes on many forms.

In general this chapter covers methods of rating cables and the conditions and problems listed above. It also provides a starting point from which further refinements may be made and other features added for improved power cable protection.

11.3 Glossary. The following symbols are used in cable protection technology.

(1) *Cable current, in amperes*

I = Current flowing in cable
I_0 = Initial current prior to a current change
I_F = Final current after a current change
I_N = Normal loading current on base ambient temperature

I_{N1} = Normal loading current on non-base ambient temperature
I_E = Emergency loading current on base ambient temperature
I_X = Current at values other than normal or emergency loading
I_{sc} = Three-phase short-circuit current

(2) *Cable temperature, in degrees Celsius*

T = Temperature in general
T_0 = Initial temperature prior to a current change
T_f = Final temperature after a current change
T_N = Normal loading temperature
T_E = Emergency loading temperature
T_X = Temperature at any loading current
T_t = Temperature at time t after a current change
T_a = Base ambient temperature
T_{a1} = Nonbase ambient temperature

(3) *Miscellaneous*

t = Time in units as noted
CM = Conductor size in circular mils
F_{ac} = Skin effect ratio or alternating-to-direct current ratio; NEC Table 9
K = Time constant or geometric factor of cable heat flow
K_t = Correction factor for initial and final short-circuit temperature

(4) *Reactances, in percent*

X_T = Transformer reactance
X_d'' = Subtransient reactance of a rotating machine
X_d' = Transient reactance of a synchronous machine

11.4 Short-Circuit-Current Protection of Cables. A cable must be protected from overheating due to excessive short-circuit current flowing in its conductor. The fault point may be on a section of the protected cable or on any other part of the electric system. The faulted cable section is, of course, to be replaced after the fault has been cleared.

During a phase fault the I^2R losses in the phase conductors elevate first the temperature of the conductor, followed by the insulation materials, protective jacket, raceway, and surroundings. During a ground fault the I^2R losses in both

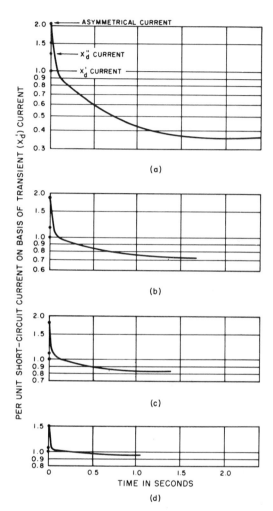

Fig 148
Typical Rate of Short-Circuit Current Decay
(a) Plant Generator System, Medium-Voltage.
(b) Utility-Power Supplied
System, Medium-Voltage, with
Large Synchronous Motors. (c) Utility-Power
Supplied System, Medium-Voltage,
No Synchronous Motor. (d) Utility or Plant
Generation, Low-Voltage 240 or 480 V
Load Centers

phase conductor and metallic shield or sheath elevate the temperature in a manner similar to that of phase faults.

Since the short-circuit current is to be interrupted either instantaneously or in a very short time by the protective device, the amount of heat transferred from the metallic conductors outward to the insulation and other materials is very small. Therefore the heat from $I^2 R$ losses is almost entirely in the conductors, and for practical purposes it can be assumed that 100 percent of the $I^2 R$ losses is consumed to elevate the conductor temperature. During the period that the short-circuit current is flowing, the conductor temperature should not be permitted to rise to the point where it may damage the insulating materials. The task of providing cable protection during a short-circuit condition involves determining the following:

(1) Maximum available short-circuit currents

(2) Maximum conductor temeperature that will not damage the insulation

(3) Cable conductor size that affects *the $I^2 R$* value and its capability to contain the heat

(4) Longest time that the fault will exist and the fault current will flow

11.4.1 *Short-Circuit Current*

(1) *Phase-Fault Current and Rates of Decay.* The fundamentals of short-circuit current behavior and the calculation of short-circuit currents are described in Chapter 2. The magnitude of short-circuit current must be properly determined. As illustrated in Fig 148 the initial peak current is called asymmetrical current (or current for momentary duty). This then decays in sequence to the subtransient current, transient current, and synchronous current or sustained short-circuit current. The short-circuit current in the subtransient period, transient period, or synchronous period decays exponentially. Fig 148 shows the approximate rate of decay of the total current. Four typical systems are illustrated here to give the reader a general picture of the fault-current behavior. The decay rate in each system depends on the X/R ratio of the system.

(2) *Maximum Short-Circuit Currents.* Generally the subtransient current of a system is used to designate the maximum available short-circuit current in the cables protected by the instantaneous overcurrent relays and medium-voltage switchgear circuit breakers. For cables protected

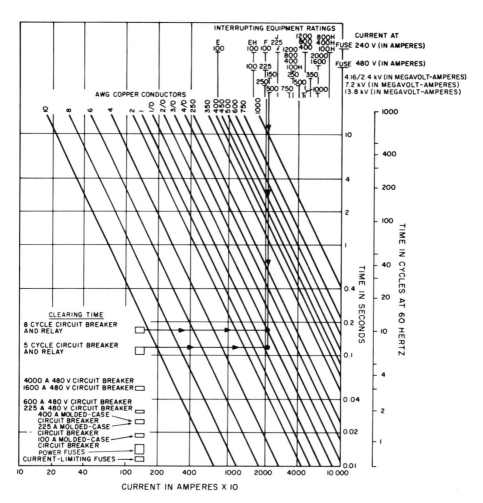

Fig 149
Maximum Short-Circuit Current for Insulated
Copper Conductors. Initial Temperature 75°C;
Final Temperature 200°C; for Other Temperatures
Use Correction Factors of Fig 151

by fuses, cable limiters and protectors (Chapter 5), or low-voltage and instantaneous trip circuit breakers, the asymmetrical current value is used. For delayed tripping at 0.2 s or longer, the rms value of the decayed current over the flow period of fault current should be used. If the latter is not possible, a reduced value of the subtransient current may be estimated. Fig 148 gives the rate of decay from which the subtransient current might be reduced to obtain a realistic value of total fault current over the entire delayed period.

(3) *Short-Circuit Currents Based on Equipment Ratings.* For liberal design margins where economic considerations are not critical, the momentary and interrupting current ratings of the switchgear, circuit breakers, or fuses may be used as the basis for cable selection and protection. The current ratings of most commonly used interrupting equipment are indicated in the upper right-hand corner of Figs 149 and 150 and can be used as maximum short-circuit currents for the selection of cable size.

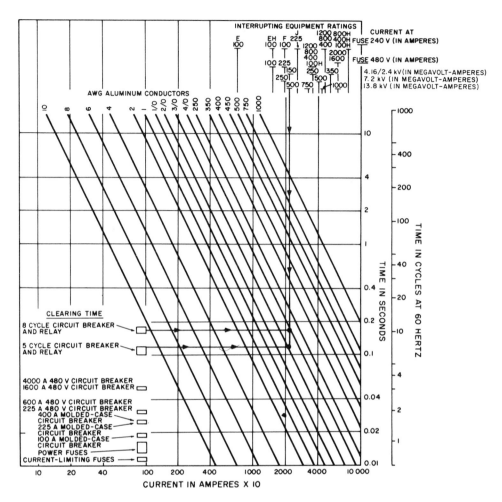

Fig 150
Maximum Short-Circuit Current for Insulated
Aluminum Conductors. Initial Temperature 75°C;
Final Temperature 200°C; for Other
Temperatures Use Correction Factors of Fig 151

(4) *Ground-Fault Currents and Rates of Decay.* The fundamentals of ground short-circuit current behavior are similar to those of phase-fault current, but the calculations are different, as described in Chapters 2 and 8. For a solidly grounded system, the ground-fault current is of about the same magnitude as the phase-fault current. For a low-resistance grounded system, the magnitude of the ground-fault current is limited to a sustained value determined by the resistor's current rating. The decay of the direct-current component occurs so rapidly that the asymmetry effect in the current wave shape can be ignored. For a high-resistance grounded or ungrounded system, the ground-fault current is small but should be immediately detected and cleared to prevent persistant arcing and the occurrence of a more serious fault involving other conductors or circuits.

11.4.2 *Conductor Temperature*

(1) *Temperature Rise of Phase Conductors.* On the basis that all heat is absorbed by the

Table 31
Temperature Rise of Shield and Sheath
Due to Ground-Fault Current

Material	Initial/Final Temperatures 65/200° C	65/150° C
Copper	$I = 0.0694 \dfrac{CM}{\sqrt{t}}$	$I = 0.0568 \dfrac{CM}{\sqrt{t}}$
Aluminum	$I = 0.0453 \dfrac{CM}{\sqrt{t}}$	$I = 0.0371 \dfrac{CM}{\sqrt{t}}$
Lead	$I = 0.0124 \dfrac{CM}{\sqrt{t}}$	$I = 0.0103 \dfrac{CM}{\sqrt{t}}$
Steel	$I = 0.0249 \dfrac{CM}{\sqrt{t}}$	$I = 0.0205 \dfrac{CM}{\sqrt{t}}$

Table 32
Maximum Short-Circuit Temperatures

Type of Insulation	Continuous Temperature Rating T_0 (°C)	Short-Circuit-Current Temperature Rating T_f (°C)
Rubber	75	200
Rubber	90	200+
Silicone rubber	125	250
Thermoplastic	75	150
Thermoplastic	85	200
Paper	85	200
Varnished cloth	85	200

conductor metal and there is no heat transmitted from the conductor to the insulation material, the temperature rise is a function of the size of the metallic conductor, the magnitude of the fault current, and the time of the current flow. These variables are related by the following formula [IPCEA P-32-382, Short-Circuit Characteristics of Insulated Cable (Rev Mar 1969)]:

for copper,

$$\left(\frac{I}{CM}\right)^2 \times t \times F_{ac} = 0.0297 \log_{10} \frac{T_f + 234}{T_0 + 234}$$

for aluminum,

$$\left(\frac{I}{CM}\right)^2 \times t \times F_{ac} = 0.0125 \log_{10} \frac{T_f + 228}{T_0 + 228}$$

If the initial temperature T_0 and final temperature T_f are predetermined on the basis of continuous-current rating and insulation material, respectively, the current I versus time t relation of current flow can be plotted for each conductor size (CM).

(2) *Temperature Rise of Shield and Sheath.* On the same basis as for phase conductors, the temperature rise on the shield or sheath due to ground-fault current can be related to the magnitude of the fault current I, the cross section (CM) of the shield and sheath, and the time t of current flow as shown in Table 31 [IPCEA P-45-482, Short-Circuit Performance of Metallic Shielding and Sheaths of Insulated Cable (Jun 1963)].

(3) *Maximum Short-Circuit Temperature Ratings.* IPCEA P-32-382 established a guideline

for short-circuit temperatures for various types of insulation as shown in Table 32. The short-circuit temperature ratings are considered the maximum temperatures and are not to be exceeded in order to protect the cable insulation from damage.

(4) *Temperature—Current—Time Curves.* For convenience in determining the cable size, the curves depicting the relationship of temperature—current—time are prepared from the "temperature rise" formula and are based on the temperature rise from the continuous to short-circuit temperature presented above. Figs 149 and 150 show the curves for copper and aluminum conductors from 75 to 200°C. They also incorporate the total fault clearing times of instantaneous trip devices and the interrupting ratings of various types of switching equipment. For safe design at present and in the future, a cable may be selected on the basis of total clearing time and equipment interrupting rating. For example, AWG No 2/0 copper cable may be selected for connection to 500 MVA, 13.8 kV, 8 cycle circuit breakers, and No 4/0 aluminum cable may be selected for connection to the same circuit breaker. However, a smaller cable can be selected if the maximum short-circuit current is less than the equipment interrupting rating.

(5) *Initial Temperature.* When the initial temperature of the conductor during maximum loading is less than the continuous temperature of rated current, the conductor can sustain a longer fault current flow or can carry more current than is allowed by Figs 149 and 150. However, the accurate temperature of the conductor can never be known, since it depends upon the daily loading of the cable and the daily ambient temperature. For a conservative approach it is generally not recommended to reduce the cable size. For an economical design, the reduction of cable size should be determined on the basis of estimated load and the expected ambient temperature, which may result in lower initial conductor temperatures. Fig 151 shows the correction factor to be used with Figs 149 and 150. This correction is mainly used to either increase or decrease the maximum short-circuit value by a factor of K_t. The K_t factor is based on initial temperature or final temperature, or both, which are above or below 75 and 200°C, respectively.

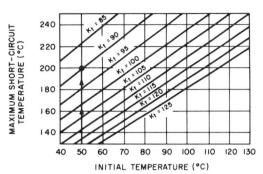

Examples of Use of Correction Factors:

(1)   Initial temperature 75°C
      Maximum transient temperature 200°C
      K_t = 1.00
      No correction on Figs 149 and 150

(2)   Initial temperature 50°C
      Maximum transient temperature 200°C
      K_t = 0.899
      Current = 0.899 $\times$ I_{sc} on Figs 149 and 150

Fig 151
Correction Factors K_t for Initial and
Maximum Short-Circuit Temperatures

11.4.3 *Protective Devices*

(1) *Total Fault-Clearing Time.* Devices to protect cables against short-circuit damage should have high reliability and fast tripping speed. In the protective scheme, primary protection is the first line and backup protection the second line of defense. The primary protection normally provides instantaneous tripping, while backup protection provides delayed tripping. Relay time plus circuit breaker time equals total clearing time t:

(a) Relayed circuit breaker:
     Total fault clearing time equals current relay time plus auxiliary relay time (if used) plus circuit breaker interrupting time

(b) Direct tripping circuit breaker:
     Total fault clearing time equals circuit breaker clearing time

(c) Fuses:
     Total fault clearing time equals fuse total clearing time

(2) *Protective Devices and Clearing Time.* The total clearing time of various types of protective devices depends on the type of relay and circuit breakers or fuses used. Table 33 estimates the

Table 33
Estimated Clearing Times of Protective Devices

1. *Relayed Circuit Breakers, 2.4—13.8 kV*

| | Type of Relay | | |
	Plunger, Instantaneous	Induction, Instantaneous	Induction, Inverse-Time
Relay times, cycles	0.25—1	0.5—2	6—6000
Circuit breaker interrupting time, cycles	3—8	3—8	3—8
Total time, cycles	3.25—9	3.5—10	9—6000

2. *Large Air Power Circuit Breakers, Below 600 V*

| | Frame Size | |
	225—600 A	1600—4000 A
Instantaneous, cycles	2—3	3
Short time, cycles	10—30	10—30
Long time, cycles	More than 60	More than 60
Ground fault, cycles	10—30	10—30

3. *Molded-Case Circuit Breakers, Below 600 V*

| | Frame Size | |
	100 A	225—1200 A
Instantaneous, cycles	1.1	1.5
Long time, cycles	More than 60	More than 60

4. *Fuses*
Total clearing time varies from about 0.25 cycle to many minutes, depending on fault current magnitude

total clearing times of various types of protective devices.

For convenience, the total clearing time of instantaneous trip devices is shown in the lower left-hand corner of Figs 149 and 150. These data can be used together with maximum short-circuit current for proper selection of cable sizes.

(3) *Time—Current Characteristics of Protective Devices.* A protective device will provide maximum protection if its time—current characteristics closely match those of the cable short-circuit current versus time curves shown in Figs 149 and 150. Thus the selection of overcurrent relays or devices is vitally important to the protection of cables. Figs 152—154 illustrate the characteristics of relays and devices commonly used in feeder circuits. Shown also are the maximum available short-circuit currents of the system and the maximum short-circuit current curve of the cable.

11.4.4 *Application of Short-Circuit-Current Protective Devices*

(1) *Typical Cases of Cable Protection.* Power cables are used for transmission or distribution, or as feeders to utilization equipment. The following are typical cases in industrial and commercial power systems.

(a) A single cable feeder through a pull box or splice joint or with taps should be protected in the same manner as feeders to panels without transformers [Fig 155 (a), (c), and (d)].

(b) A single or multiple cable feeder without a pull box or taps should be protected from the maximum short-circuit current which can occur at any point in the feeder circuit.

A multiple cable feeder with or without pull box or splice joint should be protected from the maximum short-circuit current caused by a fault on one cable. The short-circuit current in each cable is not equally distributed, since the maxi-

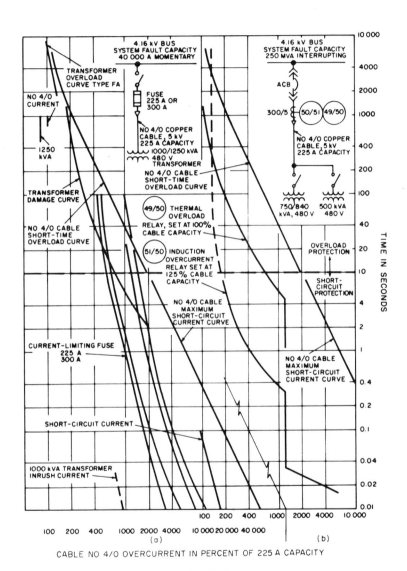

Fig 152
Short-Circuit and Overload Protection of 5 kV Cables
(a) Power or Current-Limiting Fuses. (b) Overcurrent Relays

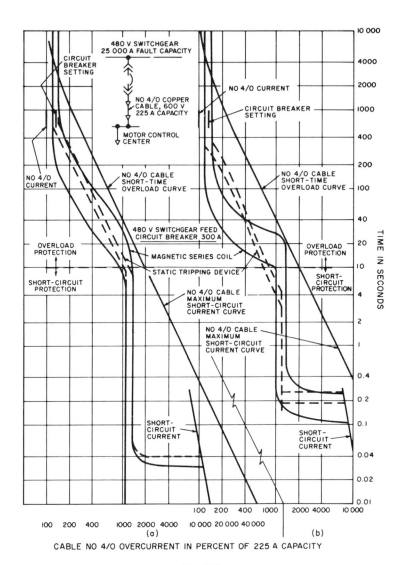

Fig 153
Short-Circuit and Overload Protection of 600 V Cables
(a) Long-Time and Instantaneous Equipped Circuit Breakers
(b) Long-Time and Short-Time Equipped Circuit Breakers

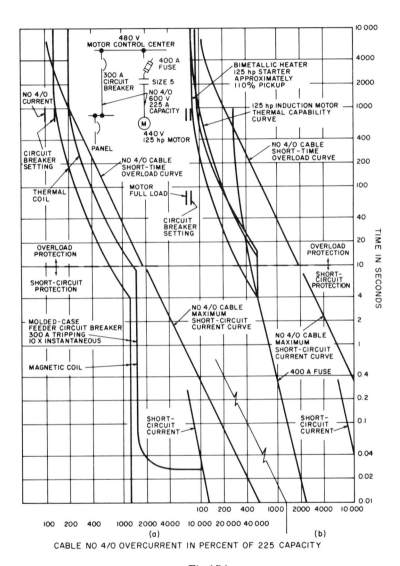

Fig 154
Short-Circuit and Overload Protection of 600 V Cables
(a) Thermal Magnetic Circuit Breakers. (b) Heaters and Fuses

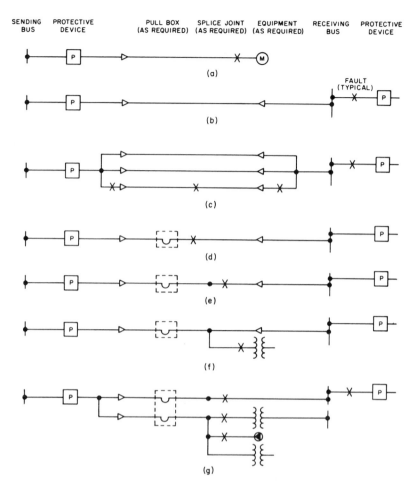

Fig 155
Application of Protective Devices
(a) Single Feeder to Utilization Equipment. (b) Single Feeder to Panel
(c) Multiple Cable Feeder to Panel. (d) Single Feeder with Pull Box
(e) Single Feeder with Pull Box and Splice. (f) Single Feeder with Pull
Box and Tap. (g) Multiple Feeder with Pull Box and Taps

mum current on the faulted cable is greater than
the total current divided by the number of cables
[Fig 155 (b) and (c)].

(c) A single or multiple cable feeder through
a pull box or splice joint, or with taps, should be
protected from the maximum short-circuit current caused by a fault on the tapped cables or end
section of spliced cable. A cable fault requires the
replacement of the faulted cable section only
[Fig 155 (d) and (e)].

(d) A single cable feeder with pull box, a
splice joint, and transformer tap should be protected from the maximum short-circuit current
caused by a fault on the tap or after the splice
joint [Fig 155 (f)].

(e) Multiple-feeder circuits should be protected in a manner similar to that of each individual circuit [Fig 155 (g)].

(2) *Protection and Coordination.* The protective device should be selected and coordinated to

give the cable sufficient short-circuit protection. This can be done easily by plotting the time—current curves of the protected cable and the protective device on the same log—log graph paper. The time—current curve of the protective device should always be below and to the left of the maximum short-circuit-current—time curve (Figs 149 and 150) of the protected cable. Figs 152—154 illustrate that a No 4/0 copper insulated cable may be protected by various protective devices as follows. A 5 kV No 4/0 feeder is protected by a current-limiting fuse [Fig 152 (a)] or a 51 or 49 relay [Fig 152 (b)]. A 600 V No 4/0 feeder is protected by instantaneous tripping [Fig 153 (a)], by instantaneous and short-time tripping [Fig 153 (b)], or by an instantaneous molded-case circuit breaker [Fig 154 (a)]. A 600 V No 4/0 motor circuit is protected by an NEC fuse [Fig 154 (b)].

11.5 Overload Protection of Cables. Overload protection cannot be applied until the current-carrying capacity of a cable is determined. Protective devices can then be selected to coordinate cable rating and load characteristics.

11.5.1 *Normal Current-Carrying Capacity*

(1) *Heat Flow and Thermal Resistance.* Heat is generated in conductors by I^2R losses. It must flow outward through the cable insulation, sheath (if any), the air surrounding the cable, the raceway structure, and the surrounding earth in accordance with the following thermal principle [1]—[4]:

$$\text{heat flow} = \frac{\text{difference between conductor and ambient temperature}}{\text{thermal resistance from materials}}$$

The conductor temperature resulting from heat generated in the conductor varies with the load. The thermal resistance of the cable insulation may be estimated with a reasonable degree of accuracy, but the thermal resistance of the raceway structure and surrounding earth depends on the size of the raceway, the number of ducts, the number of power cables, the raceway structure material, the coverage of the underground duct, the type of soil, and the amount of moisture in the soil. These are important considerations in the selection of cables.

(2) *Current-Carrying Capacity.* The current-carrying capacity of each cable is calculated on the basis of fundamental thermal laws incorporating specific conditions, including (1) type of conductor, (2) alternating-current—direct-current resistance of the conductor, (3) thermal resistance and dielectric losses of the insulation, (4) thermal resistance and inductive alternating-current losses of sheath and jacket, (5) geometry of the cable, (6) thermal resistance of the surrounding air or earth and duct or conduits, (7) ambient temperature, and (8) load factor. The current-carrying capacities of the cable under the jurisdiction of the NEC are tabulated in its current issue or amendments thereto. The current-carrying capacity of cables under general operating conditions that may not come under the jurisdiction of the NEC are published by the Insulated Power Cable Engineers Association (IPCEA). In their publications they describe methods of calculation and tabulate the current-carrying capacity for 1, 8, 15, and 25 kV cables [IPCEA S-19-81, Rubber-Insulated Wire and Cable for the Transmission and Distribution of Electrical Energy (1969), (NEMA WC3-1969); IPCEA S-61-402, Thermoplastic-Insulated Wire and Cable for the Transmission and Distribution of Electrical Energy (1973), (NEMA WC5-1973); IPCEA S-65-375, Varnished-Cloth-Insulated Cables (1966), (NEMA WC4-1966)]. The current-carrying capacities of specific types of cables are calculated and tabulated by manufacturers. Their methods of calculation generally conform to IPCEA P-54-440, Ampacities of Cables in Open-Top Cable Trays (1972), (NEMA WC51-1972).

(3) *Temperature Derating Factor.* The current-carrying capacity of a cable is based on a set of physical and electrical conditions and a base ambient temperature defined as the no-load temperature of a cable, duct, or conduit. The base temperature generally used is 20°C for underground installation and 30°C for exposed conduits or trays.

Temperature derating factors (TDF) for ambient temperatures and other than base temperatures are based on the maximum operating temperature of the cable and are proportional to the square root of the ratio of temperature rise, that is,

$$TDF = \frac{I_N}{I_{N1}} = \frac{\begin{array}{c}\text{current capacity}\\\text{at base ambient temperature}\end{array}}{\begin{array}{c}\text{current capacity}\\\text{at other ambient temperature}\end{array}}$$

$$= \sqrt{\frac{T_N - T_a}{T_N - T_{a1}}}$$

$$= \sqrt{\frac{\begin{array}{c}\text{temperature rise above}\\\text{base ambient temperature}\end{array}}{\begin{array}{c}\text{temperature rise above}\\\text{other ambient temperature}\end{array}}}$$

(4) *Grouping Derating Factor.* The no-load temperature of a cable in a group of loaded cables is higher than the base ambient temperature. To maintain the same maximum operating temperature, the current-carrying capacity of the cable must be derated by a factor of less than 1. Grouping derating factors (GDF) are different for each installation and environment. Generally they can be classified as follows:

(a) For cable in free air with maintained space

(b) For cable in free air without maintained space

(c) For cable in exposed conduits

(d) For cable in underground ducts

The NEC Table 318-8 lists the grouping derating factor for cables with maintained space and manufacturers generally provide recommended grouping derating factors for their cable under various environmental conditions.

11.5.2 *Overload Capacity*

(1) *Normal Loading Temperature.* Cable manufacturers specify the normal loading temperature for their products which results in the most economical and useful life of the cables. Based on the normal rate of deterioration, the insulation can be expected to have a useful life of about 20 to 30 years. Normal loading temperature of a cable determines its current-carrying capacity under given conditions. In regular service, rated loads or normal loading temperatures are reached only occasionally because cable sizes are generally selected conservatively in order to cover the uncertainties of load variations. Table 34 shows the maximum operating temperatures of various types of insulated cables.

(2) *Cable Current and Temperature.* The temperature of a cable rises as the square of its current. The cable temperature for a given steady load may be expressed as a function of percent full load by the formula

$$T_X = T_a + (T_N - T_a)(I_X/I_N)^2$$

Fig 156 shows this relation for cables rated at normal loading temperatures of 60, 75, 85, and 90°C.

(3) *Emergency Loading Temperature.* An overload may increase the cable temperature above its allowable normal limit. The life of cable insulation is about halved, and the average rate of service failures about doubled for each 10°C increase in operating temperature. It is obvious that the shorter the time and the less frequently such loads are applied, the longer the cable life will be. Emergency overload time should not exceed 100 h a year and 5 h per overload. On this basis, the recommended emergency overload temperature limits are listed in Table 34 and should be considered as the absolute maximum for all circumstances. The emergency overload with the above limitation should not materially increase the rate of service failures.

(4) *Emergency Overload Current.* The emergency overload current can be determined on the same basis as the temperature derating factor, namely, the current is proportional to the square root of the ratio of temperature rises as follows:

$$\text{percent loading} = \frac{I_E}{I_N} \times 100$$

$$= \frac{\begin{array}{c}\text{emergency}\\\text{loading current}\end{array}}{\begin{array}{c}\text{normal}\\\text{loading current}\end{array}} \times 100$$

$$= \sqrt{\frac{T_E - T_a}{T_N - T_a}} \times 100$$

$$= \sqrt{\frac{\begin{array}{c}\text{emergency temperature}\\\text{rise above ambient}\end{array}}{\begin{array}{c}\text{normal temperature}\\\text{rise above ambient}\end{array}}}$$

$$\times 100$$

As an example, the emergency overload current for three single-conductor 5 kV rubber-insulated cables in an underground duct can be calculated using the following data: $I_N = 210$ A for AWG No

Table 34
Typical Normal and Emergency Loading of Insulated Cables

Insulation	Cable Type	Normal Voltage	Normal Loading (°C)	Emergency Loading (°C)
Thermoplastic	T, TW	600 V	60	85
	THW	600 V	75	90
	Polyethylene	0—15 kV	75	95
		15 kV	75	90
Thermal setting	R, RW, RU	600 V	60	85
	XHHW	600 V	75	90
	RHW, RH-RW	0—2 kV	75	95
	Cross-linked polyethylene	5—15 kV	90	130
	Ethylene-propylene	5—15 kV	90	130
Varnished polyester		15 kV	85	105
Varnished cambric		0—5kV	85	102
		15 kV	77	85
Paper lead		15 kV	80	95
Silicone rubber		15 kV	125	150

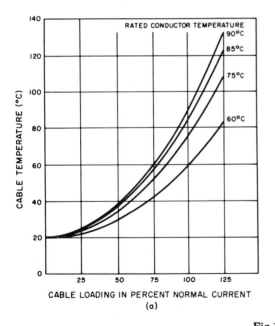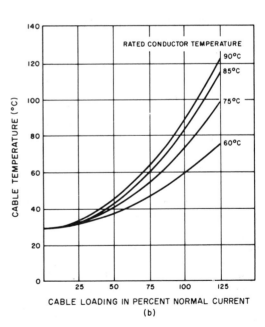

Fig 156
Cable Loading and Temperature Rise
(a) Ambient Temperature at 20°C. (b) Ambient Temperature at 30°C

Table 35
Values of K

Size of Cable	Cable in Air	Cable in Underground Duct	Direct Buried Cable
Small	0.5	1	1.5
Intermediate	1	2.5	3
Large	1.5	4	6

4/0, 100 percent load factor, $T_N = 75°C$, $T_a = 20°C$, and $T_E = 95°C$.

$$\text{percent emergency loading} = \sqrt{\frac{95-20}{75-20}} \times 100$$

$$= 117 \text{ percent}$$

$I_E = 1.17 \times 210$ A
= 246 A continuous for a maximum period of 5 h per overload

If the initial loading is only 50 percent of normal rated current, the initial temperature from Fig 156 (a) is only 33.8°C. Then $I_N = 105$ A and $T_N = 33.8°C$, and

$$\text{percent emergency loading} = \sqrt{\frac{95-20}{33.8-20}}$$

$$\times 100 = 234 \text{ percent}$$

$I_E = 2.34 \times 105$ A
= 246 A continuous for a maximum period of 5 h per overload

The rate of temperature change in the cable is affected mainly by (1) the size of the cable, (2) the geometric factor of the cable, (3) the type of installation (such as exposed cable in air, conduit, underground duct, or buried), and (4) the condition of the surrounding environment. These factors can be expressed mathematically by a time constant K in the exponential function of the temperature rise as shown in Table 35.

Thus the temperature curve of a cable for a given current increase may be expressed by the formula [4]

T_t = initial temperature + total temperature rise × per unit rise

$$= T_0 + (T_f - T_0) \times (1 - e^{-t/K})$$

Fig 157 shows the rate of temperature rise curves for K = 0.5, 1, 1.5, 2.5, and 4. It is interesting to note that the final temperature is reached in a much longer time in large cables than in smaller ones.

(5) *Short-Time Overload.* Since temperature rises exponentially, the shorter the overload period, the greater the overload that can be applied to the cable without exceeding the allowable limit. The permissible overload current for a duration time t may be expressed by the formula

percent emergency loading

$$= \frac{I_E}{I_N} \times 100$$

$$= \frac{\text{emergency loading current}}{\text{normal loading current}} \times 100$$

$$= \sqrt{\frac{T_E}{T_N}} \times 100$$

percent emergency overload (at time t)

$$= \frac{1 - (I_0/I_N)^2 \times e^{-t/K}}{1 - e^{-t/K}} \times 100$$

In Fig 158 the overload curves show possible overload conditions for a given duration with initial loading I_0 at 0, 25, 50, or 75 percent of the normal current I_N of the cable. The normal current means the maximum continuous current

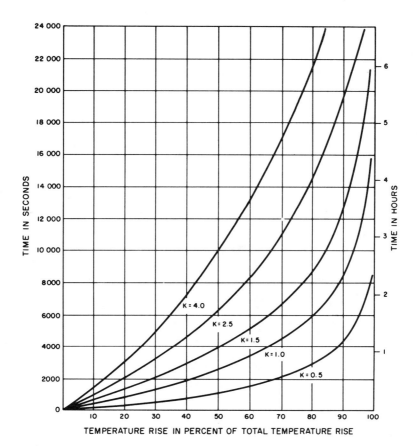

Fig 157
Rate of Temperature Rise Due to Current Increase

Temperature Rise   Percent Total Temperature Rise
Temperature at any Time Initial Temperature and Temperature Rise
Final Temperature   Initial Temperature and Total Temperature Rise

$K = 0.5$	Small Cable in Air
$K = 1.0$	Medium Size Cable in Air
	Small Cable Underground
$K = 1.5$	Large Cable in Air
	Small Cable Direct Burial
$K = 2.5$	Medium Cable Underground
	Medium Cable Direct Burial
$K = 4.0$	Large Cable Underground

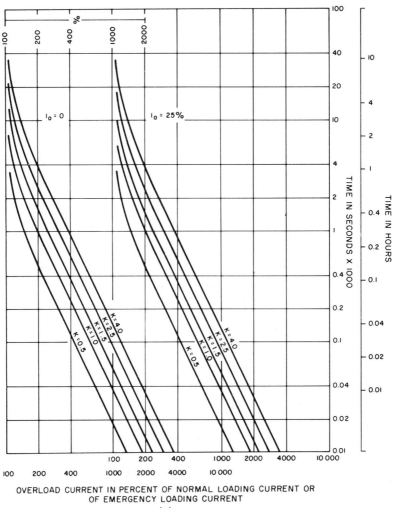

Fig 158
(a) Short-Time Overloading Curves of Insulated Cables

IEEE Std
242-1975

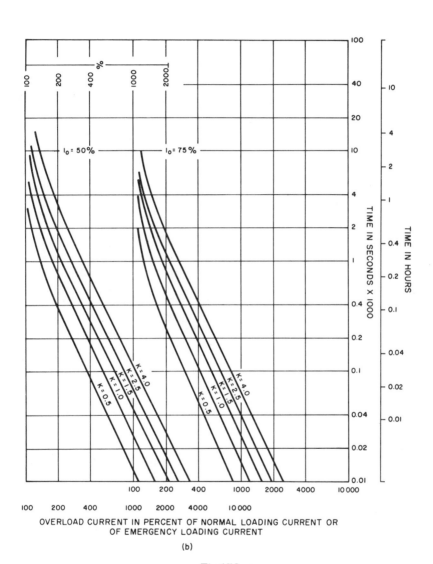

(b)

Fig 158
(b) Short-Time Overloading Curves of Insulated Cables

derated for actual operating conditions. These overload curves are calculated on a different basis than the short-circuit-current curves. In the short-circuit-current curves an initial temperature of 75°C is assumed and heat flow is neglected during the entire 15 s period; in the overload curves heat flow is considered, and an assumed initial temperature of less than 75°C or the actual initial temperature is used. Thus with less initial load, the cable overload capability will be greater than its short-circuit capability for a given period of time t.

11.5.3 Overload Protective Devices

(1) *Time—Current Characteristics.* The time—current overload characteristics (Fig 158) of the cables differ from the short-circuit-current characteristic (Figs 149 and 150). The overloads can be sustained for a much longer time than the short-circuit current, but the principle of protection is the same. A protective device will provide maximum protection if its time—current characteristic closely matches that of the cable overload characteristic. Thermal overcurrent relays generally offer better protection than do induction overcurrent relays because thermal relays operate on a long-time basis and their response time is proportional to the temperature of the cable or the square of its current.

(2) *Overcurrent Relays.* Very inverse or extreme inverse relays of the induction-disk type provide better protection than do the moderate inverse relays. However, all induction-type overcurrent relays afford the cables sufficient protection. Fig 152 (b) shows the cable protection given by overcurrent relays (device 51) and by thermal overcurrent relays (device 49).

(3) *Thermal Overcurrent Relays or Bimetallic Devices.* Thermal overload relays or bimetallic devices more closely resemble the cable's heating characteristic, but they are generally not as accurate as an overcurrent relay. Fig 152 (b) shows the cable protection given by thermal overcurrent relays (device 49), and Fig 153 (b) protection given by bimetallic heaters.

(4) *Fuses.* Power fuses or current-limiting fuses provide excellent cable protection from high-magnitude short-circuit currents but very little protection from short-time overload, or low-magnitude faults. Figs 152 (a) and 154 (b) show that the fuse curve intersects the cable overload curve. For currents less than that at the

point of intersection there is no protection. Detailed treatment of fuses is given in Chapter 5 of this standard as well as in Chapter 3 of IEEE Std 141-1969, Electric Power Distribution for Industrial Plants, Chapter 5 of IEEE Std 241-1974, Electric Power Systems in Commercial Buildings, and IEEE Committee Report [5].

(5) *Magnetic Trip Coil or Static Sensor on 480 V Switchgear.* The magnetic trip coils have a wide range of tripping tolerances. Their long-time characteristics match the cable overload curves for almost three quarters of an hour (Fig 153). Static trip devices provide better protection than magnetic direct-acting coils. However, for safe cable protection, the coils should be set below the heating curves of the cable by sizing the cable with normal loading current slightly greater than the coil pickup current.

(6) *Thermal Magnetic Coil on Molded-Case Circuit Breakers.* The characteristics of the thermal magnetic coils resemble those of magnetic trip coils. They do not provide adequate thermal protection to cables during the long-term overloads [Fig 154 (a)]. The cable should be selected and protected in the same manner as described in the preceding paragraph.

11.5.4 Application of Overload Protective Devices

(1) *Feeder Circuits to Panels.* A single or multiple cable feeder leading to a panel with or without an intermediate pull box should be protected from excessive overload by a thermal overcurrent device. If there are splice joints and a different type of installation, such as from an exposed conduit to an underground duct, the cable segment with the lowest current-carrying capacity should be used as the basis for protection [Fig 155 (b) and (c)].

A single cable feeder with tap to individual panels cannot be protected from excessive overload by a single protective device at the sending end, unless the cable is oversized. Therefore, overload protection should be provided at the receiving end. The protection should be based on the current-carrying capacity of the cable supplying power to the panel [Fig 155 (f)]. A multiple feeder with only a common protective device does not have overload protection for each cable feeder. In this case, overload protection should be provided at the receiving end [Fig 155 (g)].

(2) *Feeder Circuit to Transformers*. A feeder circuit to one or more transformers should be protected in a manner similar to that used for feeder circuits to the panels. However, a protective device selected and sized for transformer protection also provides protection for the cable. This is due to the fact that the cables sized for a full transformer load have higher overload capability than the transformer. [See Fig 152 (a) for a comparison of the time—current curves between cable and transformer.]

(3) *Cable Circuit to Motors*. A cable circuit to one or more motors should be protected in a manner similar to that for cable circuits to panels. Again, a protective device selected and sized for motor protection also provides cable protection because the overload capability of the cable is higher than that of the motor [Fig 154 (b)].

(4) *Protection and Coordination*. Protective devices should be selected and cables sized for coordinated protection from short-time overload. The method of coordination is the same as for the short-circuit protection, that is, the time—current curve of the protective device should be below and to the left of the cable-overload curve (Fig 158). Figs 152—154 illustrate the protective characteristics of relays and devices commonly used in cable circuits for overload protection.

11.6 Physical Protection of Cables. Cables require protection against physical damage as well as from electrical overload and short-circuit conditions. The physical conditions that must be considered are divided into three categories: mechanical hazards, adverse ambient conditions (excluding high temperatures), and attack by foreign elements. Cables can also be damaged (and frequently are) by improper handling during installation.

11.6.1 *Mechanical Hazards*. Electric cables can be damaged mechanically by vehicles, falling objects, misdirected excavation, or failure of adjacent circuits. Mechanical protection must serve the dual function of protecting cables and limiting the spread of damage in the event of an electrical failure.

Isolation is one of the most effective forms of mechanical protection. Conduit, tray, and duct systems are more effective if they are physically out of the way of probable accidents. A highly elevated cable is adequately protected against vehicles and falling objects. Where conduits or other enclosures must be run adjacent to roadways, large steel or concrete barriers provide adequate protection.

11.6.2 *Exposed Raceways*. The most popular form of mechanical cable protection is the use of conduits or metallic raceways. In addition to the electrical benefits of the grounded enclosure, the metallic conduit or raceway protects the cable against most types of mechanical damage. Cable trays are also becoming popular because they are economical and convenient for power and control cable systems. Cables may have increased protection from mechanical damage through the use of solid metal tray covers and metal barriers in the trays between different circuits.

11.6.3 *Underground Systems*. Underground ducts or embedded conduits provide similar mechanical protection. Ducts should be concrete encased for best results. Where they are subject to heavy traffic or poor soil conditions, reinforcement of the concrete envelope is desirable. Since excavation near underground cable runs is always a problem, it has been found advisable to color the concrete around electrical ducts. The addition of approximately three pounds of iron oxide per sack of cement provides a readily identifiable red color which is meant as a warning to anyone digging into the run. The color is effective even in mud or similar colored soil since it is conspicuous as soon as the concrete is chipped.

11.6.4 *Direct-Buried Cables*. The cables must be carefully routed to minimize damage from traffic and digging and to avoid areas where plant expansion is predicted. Cables should be covered with some type of special material such as a brightly colored plastic strip or a wooden or concrete plank. Warning signs should also be placed above ground at frequent intervals along the cable route. Additionally, plant drawings accurately locating the buried cable run may also prevent accidental dig-ins.

11.6.5 *Aerial Cable Systems*. Insulated cables on a messenger require special care. These are especially susceptible to installation damage also. They must be located away from possible interference from portable cranes and support systems and protected from vehicle damage. Space or solid barriers provide reasonable protection for supports, whereas warning signs and nonelectric cables strung between electric cables and roadways offer protection against cranes and high vehicles.

11.6.6 *Portable Cables.* Exposed portable cables require extra consideration from a mechanical standpoint. Since they must remain portable, enclosures are not practical. The proper selection of a portable cable type provides one of the best methods of protection. It must be selected to match operating conditions. Moisture resistance, resistance to cutting or abrasion, and type of armor are all considerations that influence cable life. However, even the best cables require mechanical consideration in service. They should not be subjected to vehicle traffic. Means should be arranged to allow traffic to pass over or under cables without contacting the cable. Care should also be taken in moving portable cables to avoid snags or cuts. They should be located so that they are clear of welding and placed where falling objects are not a serious hazard. A conspicuous color on the jacket is beneficial in warning personnel of the location of a portable cable.

11.6.7 *Adverse Ambient Conditions.* In the previous paragraphs, protection from overtemperature caused by short-circuit current or overload conditions has been discussed. There are other ambient conditions, however, that are not responsive to overcurrent devices or to compensation for elevated ambient temperatures.

In any type of cable enclosure, water or dampness must be considered, although underground installations are the most susceptible. Repeated cycles of high and low temperature, combined with humid air, can fill conduits or enclosures with water produced by "breathing" and condensation. It is almost impossible to stop the breathing, but suitable drains at low points will remove water as it collects. It is always desirable to prevent immersion, and duct systems and other raceways should be designed to slope so that the water can be removed.

Many of the available cable insulations are highly resistant to moisture, but where moisture is expected, extra care should be taken in selecting the insulation appropriate for that application.

The moisture problem may be amplified in industrial plants by the presence of various chemicals, and the possibility of chemical contamination must be considered for cables run through any process facility. Chemical seepage into ground water, or direct contact due to process misoperation, may result in chemicals coming into contact with a cable system. The enclosure, insulation, and conductor must all be tested to determine the effect of any particular chemical, and selected to be most resistant to the chemical involved. Where chemical contamination is severe, rerouting of the system should be considered.

Fires, which may result directly from cable failure or from unrelated external conditions, can cripple almost any cable system. It is easier to protect against damage to one cable caused by the failure of an adjacent cable, than an external fire. The enclosure of individual power circuits and fireproof coatings are the most effective means of limiting this type of damage, since it is unusual for a cable with proper electrical protection to burn through its individual conduit or raceway.

Pullboxes, pits, and manholes used as pulling points or sorting areas are the greatest fire hazards with respect to fault conditions, and elimination of the common enclosure for several circuits, where possible, offers the best protection.

The combustion of materials adjacent to a cable system is a difficult condition to protect against. The obvious solution is to remove all combustibles from the vicinity of the cables. As much as possible, this should be the practice. Critical circuits should be separated to lessen the extent of fire damage, and the use of multiple circuits following different routes can assure continuous service.

In all cases the selection of proper enclosures can minimize fire damage; however, the method chosen must be based upon the possible hazards involved. For underground installation, heavier enclosures, higher racks, and overinsulation using materials with greater temperature resistance are all considerations, and under severe conditions, the use of MI cables may help.

11.6.8 *Attack by Foreign Elements.* In some environments, cable systems may be subject to attack by animals, insects, plants, and fungus, all of which may possibly cause cable failure. Small gnawing-type animals have been known to chew through cables, and insects and small animals such as lizards and snakes can cause difficulties at terminations where they or their nests may bridge the gap between terminals. The use of more resistant enclosures, armor, or indigestible cable materials are effective protection.

In tropical atmospheres, fungus may grow on cable and wire systems. The creation of a dry atmosphere is an effective deterrent, although fungus-resistant coatings and insulations are protection methods most often applied.

11.7 Code Requirements for Cable Protection. Codes and regulations are established to control the installation and operation of electric cable systems. Although there are many different codes and regulations that may be applied, depending on governmental, geographical, or company requirements, the NEC is most often quoted and as a result of the Federal Occupational Safety and Health Act is now mandatory. It is the responsibility of the engineer involved to determine which codes are applicable to the particular project. This discussion is limited to the NEC. The NEC is principally concerned with overtemperature (overcurrent), short-circuit, and mechanical protection in regard to cable applications.

Overcurrent protection is covered in Article 240 under the provision requiring all conductors to be protected in accordance with their current-carrying capacity. In general, the current-carrying capacity of cables is determined from the tables contained in Article 310, which concerns the installation of conductors.

In Article 240, short-circuit currents and overload currents are considered generally in the same manner, and rules are presented for the selection and setting of these devices. This article allows the use of fused or circuit breakers for overcurrent protection. The tables in Article 310 also offer rules for derating cables for elevated ambient conditions. In general, no specific rules are presented to coordinate insulation heating characteristics with the overcurrent devices concerning temperature versus time.

Motor feeders receive particular attention in Article 430. In general, Article 430 governs the selection of the current-carrying capacity of cables used for motor feeders. After the cable size is selected in accordance with this article, the actual protection is applied in accordance with Article 240. It should be noted, however, that Article 430 provides rules for the overcurrent protection of the motors themselves, and although this discussion concerns only cables, it is possible that motor protective devices may also provide the required cable protection, or vice versa.

Article 310 insures that cables are adequate for their service applications by specifying currents which may be carried by particular conductors with specific insulation classifications and under specific governing conditions. It also requires the selection of cable materials that are suitable for application conditions, including moisture, chemicals, and nonstandard temperatures. This article permits the use of multiple cables, providing that means are provided to ensure the equal division of current, and that essentially identical conditions and materials are used for each of the parallel paths. Article 310 also covers installation methods designed to ensure the installation of cables without damage and with adequate working space.

Article 300 specifies wiring methods and protection required for cables subject to physical damage.

These articles pertain specifically to cable protection, but are not the only provisions of the NEC that deal with the subject. Any specific cable or cable system will come under the provisions of one or more sections of the NEC, and responsible parties must ensure that the protective methods they have selected comply with both the relevant provisions and any special requirements that they may impose.

B. Busway Protection

11.8 Busway Protection. Due to their economies, convenience, and excellent electrical characteristics, 600 V busway systems have gradually assumed a role of greater importance in today's industrial and commercial buildings. It is because of this consolidation of numerous small cable runs into a single large bus duct run that the reliability of duct runs has become a critical factor in building design. Today's busways are very well designed for their intended use but, because of the critical nature of their purpose, they not only must suffer fewer outages, they must also be returned to service with a minimum of downtime. Thus, while the duct manufacturers can incorporate improved design concepts, better insulation, etc, it behooves the system

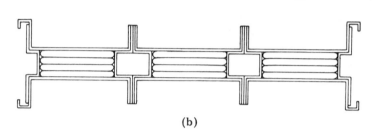

(a)   (b)

Fig 159
Low-Impedance Busway
(a) Sandwich-Type Construction for Heat Dissipation;
Close Proximity of Conductors Reduces Reactance
(b) Higher Current Ratings often Employ Multiple Sets of Bars

designer to spend an extra amount of time on incorporating the best possible protection into the integrated system so that outages due to factors beyond his control are minimized in duration. This is an important concept in that it does not suggest that the number of outages can be controlled. It does, however, suggest that by minimizing the duration of the outage, disruption of the normal activities can also be minimized. The duration of the outage is to a certain degree directly proportional to the amount of damage suffered by the busway during the fault, and this amount of damage is determined by the protective elements in the circuit.

11.8.1 *Types of Busways.* There are several different designs of busways available, and each offers certain features which are significant when considered in an integrated building plan. These can be identified as follows:

    (1) Low-impedance busways
        (a) feeder type
        (b) plug-in type
        (c) high-frequency type
    (2) High-impedance busways
        (a) service entrance
        (b) current limiting
    (3) Simple plug-in busways

(1) *Low-Impedance Busways.* Low-impedance busway designs (Fig 159) achieve their low-reactance characteristics by a careful posi-

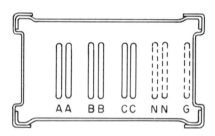

Fig 160
High-Impedance Busway;
Wide Phase-to-Phase Spacings Result
in Increased Reactance

tioning of each bus bar in close proximity to other bars of an opposite polarity. This close physical spacing (ranging from 0.050 to 0.250 in) demands that each bar be coated with some form of insulation to maintain satisfactory protection from accidental bridging. The losses in such designs are very low. Low-impedance busways are offered in feeder construction for the purpose of transmitting substantial blocks of power to a specific location or in plug-in construction. Plug-in designs feature door-like provisions at approximately 24 in increments along the length of the busway which, when opened, expose the bus bars so that plug-in taps may be made with minimum effort. Although most low-impedance designs are intended for use on

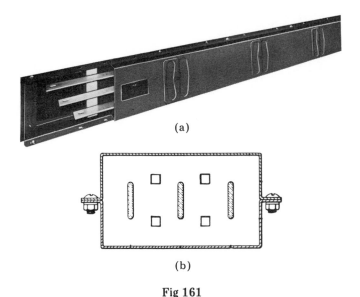

Fig 161
Simple Plug-in Busway
(a) Bare Bars Spaced Far Apart Offer Generous Electrical Clearances and Low Cost
(b) Efficiency Is Low but Acceptable for Loads of 1000 A or Less

60 Hz applications, there are some that may be used at higher frequencies. Low-impedance designs are offered in voltage ratings of 600 V or less and current ratings up to 4000 A or more. Low-impedance busways may be ventilated or nonventilated and offered in indoor or outdoor construction, except that plug-in is available for indoor use only.

(2) *High-Impedance Busways.* High-impedance busways (Fig 160) are of two general types, (a) those with deliberate impedance introduced to minimize fault current levels, and (b) those which achieve high-impedance characteristics as an incidental by-product of their construction.

In the first type, high reactance is obtained just as low reactance was, by a careful placement of each bar relative to every other bar. In this case, however, the goal is to maximize the spacing between pairs of bars of opposite polarity. Since these high-impedance designs experience high losses and since these losses appear as heat, ventilated construction and insulated conductors are frequently employed. They are offered in generally the same ratings as low-impedance designs.

Under the provisions of some standards a special-purpose bus duct design may be built in total

installed lengths of 30 ft or less for the purpose of connecting between an incoming service and a switchboard. This duct generally is constructed with a nonventilated enclosure and bars that may or may not be insulated. The bars are physically separated, and it is this large separation, introduced for safety, that results in high reactance characteristics. Short-run busways are limited to 2000 A and 600 V.

(3) *Simple Plug-in Busways.* Among the first busway designs (Fig 161), introduced in the mid 1930s, was a simple construction which supported bare conductors on insulators inside a nonventilated casing with periodic plug-in access doors. Like the short-run busway, this type of duct offered generous bar-to-bar spacings, but because it was not used to carry large quantities of current for lengthy runs, its losses were not objectionable. Short-circuit ratings are usually modest. It is still widely used today and is available in ratings of 100 to 1000 A and voltages of 600 V or less.

(4) *Bolted Faults.* Faults associated with busways are either of the bolted type or the arcing type. Due to the prefabricated nature of busways, bolted faults are rare. Bolted fault, in this context, refers to the inadvertent fastening to-

Fig 162
Simple Plug-In Busways Subjected to
Fault Currents above Their Ratings. Busway
on Left was Protected by Current-Limiting
Fuses, while that on Right Was Protected
by a Circuit Breaker

Table 36
Busway Minimum
Short-Circuit-Current Ratings

Continuous Current Rating of Busway (amperes)		Minimum Short-Circuit-Current Ratings (amperes)	
Plug-in	Feeder	Symmetrical	Asymmetrical
100		10 000	10 000
225		14 000	15 000
400		22 000	25 000
600		22 000	25 000
	600	42 000	50 000
800		22 000	25 000
	800	42 000	50 000
1000		42 000	50 000
	1000	75 000	85 000
1350		42 000	50 000
	1350	75 000	85 000
1600		65 000	75 000
	1600	100 000	110 000
2000		65 000	75 000
	2000	100 000	110 000
2500		65 000	75 000
	2500	150 000	165 000
3000		85 000	100 000
	3000	150 000	165 000
	4000	200 000	225 000
	5000	200 000	225 000

gether of bus bars in a solid fashion resulting in an unintended connection between phases. Bolted faults can occur during the initial installation or at a later date when modifications are made to the system. The actual offending connection might be found in a bus duct cubicle, but it will more often be found in some pieces of equipment connected to the busway, such as a switchboard connection or a load served from a bus plug. Because a bolted connection implies a low-resistance connection, we may expect the maximum level of fault current to flow, and circuit protective elements must, therefore, be sized in accord with this maximum fault level. Bolted faults result in a distribution of energy through the entire length of the bus duct circuit. This energy flow results in an intense magnetic field around each conductor which opposes or attracts fields around adjacent conductors. The mechanical forces thus created are very high and are quite capable of bending bus bars (Fig 162), tearing duct casings apart, or shattering insulation. For this reason, the busway must have a short-circuit withstand rating that is greater than the maximum available fault current. Such ratings are

published by the various busway manufacturers and are based on a 3 cycle duration. Table 36 reflects the NEMA standard ratings.

(5) *Arcing Faults.* In contrast to the bolted fault, arcing faults can occur at any time in the life of a system. Although there can be many individual agencies initiating an arcing fault, they generally involve one or more of the following: loose connections, foreign objects, insulation failure, voltage spikes, water entrance. Because of the resistance of the arc and the impedance of the return path, current values are substantially reduced from the bolted fault level.

The interaction of the magnetic fields around the conductors and around the arc results in an unbalanced force which causes the arc to try to move away from the power source. If the path is unobstructed, the arc will accelerate and move very quickly toward the remote end of the run. The only mark of its passage may be a scarcely noticeable pinhead size pit every several inches

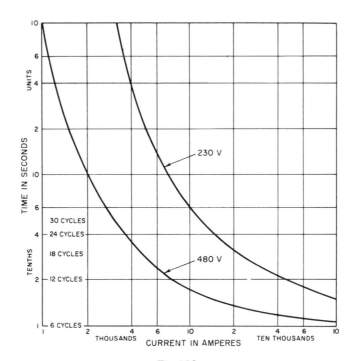

Fig 163
Time—Current Curves for a Power Arc to Burn
a ¼ by 2 in Copper Bar Halfway Through.
Bars Are Spaced on 2¼ in Centers in
Standard Plug-in Bus Duct

along the edge of the bus bar. These tiny marks, however, provide a clear trail for the investigator and can often lead him back to the origin of the fault. As the arc travels away from the power source, the length of the circuit becomes greater and the forces causing movement become smaller. Eventually, the arc reaches some obstruction which causes it to hesitate long enough to cause serious burning or even hang up until the bus duct is burned open. Busways employing insulated conductors, of course, do not permit traveling arcs. Arcing, therefore, remains at the point of initiation, or may burn slowly toward the source.

Although the magnitude of current present in an arcing fault is usually less than that present in a bolted fault, the entire thermal effect is concentrated at the arc location and results in major damage at that point. Fig 163 indicates the damage anticipated in terms of the quantity of conductor material vaporized by a phase-to-phase arc at 480 V. A 15 000 A arc persisting at one

location for 9 cycles would remove about half of the ¼ by 2 in copper conductor. These charts are based on a simple plug-in busway design with bars on 2¼ in centers. Designs with bars closer together or designs employing aluminum conductors may be expected to show much more extensive damage. If the designer intends to minimize the duration of system outages, it is toward the arcing fault that he must direct his concern.

11.8.2 *Types of Protection.* Like any other circuit, busways are subject to overloads, arcing faults, and bolted faults. Each of these is characterized by an entirely different set of parameters and, therefore, requires an entirely different set of protective concepts. No single protective element suits all requirements. An examination of the protection required will suggest the need for several protective devices.

(1) *Overload Protection.* Overloads are, of course, those temporary conditions that cause the busway to carry currents greater than its

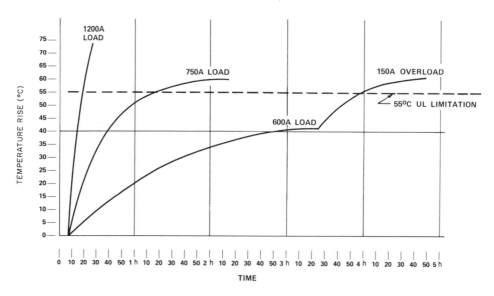

Fig 164
Time—Temperature Curves of 100, 125, and 200 Percent Loads
on a 600 A Rated Low-Impedance Busway; One ¼ by 2½ in Aluminum Bar per Phase

continuous current rating. Overloads such as stalled rotor currents or motor starting currents generally are not harmful to the busway because each motor served is usually small in comparison to the capacity of the busway. Overloads are more likely to occur as a result of adding more or larger pieces of equipment, over a period of years, to an existing busway circuit until its capacity is exceeded. Since busways tend to use large conductors, they exhibit considerable thermal inertia. For this reason, overloads of a temporary nature require a substantial time before their effect is noticed. Fig 164 displays time versus temperature rise for three different loading conditions. Note that this particular busway required only 12 min to reach 55°C rise at 200 percent current. The same busway took over an hour to reach 55°C at 125 percent loading. When operating at 100 percent loading, it required 25 min to raise the temperature up to the 55°C limit with a 25 percent overload. Naturally, this time will vary from size to size depending on the stable temperature produced by 100 percent loading. Most busway sizes are designed to operate at close to a 55°C rise at full current. The value of 55°C was selected because this is the maximum

rise allowed by Underwriters Laboratories on plated bus bar joints. It was generally assumed that the busway will operate in a 30°C (86°F) ambient, and for this reason, busway manufacturers have employed 85°C insulation for many years. (30°C ambient + 55°C rise = 85°C operating temperature.) Newer designs of busway employ higher temperature insulation, although they are still limited to a 55°C rise at the hottest spot. This suggests that busways operated in a 50°C ambient could still carry full load (producing a 55°C rise) without exceeding the 105°C total temperature limit of the newer insulations. From the foregoing it is apparent that the older busways could easily suffer insulation damage should they be subject to a high ambient or a moderate overload, or both. For this reason, any protective device must be sensitive to overload conditions.

The new 105°C or 130°C insulations are intended to provide increased protection from the danger of high ambients or temporary overloads. The designer cannot apply long-duration overloads to even these newer insulations without eliminating all the extended life factors that they provide.

(2) *Arcing-Fault Protection*. Arcing-fault currents are found to be as low as 38 percent of the bolted-fault current calculated for the same circuit (arcing ground-fault currents can be much smaller). Such faults, because of their destructive nature, must be removed with no intentional time delay. Unfortunately, the magnitude of this current may be so low that the time-delay characteristics of the overload protective device confuses the low-magnitude arcing fault with the moderate overload or temporary inrush current and allow it to persist for lengthy periods. For example, a 200 percent overcurrent on a fuse might require 200 s before the fuse functions.

Since it is the circuit impedance which limits the current flowing in an arcing fault, most busway manufacturers offer an optional ground conductor, located inside the busway casing, to provide a low-reactance ground path for arcing current. Without this conductor, the arcing current to ground would be forced to travel on the high-impedance steel enclosure, including its many painted joints. This would have a tendency to reduce the already low current to an even lower level and further confuse the low-level protective device.

The best protection against this type of fault is found to be ground-fault protection and the second most suitable is a circuit breaker with a low-range instantaneous trip. Chapter 8 gives a more thorough discussion of this problem.

(3) *Bolted-Fault Protection*. Bolted-fault currents can approach the maximum calculated available fault levels, and therefore protective elements must be capable of interrupting these maximum values. Circuit breakers and current-limiting fuses are most suitable here.

Under certain conditions busways may be applied on circuits capable of delivering fault currents substantially above the busway's short-circuit rating. While it is true that electromagnetic forces increase as the square of the current, the use of very fast fuses permits busways to be applied on circuits having available fault currents higher than the busway short-circuit rating. The reason for this is that the busway rating is based on a 3 cycle duration while class J, K1, or L current-limiting fuses function in much less time, perhaps as low as 1 half-cycle, during high-level faults. The property of inertia, exhibited by the heavy bus bars, causes the bars to resist movement during the very short period of time that the fuse allows current to flow. The fuse limits the magnitude of the current to its let-through value.

In general, a busway may be protected by a class J, K1, or L fuse against the mechanical or thermal effects of the maximum energy the fuse will allow to flow, providing the fuse continuous current rating is equal to the bus-duct continuous current rating. Most manufacturers have conducted tests and will certify that their designs are satisfactory for use with fuses at least one rating larger than the busway. These higher fuse ratings are often needed for coordination with a circuit breaker in series with the fuse.

(4) *Typical Busway Protective Device*. While no single element incorporates all the necessary characteristics, it is possible to assemble several elements into a single device. A particularly effective device is the fused circuit breaker equipped with ground-fault protection. In such a device, the circuit breaker elements provide operation in the overload or low-fault range while the coordinated current-limiting fuse functions during high-level faults. The ground-fault sensor detects those arcing faults which go to ground and, regardless of their low magnitude, signals the circuit breaker to open.

11.8.3 *Busway Testing and Maintenance*. There are several well-known tests which should be performed before any busway is energized: (1) continuity check, (2) voltage test, and (3) high-potential test. These are conducted with the busway disconnected from the supply source and without bus plugs attached.

(1) *Continuity Check*. By using a low-voltage source and a bell, the system should be checked to be sure that there is no accidental solid connection between phases or from phase to ground.

(2) *Voltage Test*. Application of a 500 V megohmmeter test to the system will indicate the insulation resistance values between phases and from phase to ground. While it is not practical to assign specific acceptability limits to meter readings, any reading of less than 1 MΩ for a 100 ft run should be investigated.

(3) *High-Potential Test*. Application of 2200 V (two times maximum design voltage plus 1000 V for 600 V equipment) between phases for 1 min while measuring the leakage current should disclose incipient insulation failures.

(4) *Visual Inspection and Joint Tightening.* None of the preceding tests will disclose the presence of loose joints in the system. It is therefore essential that one individual be assigned the task of inspecting each joint to see that it is bolted properly to the manufacturer's recommended torque values. The importance of this step cannot be overemphasized. Periodic inspections to ensure that the joint integrity has not been affected by vibration, heat cycling, or accumulation of dust or foreign matter should be performed as part of a normal preventive maintenance program. The frequency of such inspections will be determined by the nature of the installation, but ideally it might take place after three months, six months, and one year to build a history and provide a basis for scheduling future inspections.

11.9 Standards References. The following standards publications were used as references in preparing this chapter.

IEEE S-135-1962, Power Cable Ampacities; vol 1, Copper Conductors; vol 2, Aluminum Conductors (IPCEA P-46-426, 1962)

IEEE Std 141-1969, Electric Power Distribution for Industrial Plants

IEEE Std 241-1974, Electric Power Systems in Commercial Buildings

IPCEA P-32-382, Short-Circuit Characteristics of Insulated Cable (Rev Mar 1969)

IPCEA P-45-482, Short-Circuit Performance of Metallic Shielding and Sheaths of Insulated Cable (Jun 1963)

IPCEA P-54-440, Ampacities of Cables in Open-Top Cable Trays (1972) (NEMA WC51-1972)

IPCEA S-19-81, Rubber-Insulated Wire and Cable for the Transmission and Distribution of Electrical Energy (1969) (NEMA WC3-1969)

IPCEA S-61-402, Thermoplastic-Insulated Wire and Cable for the Transmission and Distribution of Electrical Energy (1973) (NEMA WC5-1973)

IPCEA S-65-375, Varnished-Cloth-Insulated Cables (1966) (NEMA WC4-1966)

NFPA No 70, National Electrical Code (1975) (ANSI C1-1975)

11.10 References

[1] WISEMAN, R.J. An Empirical Method for Determining Transient Temperatures of Buried Cable Systems. *AIEE Transactions (Power Apparatus and Systems)*, pt III, vol 72, Jun 1953, pp 545-562.

[2] AIEE Committee Report. The Effect of Loss Factor on Temoerature Rise of Pipe Cable and Buried Cables, Symposium on Temperature Rise of Cables. *AIEE Transactions (Power Apparatus and Systems)*, pt III, vol 72, Jun 1953, pp 530-535.

[3] SHANKLIN, G.B., and BULLER, F.H. Cyclical Loading of Buried Cable and Pipe Cable. *AIEE Transactions (Power Apparatus and Systems)*, pt III, vol 72, Jun 1953, pp 535-541.

[4] NEHER, J.H., and McGRATH, M.H. The Calculation of the Temperature Rise and Load Capability of Cable Systems. *AIEE transactions (Power Apparatus and Systems)*, pt III, vol 76, Oct 1957, pp 752-772.

[5] IEEE Committee Report. Protection Fundamentals for Low-Voltage Electrical Distribution Systems in Commercial Buildings. IEEE JH 2112-1, 1974.

12. Bus and Switchgear Protection

12.1 General Discussion. The substation bus and switchgear is that part of the power system that is used to direct the flow of power and to isolate apparatus and circuits from the power system. It includes the bus, circuit breakers, fuses, disconnection devices, current and potential transformers, and the structure on or in which they are mounted.

To isolate faults in buses all power source circuits connected to the bus are opened electrically by relay action, by direct trip device action on circuit breakers, or by fuses. This disconnection shuts down all loads and associated processes supplied by the bus and may affect other parts of the power system. It is therefore essential that bus protective relaying operate for bus or switchgear faults only. False tripping on external faults is intolerable.

In view of the disastrous effects of a bus fault, the equipment should be designed to be as nearly "fault proof" as practicable, with high-speed protective relaying to keep the duration of the fault to a minimum, which limits the damage and minimizes the effects on other parts of the power system. Since industrial power systems are usually grounded through resistance to limit fault damage, the current available to detect a ground fault is small; thus the protective relaying must be very sensitive.

To prevent faults the bus and associated equipment should be installed in a location where it is least subjected to deteriorating environmental conditions. A preventive maintenance program is a must to detect deterioration, to make repairs,

and to check and test relay performance before a fault occurs [1].

12.2 Types of Buses and Arrangements. The methods of protecting substation buses and switchgear will vary depending on voltage and the arrangement of the buses. The substation bus may have many different arrangements depending on the continuity of service requirements for the bus or for essential feeders supplied from the bus.[2] The bus arrangements most applicable to industrial power systems are shown in Figs 165–168.

Industrial power system voltages fall into three categories, above 15 000 V, from 15 000 to 601 V, and below 600 V. The industrial power system usually includes only the distribution buses 15 000 V and below. However, it may include the subtransmission substation bus at a higher voltage level. Bus protective relaying at this level may overload sections of equipment supplied by the electric utility. For this reason the high-voltage relaying is usually specified by the utility, and compliance with utility practice is mandatory in most cases. Chapter 13 gives further information on utility-service supply line requirements [2].

12.3 Bus Overcurrent Protection. If the system design and operation and the function of the process served do not require fast bus-fault clear-

---

[2] See forthcoming IEEE Std 520, Protective Relay Applications to Power System Buses.

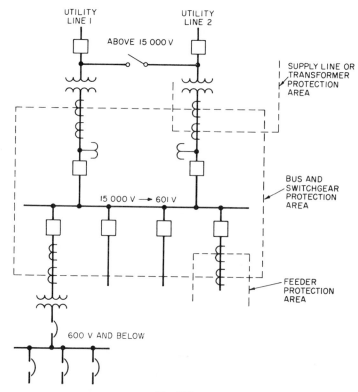

Fig 165
Single Bus Scheme Showing Zones of Protection

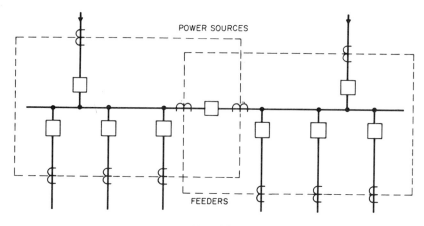

Fig 166
Sectionalized Bus Scheme with Bus Differential Relaying

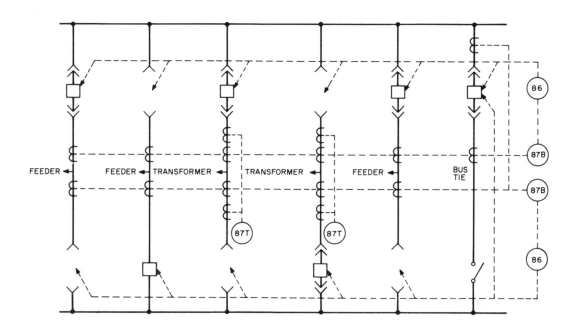

86	Auxiliary Tripping Relay
87B	Bus Differential Relay
87T	Transformer Differential Relay

Fig 167
Double Bus Scheme with Bus Differential Relaying

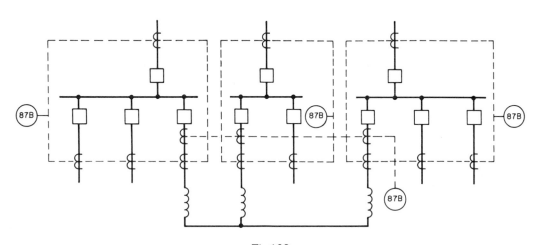

Fig 168
Synchronizing Bus Scheme with Bus Differential Relaying

ance, overcurrent protection is used on each incoming power source circuit. On medium- and high-voltage systems overcurrent relays are used. On low-voltage systems circuit breaker trip devices or fuses are used in most applications. Relays require the use of current transformers for fault sensing, and their use in low-voltage switchgear equipment is often not practicable because of physical and equipment limitations. The introduction of solid-state circuitry to perform the sensing and timing functions has provided significant improvements in the quality of protection for low-voltage circuits and apparatus.

Separate circuitry for ground faults detects the faults at much lower levels and clears them much faster than it is possible with direct-acting electromechanical phase-overcurrent devices alone. NFPA No 70, National Electrical Code (1975) (ANSI C1-1975), requires in Section 230.95 ground-fault protection on solidly grounded wye-connected electric services of more than 150 V to ground but not exceeding 600 V phase to phase for any service-disconnecting means rated 1000 A or more. Chapter 6 of this standard describes how to use low-voltage circuit breakers to their best advantage, Chapter 5 covers the application of fuses for protection, and Chapter 3 gives details on relays and procedures for proper settings.

Overcurrent relays and trip devices must have time-delay and high-current settings to prevent opening the source circuit breakers upon the occurrence of a feeder fault. Therefore, overcurrent devices cannot provide sensitive high-speed bus and switchgear protection.

An induction overcurrent relay connected to a current transformer in the power-transformer neutral-to-ground circuit will provide good sensitivity for ground faults, but it must be set to be selective for feeder faults. If the feeders have ground-sensor instantaneous protection, only a short-time delay is needed on the relay in the transformer grounding circuit. Since most faults are ground faults or eventually become ground faults, good ground-fault protection greatly improves bus overcurrent protection.

12.4 Differential Protection. Differential relaying provides the best protection for buses and switchgear. It is high speed, sensitive, and permits complete overlapping with the other power system relaying as indicated in Figs 165–168. The basic principle is that the vector sum of all measured currents entering and leaving the bus must be zero, unless there is a fault within the protected zone.

Differential relaying is provided to supplement the overcurrent protection. It is frequently used on 15 kV buses, sometimes on 5 kV buses, and rarely on low-voltage buses. The following factors determine whether this relaying should be provided.

(1) Degree of exposure to faults. For example, open-type outdoor buses would have a high degree of exposure, and metal-clad switchgear, properly installed and in a clean environment, would have minimum exposure. Contaminated environments increase the possibilities of faults, and equipment located in these environments needs better protection.

(2) Sectionalized bus arrangements make differential protection more useful and desirable, particularly when secondary-selective low-voltage distribution systems are used. The faulted bus can be isolated quickly and continuity of service maintained to a portion of the load by the other buses.

(3) Effects of bus failure on other parts of the power system and associated processes. If a long down-time period for repairs can be tolerated, differential protection may not be economically justified. On major plant buses the cost of differential relaying is usually insignificant when compared with the reduction in damage to the equipment and the reduced down time of important plant or process facilities.

(4) If there are problems in coordinating the system relay settings, differential relaying is effective in obtaining selectivity. An example would be a system where several buses are required at the same voltage level, with one feeding another. This generally results in unacceptably high overcurrent relay settings on the upstream circuit breakers to obtain coordination.

(5) On buses fed by a local generator, bus differential relaying is recommended to clear the bus quickly and hold the rest of the system together. The overcurrent relays used to protect generator circuits take considerable time to operate.

The differential protection methods generally used are, in the order of the quality of protection

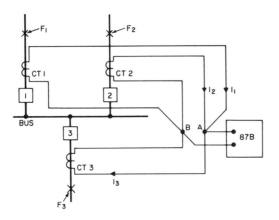

Fig 169
Voltage Differential Relaying

they provide, (1) voltage responsive and linear coupler methods, (2) percentage differential (where applicable), (3) current responsive, and (4) partial differential (sometimes not considered a differential scheme).

Since the differential relay must trip all circuit breakers connected to the bus, a multicontact auxiliary relay is needed. This auxiliary relay should be the high-speed lockout type with contacts in the circuit-breaker closing circuits to prevent "panic" manual closing of a circuit breaker on the fault. The lockout relay must be hand reset before any circuit breakers are closed.

12.5 Voltage Differential Relaying. This method uses through-type iron core current transformers. The problem of current transformer saturation is overcome by using a voltage-responsive (high-impedance) operating coil in the relay. Separate current transformers are required in each circuit connected to the bus as shown in Fig 169. Voltage differential bus protection is not limited as to the number of source and load feeders and has the following features.

(1) High-speed operation in the order of 2 to 3 cycles.

(2) High sensitivity can be set to operate on low values of phase or ground-fault currents in most installations.

(3) Relay operates from all standard bushing current transformers and from switchgear

through-type current transformers with distributed windings.

(4) Relay is not adversely affected by current transformer saturation, direct-current component of fault current, or circuit time constant.

(5) Discrimination between external and internal faults is obtained by relay settings; no restraint or time delay is required.

All current transformers should have the same ratio. Auxiliary current transformers should not be used to match ratios. Unlike current transformers can be matched by operating the higher ratio current transformers as autotransformers, using a tap to obtain the match with the lower ratio current transformers.

All current transformers must have low secondary leakage reactance; wound type are generally not suitable. Bushing-type current transformers constructed on toroidal cores with completely distributed windings will generally have negligible leakage reactance. A distributed winding is one which starts and ends at the same point on the core. Through-type current transformers having suitable characteristics are available for use in switchgear assemblies.

The relay must fulfill two requirements. First, it must not trip for any fault external to the zone of protection. Next, it should be capable of operating for all faults internal to the zones of protection. Considering the first requirement, that the relay should not operate falsely for external faults, refer to Fig 169. Assuming a three-circuit-breaker bus with a fault at the location shown, let us consider for simplicity only one of the three phases. For the fault F_3 indicated, the fault current I_3 will flow through circuit breaker 3 with the currents flowing through circuit breakers 1 and 2 each being smaller but summing up to I_3. If we assume for the moment that the current transformers behave ideally, the current transformer secondary current produced at circuit breaker 3 will balance the sum of the currents produced at circuit breakers 1 and 2. This current will circulate in the current transformer secondary circuits and produce little, if any, voltage across points A—B.

If for some reason the current transformer secondary current at circuit breaker 3 does not balance the sum of the currents produced by the current transformers at circuit breakers 1 and 2, the excess or difference current will flow in the relay circuit and produce a considerably higher voltage. It thus becomes apparent that the current transformer at circuit breaker 3 has a greater tendency to saturate than those at circuit breakers 1 and 2, for the given fault location, because it gets the total current while the other two each get only a fraction of the total. From the point of view of the relay, the worst condition would be the case where the current transformer at circuit breaker 3 saturated almost completely and hence produced no detectable secondary current while those at circuit breakers 1 and 2 did not saturate at all and hence reproduced the current faithfully. It is important to note that for the condition of complete saturation, the mutual reactance of the bushing-type current transformer approaches zero. If it has no appreciable secondary leakage reactance, then the secondary impedance of the current transformer is just its resistance. Thus for the condition of complete saturation of the current transformer at circuit breaker 3, the voltage developed between points A and B is the product of $(I_1 + I_2)$ and the total resistance in the circuit between points A and B and current transformer 3 at circuit breaker 3, including the current transformer secondary resistance. The differential relay is set so that it does not operate for this voltage. It is obvious that this voltage depends on the magnitude of the fault current, the type of fault, and the total resistance. In the case of internal faults, the secondary currents do not circulate but rather result in a high enough secondary voltage to cause the relay to operate.

A nonlinear resistor is connected in parallel with the sensitive high-impedance operating coil to limit the voltage that may be attained during high internal faults. To obtain higher speed operation for high internal faults, the unit is connected in series with the nonlinear resistor.

When offset fault current occurs or residual magnetism exists in the current transformer core, or both, an appreciable direct-current component in the secondary current is present. This has caused false tripping when simple unrestrained low-impedance relays are used for bus differential. Voltage differential relays are made insensitive to the direct-current component by connecting the relay sensitive operating coil in series with a capacitor and reactor. The circuit is resonant at the fundamental power frequency and the direct-

current component is blocked by the series capacitor [3].

12.6 Linear-Coupler Method. The linear-coupler method provides extremely reliable high-speed bus protection. It is highly flexible to future expansion and system changes. The couplers can be open circuited without any difficulties to simplify switching circuits. The operating time for one type of linear-coupler system is 1 cycle or less above 150 percent of pickup and ½–1 cycle for another type of linear-coupler system. This scheme eliminates the difficulty due to differences in the characteristics of iron-core current transformers by using air-core mutual inductances without any iron in the magnetic circuit. Therefore it is free of any direct-current or alternating-current saturation.

The linear couplers of the different circuit breakers are connected in series and produce secondary voltages that are directly proportional to the primary currents going through the circuit breakers, as shown in Fig 170.

With the simple series circuit shown in Fig 170,

$$I_R = \frac{E_{sec}}{Z_R + \Sigma Z_C}$$

$$= \frac{I_{pri}M}{Z_R + \Sigma Z_C}$$

where

E_{sec} = Voltage induced in linear-coupler secondary

I_{pri} = Primary current, symmetrical rms

I_R = Current in relay and linear-coupler secondary

M = Mutual reactance, = 0.005 Ω for 60 Hz

Z_C = Self-impedance of linear-coupler secondary

Z_R = Impedance of relay

During normal conditions or for external faults, the sum of the voltages produced by the linear couplers equals zero. During internal bus faults, however, this voltage is no longer zero, and is measured by a sensitive relay which operates to trip circuit breakers and clear the bus.

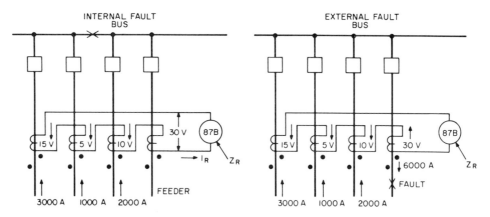

Fig 170
Linear-Coupler Bus Protective
System with Typical Values Illustrating
Its Operation on Internal and External Faults

Linear couplers are air-cored mutual reactors wound on nonmagnetic toroidal cores such that the adjacent circuits will not induce any unwanted voltage. For the conductor within the toroid, 5 V is induced per 1000 A of primary current. Therefore, by design the mutual impedance M is 0.005 Ω, 60 Hz. In other words, $E_{sec} = I_{pri}M$.

12.7 Percentage Differential Relay. Where there are relatively few circuits connected to the bus, relays using the percentage differential principle may be employed. These relays are similar to transformer differential relays which are described in Chapter 10. The problem of application of percentage differential relays for bus protection, however, increases with the number of circuits connected to the bus. It requires that all current transformers supplying the relays have the same ratio and identical characteristics. Variation in the characteristics of the current transformers, particularly the saturation phenomena under short-circuit conditions, presents the greatest problem for this type of protection and often limits it to applications where only a limited number of feeders are present.

12.8 Current Differential Relaying. When voltage or linear-coupler differential protection cannot be economically justified, a less expensive current differential scheme may be considered. This scheme utilizes simple induction-type overcurrent relays connected to respond to any difference between the currents fed into the bus and the current fed from the bus. The current transformer arrangements are the same as shown in Figs 165—168. The connections are as shown in Fig 169.

Chapter 3 gives details on these relays. A special form of overcurrent relay is available with an internally mounted auxiliary relay and connections to permit testing the integrity of the current transformer circuits for accidental ground faults and open circuits. The connections are so arranged that while checking on one phase, the relays in the other two phases are still providing protection.

12.9 Partial Differential Protection. This type of protection, sometimes called "summation relaying," is a modification where one or more of the load circuits are left uncompensated in the differential system (Fig 171). For this reason it may be a misnomer to name it a "differential" scheme. This method may be used as primary protection for buses with loads protected by fuses, as backup to a complete differential protection scheme, and to provide local backup protection for "stuck" load circuit breakers which fail to operate when they should. The phase overcurrent relays are set above the total bus load or the total rating of all loads supplied from the bus section.

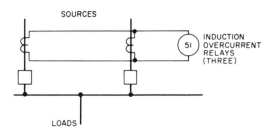

SOURCES

INDUCTION
OVERCURRENT
RELAYS
(THREE)

LOADS

Fig 171
Partial Differential System

The relays must provide enough time delay to be selective with relays on the load circuits. Consequently, the sensitivity and speed of partial differential protection is not as good as for full differential protection.

12.10 Backup Protection. In the event of a failure of the primary protective system to operate as planned, some form of backup relaying should be provided in the industrial power system or in the power supply system.

Bus backup protection is inherently provided by the primary relaying at the remote ends of the supply lines. This is known as remote backup protection. It may not be adequate because of system instability and effects on other power systems, and local backup relaying may be necessary. Kennedy and McConnell [4] analyze the performance of various remote and local backup relaying schemes. Chapter 13 gives further information on utility service supply-line requirements and the backup protection by utility relaying [2].

Circuit breaker failure can cause catastrophic results, such as complete system shutdown. Local circuit breaker failure or "stuck circuit breaker" relay schemes are available to quickly trip upstream circuit breakers if the circuit breaker on the faulted circuit fails to operate within a specified time. However, those schemes are normally applied only on buses where the extra expense can be economically justified.

12.11 Voltage Surge Protection. Protection against voltage surges due to lightning, arcing, or switching is required on all switchgear connected to exposed circuits entering or leaving the equipment. A circuit is considered exposed to voltage surges if it is connected to any kind of open-line wires, either directly or through any kind of cable, reactor, or regulator. A circuit connected to open-line wires through a power transformer is not considered exposed if adequate protection is provided on the line side of the transformer. Circuits confined entirely to the interior of a building, such as an industrial plant, are not considered exposed and ordinarily require no voltage surge protection.

The protection is provided by lightning arresters connected, without fuses or disconnecting devices, at the terminals of each exposed circuit [IEEE Std 27-1974, Switchgear Assemblies Including Metal-Enclosed Bus (ANSI C 37.20-1969 and supplements)]. Surge protection connected directly to the bus is not recommended, as the reliability of the bus will be diminished. If it becomes necessary to connect surge protection directly to the bus, it should be connected through fuses or circuit breakers. (Note that circuits will not be protected when the circuit breakers are open.)

The arresters should be of the valve type, of adequate discharge capacity, and their voltage ratings should be selected to keep the voltage surges below the insulation level of the protected equipment. When the exposed line is directly connected to the switchgear through roof entrance bushings, intermediate or station-type arresters are recommended. Where an exposed line is connected to the switchgear with a section of continous-metallic-sheath cable, the arresters should be installed at the junction of cable and overhead line. If the arresters at the overhead line are intermediate or station type, no arresters are required at the switchgear. If distribution-type arresters are used at the overhead line, another set may be required at the switchgear terminals, depending on the length of the cable. Cable without a continuous metallic sheath does not reduce the wavefront enough so that valve-type distribution arresters are required at the switchgear. In the latter case a properly installed ground wire in the duct with each three-phase cable provides very nearly the same impedance as continuous-metallic-sheath cable.

Lightning arresters may be required to protect

the switchgear at altitudes above 3300 ft, even though the circuits are not connected to exposed circuits. This is due to the voltage correction factors applicable above 3300 ft altitude (IEEE Std 27-1974). Lightning arresters are applied such that the impulse voltage protective level maintained by the arrester is about 20 percent less than the corrected impluse voltage rating of the switchgear. These arresters should be of the station type.

12.12 Conclusion. Because of its location and function in the electric power system, the bus and switchgear should be designed, located, and maintained to prevent faults. The preferred practice for bus and switchgear protection is voltage-responsive or linear-coupler differential relaying with the power system designed with sectionalized buses so that continuity of service can be maintained to a portion of the load. The best protective relaying in a single bus arrangement will operate to cut off power to all circuits supplied by the bus.

Hopefully, the relaying will never be called on to operate. Location of the equipment in a good environment and maintenance on a planned basis will help prevent the need for relays to operate [1]. If a fault does occur, high-speed sensitive relaying will limit the damage so that repairs can be made quickly and service restored in a short time. Fast clearing of faults also can save lives by minimizing explosion and fire aftermath. Fur-thermore fast clearing of human contact faults has saved lives or reduced injury.

12.13 Standards References. The following standards publications were used as references in preparing this chapter.

IEEE Std 27-1974, Switchgear Assemblies Including Metal-Enclosed Bus (ANSI C37.20-1969 and supplements)

NFPA No 70, National Electrical Code (1975) (ANSI C1-1975)

12.14 References

[1] KILLIN, A.M. How Plant Management Evaluates Electrical Performance. *IEEE Transactions on Industry and General Applications*, vol IGA-3, Mar/Apr 1967, pp 75-78.

[2] BECKMANN, J.J., DALASTA, D., HENDRON, E.W., and HIGGINS, T.D. Service Supply Line Protection. *IEEE Transactions on Industry and General Applications*, vol IGA-5, Nov/Dec 1969, pp 657-671.

[3] SEELEY, H.T., and VON ROESCHLAUB, F. Instantaneous Bus Differential Protection Using Bushing Current Transformers. *AIEE Transactions*, vol 67, 1948, pp 1709-1719.

[4] KENNEDY, L.F., and McCONNELL, A.J. An Appraisal of Remote and Local Backup Relaying. *AIEE Transactions (Power Apparatus and Systems)*, pt III, vol 76, Oct 1957, pp 735-747.

13. Service Supply Line Protection

13.1 General Discussion. This chapter discusses the interface between the supplier of electricity and the consumer. The basic desire of the power company and the consumer is to provide a reliable power supply of adequate capacity to serve the connected load. A typical sequence in designing an electrical service arrangement for a new plant could be as follows.

(1) Classify and group loads according to their characteristics:

    (a) Power required, real and reactive

    (b) Optimum nominal voltage

    (c) Sensitivity to voltage level

    (d) Sensitivity to frequency variations

    (e) Sensitivity to voltage dips

    (f) Sensitivity to interruptions

    (g) Other unusual service requirements, such as sensitivity to nonsinusoidal wave shapes

    (h) Physical location

    (i) Future load considerations

(2) Select, together with the electric utility personnel, a suitable supply service arrangement consistent with the economics of the application.

(3) Analyze the service line and equipment as to the various electric faults or system disturbances which are likely to occur. In considering service supply line protection, one must know and understand the various system disturbances in order to avoid unnecessary tripping of circuit breakers.

(4) Restructure the one-line diagram and apply equipment to minimize the effect of these faults and disturbances. This may include

    (a) Addition of power equipment, such as capacitors or voltage regulators

    (b) Surge protective equipment

    (c) Protective relaying and switching schemes

    (d) Use of auxiliary devices or stored-energy control systems

    (e) Segregation of loads capable of interruption during low-frequency and other manifestations of system trouble

13.1.1 *System Distrurbances and Circuit Arrangements.* One of the principal causes of disturbances are transient line faults. Many of these faults are cleared within a few cycles with a minimum effect on consumer loads. These disturbances normally cause only momentary voltage dips. A voltage dip is defined as a temporary reduction in voltage, but the consumer's service remains energized.

A disturbance other than on the consumer's main service is often removed within several cycles since circuit clearing time only is involved. Where a circuit breaker must reclose to restore service to the consumer, the outage is of longer duration since both the clearing time and the reclosing time are involved. Here an outage is defined as occurring when the supply circuit breaker opens to a consumer.

While it is beyond the scope of this publication to analyze supply circuit arrangements, it is obvious that the use of multiple lines, transformers, bus sections, and circuit breakers reduces outages in the supply service. It also adds to the cost and space requirements. A knowledge of the frequency of outages and the estimated cost of each outage is necessary to determine the need for

Table 37
Typical Fault Records for One Utility

| | Line Design Voltage | | |
	23—46 kV	69 kV	138 kV
Temporary line faults per 100 mi/yr	25—40	10—25	5—15
Permanent line faults per 100 mi/yr	0.1—0.2	0.2—0.3	0.1

Table 38
Outage Rates and Average Repair Times of Electric Equipment

| | Outages per Mile (or per Unit) per Year | | Average Repair Time (hours) |
	Normal	Stormy	
Semipermanent forced outages			
Secondary cable	0.005	0.005	35
Network protector	0.0038	0.0038	2
Network transformers	0.0049	0.0049	24
Primary switch	0.001	0.001	2
4 kV open wire	0.045	3.2	2
4 kV feeder circuit breaker	0.005	0.005	0.8*
4 kV distribution transformer	0.0017—0.0069	0.0017—0.079	3.5
Maintenance outage			
Network protector	0.2	—	2
Network transformer	0.2	—	10
Primary switch	0.2	—	2—10
4 kV open wire	2.0	—	4
4 kV feeder circuit breaker	0.2	—	0.8*

From Reference [2].
*Time to transfer feeder to spare position.

added equipment. These costs may include

(1) Indirect effect on safety

(2) Loss of production

(3) Production of off-tolerance products or scrap

(4) Purging and restart-up expense

(5) Additional personnel needed to be available for restart procedures if not automatic

(6) Electric and mechanical equipment damage

(7) Emotional upsets of personnel

Sample fault records for various operating voltages for one utility are given in Table 37.

Failure rates for transformers and other system components differ widely due to design and application variations. Sample supply circuit arrangements with typical protective schemes are presented in this chapter. Outage rates and average repair time of electric equipment are given in Table 38.

In order to analyze the protection required at the interconnection point between the electric utility and the consumer, it becomes desirable to first assume a typical supply operating normally. Next we might list the various abnormalities which could occur on the supply system (and also in the consumer system) and disturb this smooth flow of electric energy through the interconnection point. These possible disturbances will be analyzed as to the possible harmful effects on the various types of loads which can take the form of (1) equipment damage, (2) loss of production, (3) loss of production materials, (4) damage to plant facilities (fire), and (5) injury to personnel. Finally precautionary measures which can be applied to reduce the adverse effects of the disturbances are considered.

Some plant electric loads are not particularly critical as to momentary voltage dips or short interruptions in the plant supply. Others, however, are quite sensitive. The sensitivity may be one of the magnitude of dip, such as a contactor dropping out at about 60 percent voltage. On the other hand, the duration of the voltage dip is extremely important to other loads. Many electric motor drives are sensitive to the combination of magnitude and duration of voltage dips, both of which play an important role in whether a critical drive will ride through a disturbance. Industrial plant load tolerances to power supply variations are listed in Table 39.

13.1.2 *Types of Protection Available.* Protective relays merely initiate the proper switching at the desired time. A prerequisite to the use of protective relays is the presence of suitable sensing devices (instrument transformers) and suitable switching devices (circuit breakers, etc). Therefore, selection of a protective relaying scheme is inherently dependent on the circuit arrangement as well as the equipment and processes to be protected and the service continuity needed. The details of protecting specific devices are covered in other chapters of this publication. The discussion of protective relays in this chapter centers on the specific applications in service supply lines as indicated on the sample circuit arrangements. Lightning arresters are also important to incoming circuit protection. The application of lightning arresters for equipment protection is covered in another chapter. Typical applications are as shown on typical systems illustrated in this chapter.

13.2 Service Requirements. Consideration of the design, operation, or protection of tie lines between a consumer and his power supplier must be based on deep mutual understanding of each other's needs, limitations, and problems. Learning of the problems is a required step in arriving at this understanding. The more common ones are outlined in Table 40. It is the purpose of this chapter to also list system design or operating techniques or devices which can nullify the problems. Of the solutions presented in Tables 41 and 42, some apply mainly to tie-line operation and protection, but many will relate more generally to the power systems of the utility or consumer.

As used in this chapter, the term "load" means an electric device such as a motor, lighting fixture, computer, communication system, alarm, control system, and so on. To a consumer these loads are only a means to an end; they are the muscles and nerve systems to operate chemical processes, mines, public buildings, or manufacturing plants. The nature of consumer's operation sets the requirements of power reliability and quality.

The electric power source for an industrial or commercial power system must meet the normal peak power demands of the system and do so with minimum deviations from normal voltage and frequency. Furthermore the supplied power should be relatively free from voltage surges or distortion, must maintain normal phase rotation in multiphase systems, and must not subject loads to single phasing. Reliability of a power system — the percentage of time which the system can effectively serve the loads — is also of major importance. The simplicity of these statements of fact may tend to obscure the complexity of technical and commercial problems which sometimes arise, but the statements are the true measures of quality of service to a consumer.

Service quality involves two distinct factors, each of which must be considered separately, and each of which will have different degrees of importance among consumers. The two factors are (1) power quality and (2) power reliabilty. Together, these two factors make up service quality.

Each load device (Table 39) has specific power quality tolerances within which it will operate normally. Reliability requirements, however, may completely change service considerations. For example, an incandescent lamp will perform

Table 39
Electric Service Deviation Tolerances for Load and Control Equipment

Device	Voltage Level*	Voltage Distortion: Harmonic Content	Frequency	Comment
Alarms, systems operating on loss of voltage	Variable	—	—	
Capacitors for power factor correction	+ 16% to − 100%	**	+0% to −100%	
Communication equipment	±5%†	Variable	—	
Computers, data-processing equipment	±10% for 1 cycle†	5%	+ ½ Hz to −1½ Hz	
Contactors, motor starters				
Alternating-current coil burnout	+10% to −15%	—	—	
Alternating-current coil dropout‡	−30% to −40% for 2 cycles	—	—	
Direct-current coil dropout	−30% to −40% for 5−10 cycles	—	—	
Electronic tubes	±5%	Variable	—	
Lighting				
Fluorescent	−10% (flourescent)	—	—	Uncertain starting, shortened life.
	−25%	—	—	Lamp will extinguish.
Incandescent	+18%	—	—	10% of normal life. See Fig 173.
Mercury vapor	−50% for 2 cycles	—	—	Lamp will extinguish.
Motors, standard induction#	±10%	Variable	±5%	Sum of absolute values of voltage and frequency deviation shall be no greater than ±10%.
Resistance loads, furnaces, heaters	Variable	—	—	
Solenoids, shutoff valves for gas or oil fired furnaces, magnetic chucks, brakes, clutches	−30% to −40% for 1/2 cycle	—	—	
Transformers	+5% with rated kVA ≤ 0.80 PF +10% with no load	—	—	Voltage deviations apply at rated frequency. If frequency drops, voltage limits must reduce proportionally. Firing circuits and transformers generally determine tolerances. If supply voltage is +5%, transformer loading must be reduced by 5%.
Inverters (gaseous, thyristor)	+5% with full load +10% with no load −10% transient	2%	±2 Hz	

Table 39 (Continued)

Equipment	Voltage	Harmonic	Frequency	Remarks
Rectifiers				
Diode (gaseous)	+5% with full load, +10% with no load, −10% transient	Sensitive§	—	If supply voltage is +5%, transformer loading must be reduced by 5%.
Diode (solid state)	±10%	Sensitive§	—	Some rectifier systems rated by NEMA for voltage deviation of +5% to −10%.
Phase controlled (gaseous, thyristor)	±5% with full load, +10% with no load, −10% transient	2%	±2 Hz	Firing circuits and transformers generally determine tolerances. If supply voltage is +5%, transformer loading must be reduced by 5%.
Generators	±5%	Sensitive‖	−5%	Voltage tolerance generally a function of generator design. Surge protective devices should be applied at generator terminals.
Turbines (steam)	—	—	−1%	

*It is presumed that properly selected lightning arresters and surge protective equipments are installed throughout the system. Deviation tolerance is continuous unless specified. Percent voltage unbalance = (100 × maximum voltage deviation from average voltage)/(average voltage).

**Capacitor ratings are based on a nominal voltage distortion caused by overexcited transformers. Silicone-controlled-rectifier phase-controlled loads may cause additional and excessive distortion.

†Voltage tolerances may not be applicable for equipment with integral power supply or voltage regulator.

‡This type of contactor is not recommended where tripouts due to voltage dips are undesirable.

#Phase voltages should be balanced as closely as can be read on the usual commercial voltmeter.

§Presence of harmonics may raise or lower direct output voltage, balance between phases, or balance between rectifier units operating in parallel.

‖Systems with rectifiers or inverters may contain harmonics which cause nonsalient pole generator overheating. Generator load reduction may be required as a function of the number of phases in the rectifier system as follows: 24 phases—no reduction; 12 phases—8% reduction; 6 phases—40% reduction.

Table 40
Electric Power System Disturbances

Disturbance	Duration	Effect on System	Typical Cause
Voltage level change	Steady	± 10% voltage	Normal system voltage variation resulting from load changes.
Voltage swing	10 cycles to 5 min	± 30% voltage	Motor starting, shock loads, furnace loads, welders, planers, chippers, roughing drives.
Voltage transients*	Up to 30 cycles	+100% to −50% voltage	Remote system faults, switching surges, lightning strokes, capacitor switching.†
Voltage flicker	Variable	Voltage variations	Repetitive voltage swings or transients.
Voltage loss A	1 s maximum	Down to 0% voltage	Power transmission system or distribution system faults, network system faults.*
Voltage loss B	1 min maximum	Down to 0% voltage	Power system faults or equipment failure requiring reclosing or resynchronizing operation.‡
Voltage loss C	Extended	−100% voltage	Permanent power system faults, equipment failure; accidental opening of power circuit breaker.‡
Voltage wave-shape distortion, harmonics, noise	Variable	Fundamental or harmonic voltage up to + 200%	Arcing faults, ferroresonance, switching, transients, transformer, iron core reactor or ballast magnetizing requirements, controlled rectifiers, commutators, arc discharges, fluorescent lamps, motors.
Voltage unbalance	Steady	Up to 10% voltage variation among phases of three-phase system	Single-phase or unbalanced loads on three-phase system.
Single phasing	Extended	Down to 0% voltage on one phase of three-phase system	Open conductor, switching with single-pole devices, fuse blowing.
Power direction change, short circuits	Variable	Change of flow of current or power	Supply system faults, loss of transmission lines, synchronizing power surges, switching.
Frequency change	Variable	+ 1 to −2 Hz	Loss of generation or utility supply line.

*It is presumed that properly selected lightning arresters and surge protective equipments are installed throughout the system.
†Some types of switching transients may be amplified by coincident resonance of power factor capacitors and transformer inductances at the switching frequency.
‡Disturbance may be in either the utility or consumer system. Disturbance may be isolated in 3 to 30 cycles by circuit breakers or 35 cycles by network protectors, after which time service may be restored to disturbance-free portion of system.

Table 41
Restoration of Service after Loss of Voltage

Method	Minimum Time to Restore Service	Comments
Re-energize circuit*		
Automatic reclosing after temporary fault	25 to 70 cycles	If motor loads exist which support plant voltage after loss of system voltage, then reclosing must be delayed either for a definite time or until residual plant voltage has decayed to less than 50% normal, or as recommended by manufacturer to prevent damage to motors.
Remote-controlled reclosing of circuit breakers or switches	Up to 1 min	
Manual or remote-controlled reclosing after manual isolation of cause of disturbance; replace fuses	Up to 1 h or longer	
Transfer incoming line to alternate power source†		International time delay may eliminate unnecessary transfer under some conditions.
Automatic transfer	15 to 30 cycles	
Manual transfer	Up to 30 min	
Start generators in consumer system	Variable	Standby generation may be sufficient to supply emergency or critical loads.

*Reclosing must include resynchronizing if consumer generators are operating in parallel with utility system.
†May include transfer of emergency lighting and loads to a battery source or engine-driven generators.

Table 42
Minimizing the Effect of Power System Disturbances

Device	Minimized Disturbance	Comment*
A. Power Equipment and Power Switching Devices		
Capacitors, shunt connected	Voltage level change	Can reduce normal system voltage variations.
Reactors, shunt connected	Voltage level change	Can reduce system overvoltages during light load periods.
Voltage regulators, induction or load tap changing	Voltage level change, voltage swing, voltage unbalance	Will restore voltage level on load. Operating range usually ±10%. Response too slow to correct for voltage transients. Single-phase sensors and regulators required to correct for voltage unbalance.
Generators, synchronous motors, or condensers used with automatic voltage regulators	Voltage level change, voltage swing	Will restore system voltage within limits imposed by maximum and minimum excitation on machine field.
Circuit breakers used with protective relays, network protectors, reclosers, fuses	Voltage transients, voltage single phasing, power flow change, frequency change	Device isolates cause of disturbance. Restoration of service required. See Table 42.
B. Control Circuit Modification and Small Power Equipment		
Battery, battery charger system	All	Provides uninterrupted power for control, instrumentation, communication. Alternate power source for emergency lighting, critical loads. Limited to about 50—300 A·h capacity during loss of voltage.
Cable shielding in control or communication circuits	Voltage waveshape, harmonics, noise	Reduces or eliminates induced voltages which interfere with correct transmission of intelligence or control signals.
Circuit breaker, close/trip control circuits	Voltage loss	
Capacitor stored energy		Provides energy for releasing circuit breaker tripping mechanism for several seconds.
Compressed air storage		Provides energy for 2—3 circuit breaker closing operations.
Spring energy		Provides energy for 1 circuit breaker closing operation.
Control relays		
Time delay dropout	Voltage loss	Primarily used in motor control circuits to maintain starter circuit in "run" position for 2 or 4 s after loss of voltage.
Mechanically latched relay		Retains control circuit condition during loss of voltage.
Engine, generator set	Voltage loss B and C (Table 40)	Provides emergency power for lighting or critical loads. Normally requires minimum of 5—60 s to start if automatic.
Filters, tuned-circuits	Voltage distortion, harmonics, noise	
Motor-generator set†	All except voltage loss B and C (Table 40), extended frequency change	Loss of voltage ride-through capability with flywheel: (1) 0.3 s with deviation less than −3% voltage, −½ cycle, or (2) 1.8 s with deviation less than −3% voltage, −1½ cycles.

Table 42 (Continued)

Uninterruptible power system (UPS), static system, or motor-generator set	All	Total isolation from power system disturbance for loads up to about 1000 kVA. Provides power for 15–30 min after loss of voltage. A UPS system contains a rectifier, inverter, and battery (or motor-generator sets with battery) to supply 60 Hz power.
Voltage stabilizer‡	Voltage level change, voltage swing, voltage transient, voltage unbalance, voltage flicker	Static device. Holds constant output voltage with voltage input variations up to ± 30%. Full correction in 1/2 cycle. Limited to loads below 50 kVA.
C. System Line Diagram	All	Use of reactors, bus ties, etc, can minimize flicker on critical loads. Emergency separation of critical and noncritical loads.

*The solution to any disturbance may result in "side effects" which are equally disturbing. For example, capacitors switched without preinsertion surge-limiting resistors may create damaging transient overvoltages; improperly installed or grounded shield communications cable may cause other types of noise or burnout due to power-fault induced currents, etc.
†Improved performance results from addition of a flywheel.
‡May contain harmonic distortion in regulated voltage.

satisfactorily on voltage containing myriad abnormalities, but if it is the lamp in an EXIT fixture in a public building, or an operating lamp in a hospital operating room, then it must have power of absolute reliabilty. By contrast, a computer must be supplied with power of extremely high quality, but if it is being used to process routine business data or produce a payroll, then service reliability becomes secondary.

Lamp performance for various voltage levels is given in Figs 172-174. Table 43 shows the effect of voltage variations on standard induction motor characteristics. The borderline of visibility and irritation to the human eye resulting from flicker is given in IEEE Std 141-1969, Electric Power Distribution for Industrial Plants. A revision of this curve as used by some utilities is shown in Fig 175.

A study of a consumer's operation and loads can help the utility—consumer team arrive at the required level of service quality. Most, if not all, of the disturbances of Table 40 will eventually occur in a power system. If any of these disturbances exceed tolerances of the load devices for the equipment included in a system, then appropriate steps must be considered. The required reliability must also be kept in mind; Tables 41 and 42 (section B) will be particularly helpful here.

Reliability can be increased most effectively by designing a system so that service can be quickly restored after loss of voltage. Table 41 lists methods by which this can be done. A consumer, within his own system, can provide varying degrees of reliability by installing engine or turbine driven generators or one of the battery stored energy power systems.

Service quality can also be greatly enhanced by proper tie-line protection and operation.

13.2.1 *System Disturbances.* In an electric power system the voltage and phase angle on the lines are the driving forces for the transmission of energy to the loads. As long as the voltage is of the proper magnitude, waveshape, and frequency, the loads and control devices will operate normally.

The voltage level, however, may gradually change over a period of minutes or hours and as a result subject loads to voltages which are either too high or too low to permit continued satisfactory operation. Voltage variations of this type

Table 43
Effect of Voltage Variations on Standard Induction Motor Characteristics

Characteristic	Function of Voltage	109 V (95%)	115 V (100%)	120 V (104%)	126 V (110%)
Start and maximum running torque	V^2	90.2%	100%	108%	121%
Full-load speed	Slip $\propto 1/V^2$	98.65%	100%	100.4%	101%
Full-load current	Depending upon design	106%	100%	96%	93%
Starting current	V	95%	100%	104%	110%
Temperature rise, full load	Test	+4° C	0° C	−2° C	−4° C
Maximum overload capacity	V^2	90.2%	100%	108%	121%

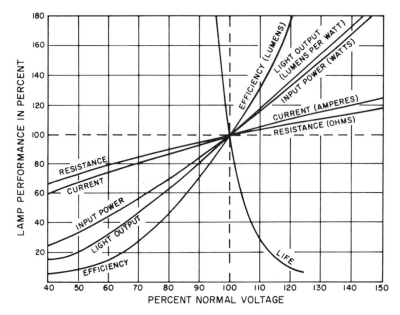

Fig 172
Incadescent Lamp Performance as
Affected by Voltage

NOTE: These characteristics are averages of many lamps.

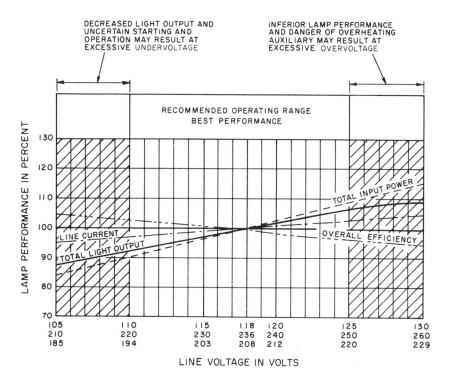

Fig 173
Fluorescent Lamp Performance as Affected by
Voltage

NOTE: Energized lamps may be extinguished if voltage
drops to approximately 75 percent of rated line voltage.

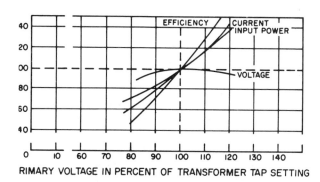

Fig 174
Mercury Lamp Performance as Affected
by Voltage; 120 V Basis

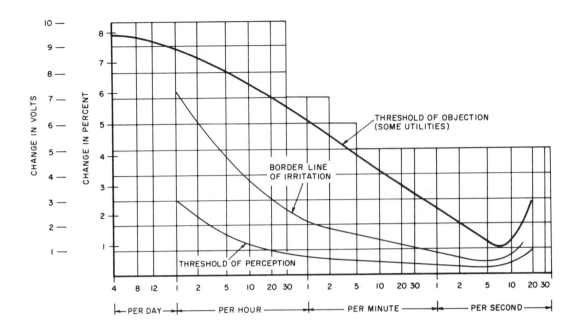

Fig 175
Typical Voltage Flicker Limits; 120 V Basis
NOTE: For more details see IEEE Std 141-1969.

result from daily changes of load on a power system. During periods of peak load the transmission lines, transformers, and distribution lines that make up the system will reduce the voltage available to the loads.

Apart from the gradual changes there will be voltage swings of much shorter duration and frequently of much greater magnitude. These swings will result from remote system faults, switching surges from operating circuit breakers in the transmission or distribution system, shock loads, open-arc or submerged-arc furnaces, starting of large motors, roughing drives, planers, chippers, and welders. Voltage swings resulting from these causes are quite transient in nature, but may be severe enough to cause malfunction of some of the load equipment. Flicker, a special case, is repeated voltage swings of such magnitude that there is a noticeable change in the amount of light produced by lamps. The effect of this problem is mainly psychological, but it is symptomatic of trouble within the power system. There is a rather wide variation between the

objection and the irritation curves. The original curves are based on low-power incandescent lamps. Higher power lamps in common use today are less sensitive to flicker due to a larger mass of the filaments.

The higher flicker curve is a calculated theoretical value which ignores usual compensating factors such as all single-phase and small-horsepower three-phase motors. This higher curve also recognizes that the flicker will not occur regularly during the critical early-evening illumination hours.

Loss of voltage is another type of power system disturbance which can vary considerably in time duration, depending upon the cause of trouble and the method used to restore service. These disturbances generally result from faults occurring on transmission lines, in transformers, or in distribution or network systems, and result from countless natural and man-made phenomena. They are quite unpredictable as to location and time. Some of the faults will be semipermanent in nature and will require isolation of a portion of the power system while repairs are made. Others are temporary and, once cleared by automatic switching devices, will permit continued normal operation of the power system.

The presence of harmonic voltages or wave-shape distortions causes problems which require consideration. Voltage levels at the loads can appear to be normal, and yet there may be severe generator overheating or interference with communication or signal systems under these conditions.

Abnormalities in current would also be considered a service disturbance, although they may be the result of and concurrent with voltage problems. During some system disturbances the flow of current or power is of unreasonable magnitude or improper direction. The problems caused by a reversal of the normal direction of the flow of power may be commercial rather than technical.

Frequency deviation from the nominal system value is a service disturbance of rather special nature. The change in frequency in itself may be undesirable for some particular loads such as computers, but is usually associated with other system troubles and a drop in system voltage and is an indication that the total load connected to the power source is greater than the capability of that system to supply the load. Many load-shedding schemes are activated by a drop in system frequency.

Any particular power system disturbance, as will be noted in Table 40, characteristically originates in a certain part of the power system. Voltage swings and flicker are most frequently conditions arising from within the loads, as are spurious noise and harmonics. However, single phasing of the power system — a condition where voltage is lost on one phase of a three-phase system — generally occurs within the power supply circuits or on the tie line and results from the use of single-pole switching devices or the blowing of a fuse. Electric faults, however, can occur anywhere within the power system. A fault may result in loss of voltage for the portion of the system within which the fault occurred, but if a selective relaying system is in operation, then the remainder of the load will function normally after seeing a voltage dip for a period of time lasting from 2 to 35 cycles.

13.2.2 *Corrective Measures.* In order to define the quality of desired service, factors in addition to the load and consumer requirements should be considered. Some consumer systems have inherent alleviating factors such as local generation to cope with certain types of disturbances. Large systems with a mixture of different types of load equipment can reduce some types of disturbances to insignificance. For example, starting a large motor can cause voltage swings detrimental to other loads, but the effect of starting the motor may be negligible if the system is large and relatively "stiff."

After a study has been made, it may be found that corrective measures will have to be considered. Tables 41 and 42 list various approaches. Those listed in Tables 41 and 42 (section A) all involve reconfiguration of the power system by means of power circuit switching. The application of each of these depends upon specific system conditions. Load shedding, for example, would probably not be considered unless there is local generation. After considering all other factors, a consumer might find some load which is so demanding in its service requirement that special corrective measures would have to be taken. These will follow one of two approaches [Table 42 (section B)].

(1) *Control Modifications.* A common approach here would be to use a time-delay dropout scheme for motor control. This time delay can be produced by different techniques, but basically it will ensure that the seal-in circuit for a motor starter will remain closed for a period up to several seconds after a loss of voltage. If service is restored within this period of time, then the motor will automatically restart and will require no attention by an operator.

Another approach is to supply the control system from a battery or other reliable direct-current source. Disturbances in the alternating-current portion of the system are thus not reflected into the control circuit, and power devices will resume their normal operating conditions after restoration of service.

Stored-energy tripping mechanisms for circuit breakers are sometimes used to provide energy for tripping a circuit breaker after alternating-current voltage has disappeared.

(2) *Power Devices.* Some processes or loads cannot tolerate a shutdown and must be provided with a source of energy to ride through a loss of voltage.

One device which provides short-time continuous power to a load during service interruptions is an isolating motor-generator set. This electrical isolation of the load from the main system will also eliminate such disturbances as noise, voltage swings, or transients, but the motor-generator set itself will be sensitive to certain disturbances. The inertia of the rotating motor-generator set provides energy to the load for a period of from several cycles to several seconds, depending upon flywheel effect. This type of approach is common in power supplies for computers where it is sometimes essential to have a programmed shutdown of a system so that the computer's memory circuits can retain vital information.

The most critical loads can be supplied by a rectifier—inverter system incorporating a battery. In this approach the alternating-current load is supplied by a static or rotating inverter which receives its power from a direct-current system. The direct-current system will consist of a battery in parallel with a rectifier or generator and will in turn be supplied by the main power system. Such an array of equipment will provide uninterrupted service to a load of up to 1000 kVA for 30 min or more.

After consideration of the individual and collective service requirements of the loads and the inherent alleviating factors within the consumer's system, it becomes possible to define the level of service quality required from the source. The problem now resolves into one of selection and application of appropriate types of equipment to minimize the effect of any remaining possible disturbances, not the least important of which is a well-designed and properly implemented and operated interconnection between consumer and utility.

13.3 Types of Protection Available. Circuit protection is applied to the interconnection circuits in the same manner and employing the same principles as in all other locations in utility and industrial systems. The basic purpose is to protect all circuits and equipment from abnormal electrical conditions. When this goal is unattainable, the aim is to minimize the effects of the fault or other abnormal condition.

This is accomplished by a division of the electric circuit into zones of protection which can be accommodated in an economical manner using available relaying schemes. These zones of protection are selected on the basis of the individual characteristics of each installation. Some of the considerations are the following:

(1) Electrical characteristics of the utility supply circuits, especially fault-current distribution
(2) Load continuity requirements and capacities
(3) Probability of system disturbances due to exposure, circuit length, type of equipment, etc
(4) Utility standard requirements established to ensure maximum quality service to users
(5) Available protective equipment with due consideration to economics
(6) Motor stability and other pertinent load characteristics
(7) Reclosing requirements
(8) Fault-locating requirements

Relay and protective device locations in typical installations may be grouped as follows:
Group A — supply circuit relaying
Group B — service entrance relaying
Group C — supply transformer protection
Group D — transformer secondary relaying
Group E — plant feeder circuit relaying
Group F — in-plant generator relaying
Group G — bus relaying (not shown on typical schemes)

These protective device locations are not always well defined due to differences between installations, and in some cases one or more groups may be nonexistent. For example, when the utility supplies power at 208Y/120 V, 480Y/277 V, or even 4.16 kV or higher, no transformer is required and group C is omitted. Group D is omitted when there is no secondary circuit breaker, and group F is not present except with in-plant generation. Relaying, which often is supplied for protection of each group, will be discussed in the following paragraphs. A portion of the relaying equipment described is shown with each of the six typical supply circuit illustrations of Section 13.4.

While protective relaying is emphasized in this chapter, it is not intended to exclude series tripping or fusible protective devices from consideration. In circuits of 600 V and less, these two types of protection are more common than protective relays. However, even at these voltages ground-fault protective relaying is now used extensively to protect against arcing ground faults.

A broader connotation of protective relaying is better expressed as system or service protection. This carries the additional responsibility of automatic service restoration when warranted, separation of the more critical loads from less critical loads which can be deliberately dropped when the main service from the utility is lost for any reason, and similar sophisticated control schemes designed to make the maximum utilization of the utility service even during periods of partial inadequacy. Identification of the general area of a fault through relay targets so as to minimize operating confusion and down time is also part of this concept.

13.3.1 *Group A: Supply Circuit Relaying.* The protective equipment at this point is on utility premises and is utility property. Its primary purpose is to protect the main utility circuit from the adverse effect of faults between the utility circuit breaker and the service-entrance equipment. The main goal is to clear faults on the feeder quickly, rather than jeopardize the service of all users supplied from the source bus. Another function is to back up the service-entrance relaying and prevent an in-plant disturbance from affecting the utility source bus. Relays commonly employed at this point are as follows.

(1) Inverse time overcurrent phase and ground fault relays (devices 50/51 and 50N/51N) are used whenever possible because of simplicity and economy. The more inverse relays are sometimes used when coordination with fuses is needed or when it is desired to ride over high inrush currents upon restoration of power after a service outage. Standard inverse or inverse definite minimum time characteristics are also used for a variety of reasons, one condition being, for example, when the short-circuit current magnitude is dependent largely on system generating capacity at the time of fault. To avoid complications in coordination it is desirable not to use a more inverse relay upstream from a less inverse relay. (Instantaneous relay applications prove this concept since these can be viewed as the most inverse of all the relay time–current characteristic types and their successful use upstream requires considerable detailed analysis.) When the utility uses a fixed-time type relay on the industrial service line, a less inverse relay characteristic on the industrial system may be preferred.

Instantaneous attachments (devices 50 and 50N) must be employed only when they allow coordination with the load-side protective devices. When instantaneous units are used, they are set high enough so that they do not detect faults beyond about 80 percent of the distance to the next load-side overcurrent protective device. In some applications they must be omitted or immobilized in order to achieve coordination.

(2) Distance relays (device 21) are needed when fast tripping is desired over most of the protected line length. These relays are considerably more expensive than overcurrent relays and can be obtained with one to three zones of operation. The first zone provides instantaneous protection for up to about 90 percent of the protected line. The second and third zones, if used, are time delayed and extend backup protection into the area protected by the service-entrance relays. Distance relays are sometimes necessary when it is impossible to get selective reasonably fast protection with overcurrent relays because of the long time needed to get selective tripping over a wide variation of fault-current magnitudes. They may also be needed for conditions where fault currents are low and are difficult to distinguish from load currents. For long lines distance impedance relays are most commonly used. Distance reactance and certain other newer relays can be applied on medium to short lines. Where very short lines are involved, pilot relay systems can be considered.

Distance-controlled overcurrent relays (devices 51 and 21) can be used in combination to provide fast tripping for faults on the primary of supply transformers, plus backup time delay for low-side faults with some limitations. This combination is useful where overcurrent relays alone cannot be set to respond to transformer low-side faults.

(3) Pilot relay systems (device 87) compare the circuit conditions at both ends of the protected line simultaneously and can thus provide fast protection over 100 percent of the line.

Additional relaying must be considered for back-up protection.

The majority of pilot relay systems used on lines to consumer plants are those of the pilot wire type (device 87L). These relays measure the current in both ends of the line differentially and detect an internal fault by differences between "in" and "out" current. A metallic telephone type communication channel is required for this relaying. Pilot wire relays are limited in distance to a maximum of roughly 10—15 mi, depending on the pilot wire used. Complications arise if the line serves more than one customer, although multiterminal installations are possible.

Other pilot channels such as powerline carrier, microwave, and audio tones utilizing either wire communication or microwave, are not covered in this publication.

When selecting the basic relay types, it is well to consider the competence of the available personnel. This includes the ability to calculate the proper settings to attain the desired results as well as the qualifications of test personnel who are going to start up and maintain the relays.

13.3.2 *Group B: Service-Entrance Relaying.* The service-entrance relaying normally operates the main interrupting device. However, when fuses are used, the fuse provides the functions of both sensing and interrupting. This protective equipment is sometimes provided by the customer, but the required characteristics and settings are often specified by the utility and, when applicable, by the pertinent governmental codes. Refer to Chapter 8 for ground relaying requirements of NFPA No 70, National Electrical Code (1975) (NEC) (ANSI C1-1975) on systems below 600 V.

The protective relaying functions to disconnect the supply from the consumer's system for certain faults within the supply transformer primary and secondary connections and serves as a backup to the protective relaying associated with the transformer secondary circuit breaker.

There are several schemes which can be used to open the utility supply circuit breaker within group A when there is no transformer primary circuit breaker and the fault currents are not sufficient to operate the relays of group A. These transferred tripping schemes are often accomplished through the use of (1) pilot wire, (2) grounding switches, or (3) power-line carrier or microwave signals.

Typical protective devices associated with group B are as follows.

(1) Overcurrent phase and ground-fault relays (devices 50/51 and 50N/51N) are applied similar to the overcurrent relays described in Section 13.3.1 (1).

(2) Directional overcurrent phase and ground relays (devices 67 and 67N) if used are located at the service-entrance location whenever there is no transformation at the consumer's plant. The directional relays are normally located at the secondary circuit breaker, but the relays for groups B and D are combined in this case.

The directional relays (used when there are parallel lines and in-plant generation) operate for fault current flowing from the consumer's bus toward the utility substation. Directional ground relays (device 67N) are usually needed when there is no transformation. However, when supply transformers are involved, these relays (device 67N) are considered optional in view of the reduced ground-fault exposure and the fact that the added cost of polarization is necessary.

(3) A ground fault detector relay (device 64) is needed to clear a supply-line ground fault by disconnecting the supply line at the consumer end. This situation can occur when there are parallel supply lines or in-plant generation, and the utility end is cleared by overcurrent relays. The system ground is thus removed, and the supply line is then operating as a normally ungrounded system. An over- and undervoltage relay connected to a line-to-ground potential transformer or an overvoltage relay connected to the broken delta secondary of three line-to-ground potential transformers can be used. Alternately a sensitive power directional relay (device 32) may accomplish the same purpose.

(4) The consumer terminal of pilot relay systems (device 87) as described in Section 13.3.1(3) is located here and functions to trip the service-entrance circuit breakers as well as the utility supply circuit breakers.

13.3.3 *Group C: Transformer Protection.* Transformer protection is listed separately even though some of the protection is provided by the circuit breakers and relays covered in groups B and D, and Chapter 10 is devoted entirely to transformer protection. Nevertheless, the basic protective devices will be listed.

(1) Overcurrent phase and ground-fault relays (devices 50/51 and 50N/51N) are as described in Section 13.3.1(1). Instantaneous ground-fault relays (device 50G) should be connected to core-balance (ground-sensor) type current transformers for best results, although they can be used with conventional current transformers in a residual connection with relay 50N with good results on solidly grounded systems.

(2) Transformer differential relays (device 87T) provide fast clearing for phase-to-phase plus phase-to-ground faults and can be obtained as regular or high-speed devices. Differential protection is almost universally applied to large (above 5000 kVA) or important transformers. Harmonic restraint relays are available to allow greater sensitivity and yet not operate on magnetizing inrush currents. These differential relays are arranged to cause both the primary and the secondary circuit breakers to trip and lock out.

(3) Pressure relays (device 63) have gained acceptance as reliable relays for large power transformers. They are applicable to all transformers which have a sealed gas chamber above the oil level. They are more sensitive than differential relays for internal faults and are particularly useful for faults in tap-changing equipment. One application problem is that of properly associating all the relaying circuitry ground faults in a facility so that a possible elevation of station ground grid potential at the transformer during a phase-to-ground fault does not cause relay contact flashovers.

(4) Transformer temperature protective devices (49) are usually provided with all power transformers. These devices are essentially for overload protection and in many applications merely indicate alarm or activate cooling apparatus. They protect against excessive transformer temperatures. These excessive temperatures reduce transformer life in relation to the exposed period and, therefore, both temperature and exposure time are evaluated in ascertaining the predicted reduction in transformer life.

13.3.4 *Group D: Transformer Secondary Relaying.* Where there is no transformation, these relays are not involved. Relaying this location also varies with the number of transformers used, whether there is in-plant generation or for other

similar considerations. Basic relaying here includes the following.

(1) Overcurrent phase relays (device 51) are required to protect against bus faults and to back up the in-plant feeder overcurrent relays. When there is a primary circuit breaker with associated overcurrent relays [see Section 13.3.2 (1)], it is possible to eliminate these overcurrent relays at the secondary circuit breaker without a great sacrifice in protection. Instantaneous attachments (device 50) cannot be used here successfully without loss of selectivity between main and feeder relaying. Therefore, if high-speed protection against bus faults is desired, bus differential relays (device 87B) should be used. Residually connected overcurrent ground relays (device 51N) are not needed because the transformer neutral ground relay (device 51G), described in the following paragraph, is more effective.

(2) A transformer neutral ground relay (device 51G) is connected in the grounded neutral lead of the transformer secondary. Since no current flows through this circuit except during a ground fault, this relay can be made very sensitive and will back up feeder ground-fault relays. It is the only protection against bus ground faults when bus differential relays are not specified. A high set device 50G can provide additional protection for turn-to-ground faults in the transformer secondary winding.

(3) Directional overcurrent phase and ground relays (devices 67/67N) can be useful in those dual-service arrangements where the dual services are exclusively used by a single consumer. Otherwise the general comments of Section 13.3.2 (2) apply to this location equally well. In addition, installations with in-plant generation utilize directional relaying at this location to clear faults in the supply circuit or supply transformer from in-plant generator and bus sources.

(4) A directional power relay (device 32) can be used to remove a ground fault on the supply circuit for parallel operation or when there is in-plant generation. For an alternate method of accomplishing this, see Section 13.3.2 (2), device 64.

13.3.5 *Group E: Feeder Relaying.* This relaying is selected on the basis of the type of load, type of circuit, and general degree of protection required. The detailed considerations for load circuits are fully covered elsewhere in this publication.

13.3.6 *Group F: In-Plant Generation Relaying.* The presence of in-plant generation adds flexibility to the consumer's electrical supply sources, but it also adds certain complexities in relaying and control. Additional relaying often needed includes standard generator protective relaying (Chapter 14) as well as certain of the following.

(1) Overcurrent relays with voltage control (device 51V) are different from standard overcurrent relays in that their operation is affected by the value of circuit voltage. Low voltage usually results from faults rather than overloads. Thus when the overcurrent is accompanied by reduced voltage, the relay will operate. However, if the circuit voltage is maintained, the relay operation is blocked (or in some designs, restrained) since the condition is probably an overload which requires that the generator remain connected. It is good practice to calculate the desired relay setting before any relay is specified. In this way inadequacies in the scheme may be identified and corrected before startup. This concept is particularly true when application of 51V type relays is being contemplated.

(2) In a generator ground relay (device 51G), if the generator is grounded, a time overcurrent relay in the neutral circuit can be used in much the same manner as the transformer neutral ground relay [see Section 13.3.4 (2)].

(3) Generator differential relays (device 87G) protect against faults in the generator or generator leads.

(4) Underfrequency relays (device 81) can be used to operate at preselected frequencies to drop load or sectionalize buses in order to keep remaining generation and load in operation during disturbances. Time delays of 6 to 30 cycles are often used to prevent operation on transient or temporary conditions. Relays which operate on a rate of change of frequency characteristic are most useful for this purpose, since the relay operates faster as the rate of change of frequency increases. Frequency relays are sometimes needed to protect some generators which may be subject to overspeeding.

Frequency relays may be used to separate local generation plus critical loads from a utility system upset. They may also be used in the most simple case to drop unessential load during system upsets. The theory in this latter case is that if each consumer drops some load, the utility will not disconnect the total feeder.

Frequency relays may also be needed to disconnect the utility service in certain cases where the supply line is equipped with automatic reclosing. In this case the consumer's main circuit breaker is tripped to protect synchronous or large induction motors and generators.

(5) Directional relays (device 67) are usually needed, but are normally located elsewhere [see Section 13.3.4 (3)].

13.3.7 *Group G: Bus Relaying.* Bus and switchgear relaying has been covered in Chapter 12.

13.4 Typical Supply-System Protective Schemes. The electric supply system for an industrial plant must be designed to fulfill the particular needs of the individual plant. For example, duplicate service equipment may be needed for an industrial process requiring a constant supply of electric power.

The voltage and other characteristics of supply systems vary depending on the serving utility. Plant designers should consult with the serving utility to determine the nature of these characteristics and the requirements for the service-entrance equipment.

Six typical schemes have been selected for detailed analysis and one-line diagrams prepared to illustrate these schemes. The protective relaying will be discussed with cross reference to protection groupings A—F which appear to the left of the diagrams.

13.4.1 *Network Supply Systems Below 600 V.* In many sections of the country, the utility tariffs permit supply of power at secondary voltages only, and all higher voltage equipment is owned and installed by the utility on the consumer's premises. An example of this type installation is shown in Fig 176. The arrangement is a network system with the sources supplying network transformers operated in parallel on a common service bus. With this type installation, network protectors are required and are supplied by the utility to segregate a faulted line or transformer from the service bus. Since the short-circuit duty may be extremely high, current-limiting fuses are necessary to limit this short-circuit duty on the customer's equipment. The engineer should review very carefully with the utility the

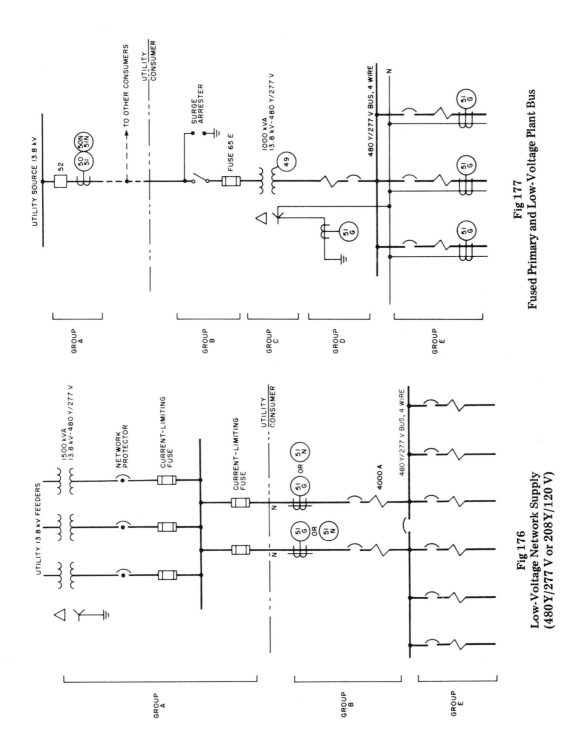

Fig 177
Fused Primary and Low-Voltage Plant Bus

Fig 176
Low-Voltage Network Supply
(480 Y/277 V or 208 Y/120 V)

type and design of approved service entrance equipment.

The protection required for the consumer's service-entrance equipment is largely regulated by governmental codes. Circuit breaker, service protector, or fuse selections are governed by the continuous-current rating and number of service main circuits allowed. In addition to phase fault protection, NFPA No 70, National Electrical Code (1975) (NEC) (ANSI C1-1975), Section 230-95, requires ground-fault protection for service entrances. Ground-fault protection is often recommended on grounded circuits since the phase protective devices usually provide little or no protection against low- to medium-magnitude arcing ground faults.

13.4.2 *Fused Primary and Low-Voltage Plant Bus.* Many utilities permit the customers to own and operate higher voltage service-entrance equipment and the associated transformers. Fig 177 shows a simple single service of this type installation adequate for small industrial customers within a metropolitan area. The fuse size on the transformer is limited to the maximum size fuse that will provide selectivity with the power company's overcurrent relays (group A, devices 50/51 and 50N/51N). This selectivity is important to the utility to prevent interruption of service to other customers on the same line.

No special relaying equipment is provided at location B. A ground relay (device 51G) is recommended at location C to protect against low-magnitude arcing ground faults. Since the transformer secondary is low voltage (480Y/277 V), the protective devices at D are integrally mounted on the low-voltage power circuit breakers. Where the transformer is not close coupled to the secondary circuit breaker, protection for this connecting circuit is dependent on fault clearing by the primary interrupting device.

The insensitivity of high-side fuses to arcing low-side faults must be evaluated before using this scheme.

13.4.3 *Single-Service Supply with Transformer Primary Circuit Breaker.* Should the transformer full-load requirements exceed the maximum permissible size fuse, it may be necessary to install a circuit breaker (device 52L) in place of the fuse (Fig 178). The high-side circuit breaker may also be justified to clear low-side ground faults, especially when lightly grounded as in this plan.

The service-entrance overcurrent relays (group B, devices 51/50 and 51N/50N) are often set and tested by the utility to coordinate with relays (group A). The service-entrance relays serve a dual purpose since they also are the protection for the transformers. Circuit breaker 52L may not be considered the service circuit breaker by the NEC, and a main secondary circuit breaker may be required if more than six feeder circuit breakers are installed.

13.4.4 *Dual Service without Transformation.* Fig 179 illustrates a method of utilizing a second service. With this arrangement, one service circuit breaker (device 52S-1) may be normally open and the other service circuit breaker (device 52S-2) normally closed. The reasons for not operating in parallel in this instance are to minimize the short-circuit duty on the plant 13 800 V bus and to eliminate the need for directional relays (device 67). Automatic transfer (device 27) must be delayed for at least 1 s to coordinate for faults on other portions of the system. The voltage will be depressed due to a fault until a circuit breaker or a fuse operates to clear it. Because of the setting requirement placed by the utility on the overcurrent relays for circuit breakers (devices 52S-1 and 52S-2), selectivity between the service circuit breakers (device 52F) may be difficult to obtain. Selectivity would depend on the instantaneous overcurrent relay (device 50) tripping only the feeder circuit breaker (device 52F) for feeder faults.

13.4.5 *Dual Service with Transformation.* Fig 180 is a preferred dual-service scheme. With this arrangement the industrial plant is protected not only from the loss of a supply line but, in addition, against the failure of a primary bus or transformer. This scheme could also be provided with automatic transfer on both the primary and secondary sides of the transformer.

The relay features discussed with Fig 179 apply also to this scheme. Transformer differential (device 87) should be considered for banks of this size. Since each installation has a wide variety of individual conditions, the relaying complement of Fig 180 is used only to illustrate certain protective schemes and is not intended to show all the devices that may be required. The transformer differential relay connection on all these low-side grounding schemes needs particular atten-

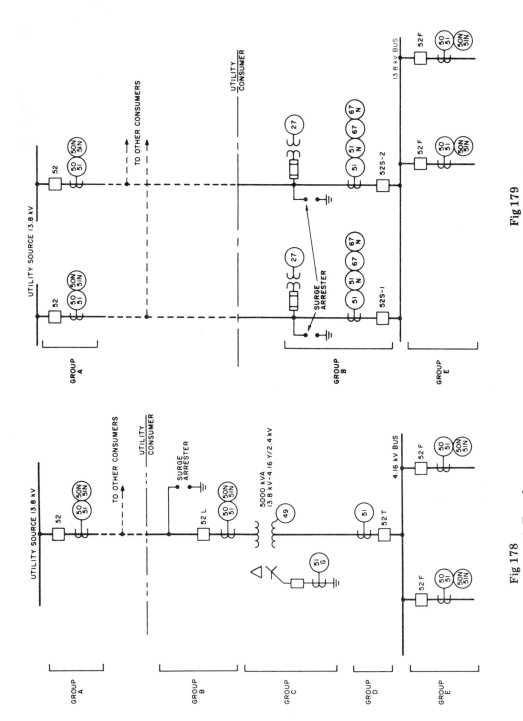

Fig 179
Dual Service without Transformation

Fig 178
Single Service Supply with Transformer
Primary Circuit Breaker

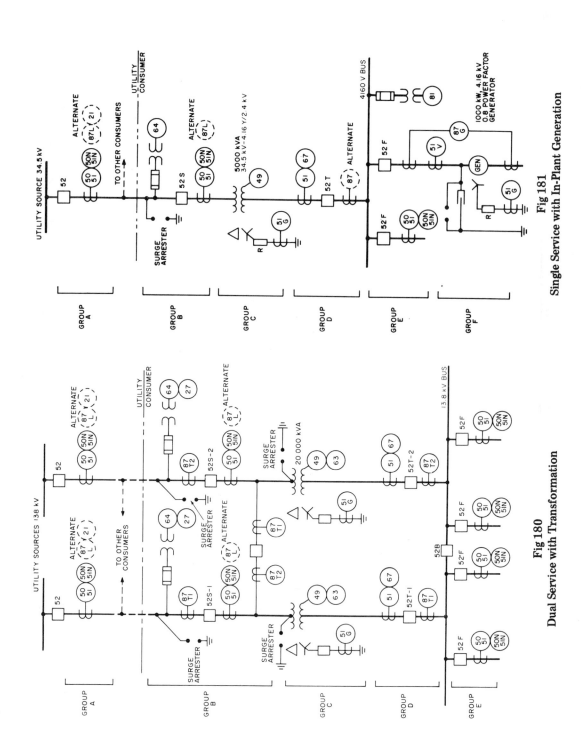

Fig 181
Single Service with In-Plant Generation

Fig 180
Dual Service with Transformation

tion for proper functioning on internal and external ground faults.

When a tie circuit breaker is normally closed, special relaying may be necessary to allow for the possibility of circulating current between bus sections.

13.4.6 *Single Service with In-Plant Generation.* An alternate to Fig 178 is shown in Fig 181. Some manufacturing plants find it economical to generate a portion of their electrical requirements in conjunction with their manufacturing process. The generator, when operated in parallel with the utility system, presents many additional protection and control problems. The system will probably have to be designed to shed load in the event the utility source is lost. This can be difficult since accurate control is needed over the generator governor to ensure that excessive loading does not occur during this load-shedding period. Another problem concerns disconnection of the generator from the utility system when the source becomes disconnected under other than fault conditions. Under these circumstances, the industrial plant's generator may attempt to pick up load normally intended to be supplied by the utility system only.

To attempt to solve these problems, a ground detection relay (device 64) would recognize a ground fault on the 34 500 V side of the transformer and would cause device 52S to trip. A directional power relay (device 32) would measure the current flow toward the utility source. The relay pickup current should be set with sufficient margin above the magnetizing current of the transformer. Calculations may show that the fault-sensing relay (device 67) can provide this function. But actual setting of device 67 for both functions should be determined before this conclusion is reached. The underfrequency relay (device 81) could be set to recognize a slowdown of the generator due to overload. The relay could trip the transformer secondary circuit breaker and preselected feeder circuit breakers to shed load to within the rating of the generator.

When the main step-down transformer is out of service, an alternate neutral ground must be established. This consideration frequently results in a separate source of neutral ground, through either a zig-zag ground tranformer or the local generators. The generators can safely carry the system ground because the main transformer del-

ta winding prevents any interaction between the utility system ground faults and the local generator neutral system.

13.4.7 *Example of Relay Coordination Plot.* Fig 182 shows a coordination plot of a portion of the supply system of Fig 179. It illustrates some of the problems involved in obtaining selectivity between protective devices.

Reactors have been indicated on the load side of the utility company's circuit breaker. Reactors are commonly installed by utilities to limit the short-circuit duty imposed upon the customer's equipment. These reactors also restrict the voltage drop imposed upon the utility bus for faults on the customer's installation.

Curve A has been selected to coordinate with the utility's 13.8 kV bus protective system and is the interface point for the consumer's relay coordination. The instantaneous relay (device 50) at circuit breaker A is often arranged to prevent reclosing, since the high fault current would indicate a local permanent type fault. The instantaneous setting has not been shown on curve A.

Curve B represents the customer's service circuit breaker protective relay setting. The setting follows the accepted practice which is to provide about 20 cycles (0.33 s) between the relay curves for the customer's circuit breaker and the utility's circuit breaker. The instantaneous device must be set less than the available short-circuit duty at point F_1. It has been selected to operate at 2000 A in order to prevent curve B from approaching curve A closer than the 20 cycles mentioned. It must be recognized that with the system shown, inselectivity will occur for high 13.8 kV faults protected by C.

In some installations it is possible to eliminate the instantaneous element on B by providing a time curve on B which allows the 20 cycle coordination interval at maximum short-circuit current. It is often useful to review coordination problems of this nature with utility engineers.

The overcurrent relays associated with circuit breaker C have been set to comply with the limitations of the NEC. They have been set high in order to obtain selectivity with the secondary circuit breaker to provide for an additional transformer or load to be connected directly to circuit breaker C. Under this arrangement it is necessary to make relay C settings as high as possible to obtain this selectivity. If provision is not made

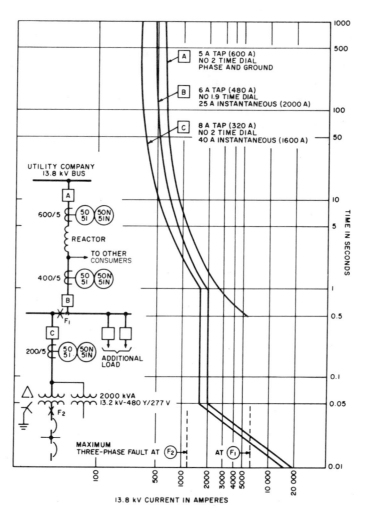

Fig 182
Example of Relay Coordination Plot

vice 50GS should be included. It will provide sensitive high-speed ground-fault protection for the generator. In the event the system is grounded at the generator neutral, relay device 51G should be furnished. This relay is of the time-delay type and will provide system backup as well as generator ground-fault protection. Device 51G will be operating more slowly than device 50GS.

If the generator is unit-connected to a transformer, high-resistance grounding is often used. Ground protection for that application is shown in Fig 118(a) of Chapter 8, Ground-Fault Protection.

Protection against motoring through a directional power relay, device 32, should be provided for all machines except those in which the prime mover provides this protection. The relay should operate for power flow into the generator. For steam-turbine driven machines, the relay setting should not be more than 3 percent of machine rating, generally as small as possible. Diesel- or gas-turbine driven machines will require less sensitive settings, generally below 10 percent of machine rating. These relays should be of the time-delay type to prevent tripping during power swings, which might occur as a result of synchronizing or of a system disturbance.

Two types of relays are available for use as device 40 to detect loss of excitation. The simplest type is an undercurrent relay. It would be connected in series with the generator field and would operate whenever current ceases to flow. The relay should be provided with a short time delay to ride through momentary interruptions of current flow which sometimes might occur due to short circuits in the power system. This type of relay, however, would not necessarily detect loss of excitation resulting from short circuits in the field circuit. Where more complete protection is required, a directional impedance type relay should be used. This relay is arranged to sense the change in impedance of the generator stator, which occurs upon loss of excitation, and thus would operate for conditions which the simpler field undercurrent scheme would fail to detect. Device 40 is normally arranged to automatically take the machine out of service.

Where it is desired to provide protection against generator rotor overheating due to unbalanced phase currents, device 46 should be furnished. A negative phase-sequence overcurrent relay is recommended for this application. This relay is sometimes provided with two contacts, each of which operates at a different level of negative phase-sequence current. The more sensitive operating contact is generally used to sound an alarm while the other is used to take the machine out of service. Where a single contact is provided, it should be arranged to trip the machine.

It is usual practice to operate generator field circuits ungrounded so that a single ground will not result in damage to the machine. However, since a second ground may cause damage, the use of a generator field ground relay, device 64F, is desirable. One type of device available for this application incorporates a separate low-voltage grounded source and a potential-sensing relay. The ungrounded side of the source is connected in series with the relay coil and to one side of the field circuit in such a manner that a ground occurring in any part of the circuit will operate the relay. Another type uses a sensitive direct-current relay connected between ground and the center of a voltage divider, which is across the field circuit. One side of the voltage divider includes a varistor (voltage-sensitive resistor) to shift the null point as the excitation changes, thus ensuring operation of the relay for ground faults anywhere in the field. The usual procedure is to arrange the field ground relay to sound an alarm when it operates. Satisfactory operation of the relay requires that the generator rotor be grounded. To accomplish this, means should be provided to bypass the bearing oil film. In some machine designs the bearing seals provide the necessary bypass, while others may require the addition of a brush on the rotor shaft to secure effective grounding of the rotor.

Satisfactory operation of the voltage regulator and certain relays, that is, devices 51V, 40, etc, require proper output from their associated potential transformers. Failures of these circuits due to fuse blowing, etc, would in the case of the regulator cause the generator excitation voltage to go to its ceiling level. In the case of device 51V, loss of restraint voltage may result in unnecessary tripping and shutdown of the generator. A type of voltage balance relay, device 60V, provides the means of monitoring these potential circuits. As long as the outputs from the two sets of potential transformers are alike, the relay will not operate. However, should there be a deviation between the outputs, the relay will operate. In case the relay potential transformer output were lower than the output from the regulator transformer, device 60V would close a set of contacts which usually would be arranged to sound an alarm and block tripping by devices 51V, 40 Etc. On the other hand, should the output from the regulator transformer be lower, device 60V would close another set of contacts which would usually be arranged to sound an alarm and switch the regulator to the manual mode of operation with "fixed" excitation.

Differential protection for generators, device 87, provides a means whereby a faulted machine may quickly be removed from service. Serious consideration should be given to its use for the following machine ratings:

(1)   Any voltage, 1000 kVA and larger
(2)   Any kilovolt-ampere rating, 5000 V and higher
(3)   2200 V and higher, 501 kVA and larger

The types of relays used for differential protection of generators are the fixed-percentage inverse time type and the variable-percentage instantaneous type. When percentage type relays are used, care should be excercised in selecting the current transformers connected to each end of the generator windings so that their performance characteristics will be closely matched. The variable-percentage relay is less sensitive to current transformer mismatch than the fixed-percentage type. When the generator is grounded through an impedance which limits the maximum ground-fault current to less than the generator full-load current, the differential relays should be supplemented with device 87G. A current-polarized directional relay may be used for this application, its operating coil being connected to the differential circuit and its polarizing coil energized from a neutral current transformer.

Where suitable current transformers are available, the self-balanced primary current differential scheme shown in Fig 184 may be used. The relays for device 87 would in this instance be instantaneous overcurrent types. The scheme can be made quite sensitive; since the load currents affect only the sizing of the primary conductor, the current transformer ratio can be kept small.

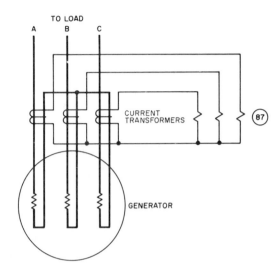

Fig 184
Self-Balanced Primary Current Differential Scheme

Generally the sensitivity of the scheme is such as not to require the use of device 87G. Since the cables connecting the machine are not usually included in the differential zone, supplemental cable protection may be required.

Both the percentage differential and the self-balancing differential schemes protect against phase-to-phase and phase-to-ground faults. Neither scheme will detect turn-to-turn faults. Additional information on rotating machine fault protection will be found in Chapter 9, Motor Protection.

The differential relays should be arranged to operate the lockout relay, device 86, which trips and locks out the main generator circuit breaker, device 52, and the generator field circuit breaker, device 41. Where automatic CO_2 fire-extinguishing equipment is present, consideration should also be given to initiating its operation by the differential relays. The use of differential relays requires that both ends of the generator windings be brought to terminals. For application of self-balancing differential schemes, both ends of each winding must be brought adjacent to one another to accommodate the current transformers in Fig 184.

14.3 Some Other Considerations. The insulation that can be economically included in the slots for the stator winding of a generator is limited and may fail when subjected to transient overvoltage. Surge protection equipment in the form of lightning arresters and capacitors is therefore generally provided. To be most effective, this equipment should be connected to the generator terminals with interconnecting leads kept as short as possible.

Where a generator is not operated in parallel with other sources, consideration should be given to the location of current transformers which energize the relays used as device 51V. Since for this arrangement there will generally be little or no backfeed from the system, these current transformers should be located as close to the generator as possible, and preferably should be connected in the generator neutral. Such an arrangement, in the absence of differential relaying, will provide protection for the connections between the generator and its circuit breaker, device 52.

The use of static exciters for generators is increasing. Generally these exciters are energized from the stator circuit of the generator. For an arrangement such as this, it is necessary that the static exciter include some means of maintaining excitation should the generator stator become short circuited. Failure to maintain excitation under these conditions may result in insufficient flow of fault current to ensure proper relay operation.

14.4 Bibliography

[1] AIEE Committee Report. Relay Protection of AC Generators. *AIEE Transactions*, pt I, vol 70, 1951, pp 275-282.

[2] *Applied Protective Relaying*. Newark, NJ: Westinghouse Electric Corporation, Relay-Instrument Division, publication number B-7235-D, 1974.

[3] BEEMAN, D.L., Ed. *Industrial Power Systems Handbook*. New York: McGraw-Hill, 1955.

[4] *Electrical Transmission and Distribution Reference Book*. East Pittsburgh, PA: Westinghouse Electric Corporation, Central Station Engineers, 1964.

[5] IEEE Committee Report. Bibliography of Relay Literature, 1965-1966. *IEEE Transactions on Power Apparatus and Systems*, vol PAS-88, Mar 1969, pp 244-250.

[6] LATHROP, C.M., and SCHLECKSER, C.E. Protective Relaying on Industrial Power Systems. *AIEE Transactions*, pt II, vol 70, 1951, pp 1341-1345.

[7] MASON, C.R. *The Art and Science of Protective Relaying*. New York: Wiley, 1956.

15. Maintenance, Testing, and Calibration

15.1 General Discussion. It is impossible to predict when an abnormal electrical condition will occur, and, therefore, the entire electric protective system must be ready to operate properly at all times. To reduce the possibility that the electric protective devices will not perform they must be maintained [1].

Maintenance should not be confused with repairs after a breakdown. The definition of maintenance is, "to keep an object, machine, or system in the state of good repair." Maintenance implies that we inspect this equipment or system to discover its weaknesses and then repair or replace them before a breakdown occurs. A maintenance program for protective devices and the electric system in general could be broken down into five simple steps.

(1) *Clean.* Eliminate all the contaminants that are not compatible with electricity. These contaminants will include dust, dirt, moisture, water, rust, or anything else that may reduce the efficiency of the electric system.

(2) *Tighten.* Loose electrical connections are heat sources which reduce the efficiency of the electric system and will ultimately result in the destruction of a portion or portions of the system. An energized alternating-current system is subjected to constant vibration, and, therefore, connections frequently become loose.

(3) *Lubricate.* Direct-acting low-voltage power circuit breakers, relay-actuated power circuit breakers, and switches contain many pivoting or sliding mechanical parts. These parts require lubrication so that they do not become inoperable because of nonuse. However, care must be exercised when lubricating so that improper types of grease and lubricants do not come in contact with current-carrying parts of the system, and lubricants which collect dust and dirt are to be avoided.

(4) *Inspect.* From time to time it becomes necessary to check the ratings and settings of all electric protective devices to assure that they are in accordance with good design. Conditions change, loads shift, and human beings tamper to "solve a problem."

(5) *Test.* This requires that the system or protective device be subjected to abnormal electrical conditions and the operation of the system or devices compared to manufacturers' specifications for these conditions.

The primary scope of this chapter will be to deal with the maintenance, testing, and calibration of the following types of electric protective devices:

(1) Motor overload relays
(2) Molded-case circuit breakers
(3) Low-voltage power circuit breakers
(4) Protective relays
(5) Medium-voltage circuit breakers using series trip coils
(6) Fuses

In addition to the electric protective devices mentioned, the protective system includes accessory items such as:

(1) Battery chargers
(2) Storage batteries
(3) Current and potential transformers
(4) Control circuitry
(5) Auxiliary or control relays

15.2 Motor Overload Relays. Motor overload protection and how it is applied are covered in Chapter 9. These principles should be understood by the persons responsible for maintaining motor overload protection.

There are four basic types of motor overload relays:

(1) Solder pot
(2) Bimetal strip
(3) Winding-temperature sensor
(4) Hydraulic—magnetic relay

Small motors may be protected by a direct-acting bimetal or solid-state temperature-detecting device adjacent to the motor winding. When such a device opens at a predetermined temperature, the motor is shut sown. These units are not adjustable.

15.2.1 *Selecting Heater Size.* The most widely used motor overload relays are the bimetal or solder-pot types, both of which incorporate heater elements. Proper heater size is essential for the effective operation of these relays and should be regularly checked.

All manufacturers of heater-type thermal motor overload relays have heater selection tables. The user of these tables should read all the relay manufacturers' instructions before selecting a heater. This is necessary so that the basis of the tables will be known. For instance, one manufacturer prepares his tables so that heater sizes may be used directly if the service factor of the motor is 1.15 with motor and controller in same ambient temperature. If these conditions are not met, appropriate steps must be taken to either uprate or derate the heater.

15.2.2 *Testing Motor Overload Relays.* The test of the overload relay should be done as efficiently as possible. Normally, only one test point is needed to establish whether the device is operating correctly, that is, whether its time of operation falls within its manufacturer's band of tripping time. It is suggested that the value of test current be equal to three or four times the normal rating of the relay. to three or four times the normal rating of the relay.

To test the motor overload relay it is necessary to disconnect the relay from the electric system, pass the test current through the relay heater circuit, and determine the time required for the device to trip. This time should be compared with the manufacturer's recommended time of operation for the value of current used in the test. Generally speaking, motor overload relays are nonadjustable, and if the test reveals an improper operation, it may be necessary to replace the relay heater with one of the proper size. Sometimes it is necessary to replace the entire relay.

15.3 Molded-Case Circuit Breakers. Molded-case circuit breakers are used in low-voltage systems to interrupt fault currents, provide overload protection, and switch electrical circuits. They are described in Chapter 6.

Molded-case circuit breakers may be equipped with time-delay and instantaneous trip devices. Generally, a thermal element composed of a bimetal strip provides the time-delay characteristics, but some manufacturers use a magnetic sensing device with a dashpot for time-delay characteristics. The instantaneous trip device is magnetic, and is adjustable in some models. When it is adjustable, the pickup of the instantaneous trip may be changed in the field from approximately five to ten times the circuit breaker trip unit continuous current rating. The factory setting of molded-case circuit breakers with adjustable magnetic trip devices should be checked to determine whether or not it is satisfactory for the installation. Circuit breakers having the same continuous rating may have different interrupting ratings. Some trip devices are of solid-state design.

Manufacturers of molded-case circuit breakers make also "magnetic only" and "nonautomatic" circuit interrupters. These devices are molded-case circuit breakers without time-delay tripping characteristics and circuit breakers without the automatic trip feature, respectively. Care must be exercised so that these three devices are not interchanged in the field.

An integrated unit composed of a current-limiting fuse and a molded-case circuit breaker connected in series is now available for installations where the available short-circuit current is higher than the interrupting capacity of the molded-case circuit breaker alone. Devices of this type incorporate an antisingle-phasing feature. Blowing of any one fuse automatically opens the circuit breaker on all poles.

15.3.1 *Maintenance and Testing of Molded-Case Circuit Breakers.* The circuit breaker should be kept clean and dry, and the line and load connections should be checked occasionally for tightness. If located in a dirty or dusty atmosphere, it should frequently be removed and blown out with clean, dry, low-pressure air. Do not blow the dirt into the recesses of the unit.

Check for excessive heating. Heating may be due to poor electrical connections, improper trip unit alignment, or improper mechanical connections within the circuit breaker housing or at the terminals.

Should heating be abnormal, as evidenced by discoloration of terminals, deterioration of molded material, or if there is nuisance tripping, measurement of load current is advised. If the current in each of the individual poles is less than the rating of the circuit breaker, all bolted connections and contacts should be examined and tightened in accordance with the manufacturer's recommendations.

Other reasons for the circuit breaker to operate improperly include

(1) Excessive friction or binding of the trip bar so that the bimetal cannot exert sufficient pressure to move the trip bar

(2) Defective bimetal or dashpot assembly disabling its operating characteristics

(3) Open turn or terminal of a trip coil

(4) Foreign matter preventing movement of one or more of the following: trip bar, armature, latch release, trigger, bimetal, bellows or dashpot, or spring assemblies

If the circuit breaker has a removable trip unit or is of the design which incorporates fuses, the trip function can be readily checked. On units with removable trip units, the loosening of the trip unit mounting hardware is usually sufficient to cause tripping. Under no circumstances should a molded-case circuit breaker remain closed and latched when the trip unit is physically removed. When replacing the trip unit, the manufacturer's torque requirements must be referred to.

Fused units have an added feature that should be checked: a spring-loaded plunger which releases when a fuse blows and strikes a trip bar extension on the trip unit. This action causes the circuit breaker to trip on any damaging fault current above the rating of the fuse. A safety feature requires that the circuit breaker be in the open position when checking or replacing the fuses. If it does not trip when the fuse blows, either the circuit breaker or the fuse is defective. A defective fuse will not eject its plunger when the link melts. The defective unit must be replaced. In no instance should it be possible to remove the fuse cover without the circuit breaker automatically tripping.

To determine if the circuit breaker mechanism allows its closure under a blown-fuse condition, one of the fuses is removed and a blown fuse substituted for the good fuse; then the cover is replaced. It should be impossible to latch and close the circuit breaker with a blown fuse in any of its fused legs. If the circuit breaker will latch and close with a blown fuse in any of the phases, it is defective and should be repaired or replaced.

If a molded-case circuit breaker does not trip under abnormal load conditions, or within the required Underwriters Laboratories specified times, and mechanical troubles are not found, the circuit breaker is either labeled incorrectly or it is out of calibration. Recalibration of molded-case circuit breakers can be accomplished only under controlled test conditions by the manufacturer. It is unwise to attempt to make changes in the calibration of a circuit breaker trip unit except for any adjustable feature such as pickup of the instantaneous unit. This adjustment is readily made with a calibrated dial which is accesible without opening the case.

Periodically these circuit breakers must be electrically tripped to assure proper operation. Experience has indicated that if they are allowed to remain in service for an extended period of time without an electrical operation the internal mechanism and joints may become stiff so that the circuit breaker operates improperly when subjected to abnormal current. Therefore, each pole of the circuit breaker should be electrically exercised.

In testing a molded-case circuit breaker, several points must be remembered.

(1) Nameplate rated voltage must be available at the input terminals throughout the test.

(2) The values of current are high and voltage is low. Therefore it is advisable to use connections having the shortest possible length and largest possible cross-sectional area between test unit and circuit breaker. In some cases, pieces of bus bar may be used for these connectors.

(3) The connection to the circuit breaker must be tight.

(4) The circuit breaker is tested one pole at a time.

(5) Trip devices must be allowed to fully reset before performing a check test.

The recommended tests for a molded-case circuit breaker are (1) timing and (2) instantaneous pickup.

It is suggested that these circuit breakers be subjected to a test current equivalent to 300 percent of the circuit breaker trip unit rating and the tripping time be measured. This time should be compared to the manufacturer's specified values or curves. Repeated testing of the time-delay element without a cooling period will raise the temperature of the thermal element, thereby decreasing the tripping time noticeably on successive tests.

Molded-case circuit breakers may be relatively precise; however, the published time-delay characteristic indicates a wide band of operation. The electrical test will reveal circuit breakers that will not trip, those that take abnormally long to trip, and those that have no time delay. If the test reveals that the circuit breaker is tripping within ±15 percent of the outside limits of its published curves and this tolerance does not affect the electric system coordination or stability, the circuit breaker should be considered satisfactory.

In some molded-case circuit breakers the instantaneous element is not adjustable and it is set and sealed at the factory. Other molded-case circuit breakers have an adjustment which permits changing the pickup of the instantaneous unit. This type of circuit breaker may be shipped from the factory with the instantaneous unit pickup set at its maximum if a setting has not been specified by the purchaser. Therefore it is necessary to check these adjustable instantaneous settings before putting the circuit breaker in service to be assured that the instantaneous pickup is not above the available short-circuit current at its location in the electric system. In many electric systems one purpose of initial startup tests is to determine optimum setting for instantaneous units.

An electrical test for pickup of the instantaneous unit should be run to verify that the circuit breaker is tripping magnetically. Testing at one of the lower calibration marks is satisfactory.

The adjustment may be set to the lowest calibration point and tests made at that point to verify that the unit will pick up. If the instantaneous unit picks up at the minimum calibration point, it may be reasonably assumed that, when reset to the desired calibration point, pickup will be within manufacturer's tolerances. Published tolerances for standard adjustable instantaneous trip units usually are as follows:

"LO" position   ±25 percent
"HI" position   ±10 percent

The NEMA publication, "Procedures for Verifying Performance of Molded-Case Circuit Breakers" [2] may be used as an alternate guide for testing molded-case circuit breakers.

15.4 Low-Voltage Power Circuit Breakers. Types of low-voltage power circuit breakers and their applications are described in Chapter 6.

15.4.1 *Series-Trip Characteristics.* The series-trip device may be equipped with any one or any combination of four different time—current characteristics.

(1) *Long-Time Delay.* Designed for overload protection, this characteristic gives time-delay tripping measured in seconds or minutes. Circuit breaker standards allow an adjustment of long time delay pickup above 100 percent of the continuous current rating of the trip coil. This feature is provided so that the device will "ride through" system transient overloads and still provide protection for prolonged overloads. This feature does not change the rating of the circuit breakers. It must be clearly understood that this circuit breaker cannot be used without trouble for any continuous duty in excess of the 100 percent continuous current rating of the trip coil. When the pickup of the long-time-delay element is set above the continuous rating of the trip coil, an electrical test using current magnitudes of 300 or 400 percent of the pickup value may result in permanent damage to the trip coil.

(2) *Short-Time Delay.* This characteristic gives a delay trip time measured in cycles for fault-current or short-circuit protection.

(3) *Instantaneous Trip.* This has no intentional time delay and gives short-circuit protection.

(4) *Ground Fault Protection.* This is optional on circuit breakers with solid-state tripping devices.

15.4.2 *Reasons for Maintenance.* Causes of malfunction of the circuit breaker are dependent upon time and severity of duty. Common causes of malfunction of low-voltage power circuit breakers generally fall into three categories.

(1) Loss of oil or air seal due to physical damage to dashpot; aging of seals or physical wear, causing incorrect operation

(2) Clogging of orifices with foreign matter or oil sludge that forms due to atmospheric and environmental conditions and aging of oil

(3) Freezing of components in plunger assembly due to corrosive atmospheres and extensive periods of inoperation

Of all the possible faults, improper delay in opening automatically under overload conditions is the most dangerous to safety. Normal operational procedures and careful maintenance inspections will reveal most of the other conditions that are likely to remove protection from the circuit. However, unless overload tests are run, there is no way to predict whether the circuit breaker will recognize an abnormal circuit condition and operate properly.

There are a number of conditions which may render a low-voltage power circuit breaker unfit for service until they are corrected. Some of the more common are listed below.

(1) Frozen contacts or mechanism (circuit breaker may not open automatically or manually)

(2) Improper calibration, that is, trips too fast or too slow

(3) Trip element improperly set

(4) High contact resistance

(5) Trip element armature fails to strike trip bar

(6) Open contacts or damaged series element

(7) High resistance or arcing fault often due to loose or improper fit between primary disconnect assembly and bus stabs

(8) Broken or cracked arc chutes

(9) Loose parts

(10) Excessive force required to operate circuit breaker

(11) Dirt

(12) Contaminated dashpot oil

These conditions can be initiated by a variety of causes including moisture, corrosion, abuse, wear, and vibration.

15.4.3 *Maintenance and Test Procedures.* A complete test program should include checks to determine the condition of the circuit breaker with respect to each of the foregoing items. The following recommendations are made concerning action to be taken on circuit breakers undergoing overhaul. Most of these conditions can be determined easily.

(1) Broken or cracked arc chutes are revealed by visual inspection.

(2) Frozen or open contacts as well as excessive force required to operate circuit breaker may be determined by manually opening and closing the circuit breaker.

(3) Determination of leakage current requires an insulation resistance tester.

(4) Improper calibration and high contact resistance can be determined by electrical tests.

Safety rules and cooperation with operating personnel must be strictly observed. This cannot be overemphasized.

When working on a low-voltage power circuit breaker, it must be de-energized and removed from the cubicle. Hereby it should be kept in mind that the stabs on the line side of the cubicle are still energized unless the entire substation or section of bus is out of service.

15.4.4 *Mechanical Inspection.* The following steps should be taken in servicing and testing these circuit breakers:

(1) Schedule work.

(2) Ascertain that the circuit is dead and the circuit breaker can be removed for servicing.

(3) Open the circuit breaker, rack it out, and remove it from the cubicle. Note that these circuit breakers are heavy and should be handled with some type of lifting device. If no lifting device is available, at least two men will be needed to lower the circuit breaker to the floor.

(4) Inspect primary disconnect assembly in back of the circuit breaker. Check that no springs are missing or broken. Check for excessive wear. Clean out dust or dirt.

(5) Remove arc chutes; inspect for cracked, broken, or burned parts. Replace defective parts and clean.

(6) Inspect and clean main and arcing contacts in accordance with manufacturer's instructions. Note that commercial products are available to resilver these contacts.

(7) With the arc chutes removed, mechanically close the circuit breaker to inspect contact action and alignement. When the arc chutes are removed, moving parts of the circuit breaker are exposed. Be extremely careful to keep all parts of the human body clear. These exposed circuit breaker parts are actuated by stored energy devices, and serious injury will result if any part of the body is caught by the contacts or moving mechanism.

(8) Open and close the circuit breaker several times to determine that operation is smooth and there is no binding.

(9) Lubricate and inspect the "racking-in" device located both in the cubicle and on the circuit breaker. Make sure that there is no evidence of binding.

(10) Lubricate in accordance with the manufacturer's instructions all mechanical joints and mechanism used to close and open the circuit breaker.

(11) Tighten all screwed or bolted connections that are not pivotal.

(12) Remove dust and dirt from the cubicle and arc chutes. Use a high-suction industrial vacuum cleaner to clean cubicle. The circuit breaker and arc chutes may be removed to a cleared area and cleaned with dry low-pressure compressed air.

(13) Determine that proper movement of the trip bar will trip the circuit breaker.

(14) Determine that the trip arms on the trip devices of each pole properly engage the trip bar.

(15) Determine that the circuit breaker position indicator is showing proper position of the circuit breaker.

(16) If any adjustments are necessary, consult the manufacturer's instruction bulletin.

15.4.5 *Electrical Test.* Since 90 percent of all the low-voltage power circuit breakers manufactured contain the long-time delay and instantaneous trip unit combination, a discussion of a typical test program covers these units.

Most of these circuit breakers are equipped with one series overcurrent trip device per phase. The electrical test must be run on each individual trip device. The operation of any one of these devices will trip all poles of the circuit breaker.

In testing a low-voltage power circuit breaker equipped with series overcurrent trip devices,

several points must be remembered.

(1) Nameplate rated voltage must be available at the input terminals of the test set throughout the test.

(2) The values of current are high and the voltage is low. Therefore it is advisable to use connections having the shortest possible length and largest possible cross-sectional area between test unit and circuit breaker. In some cases pieces of bus bar are used for these connectors.

(3) The connection to the primary disconnect assemblies of the circuit breaker must be tight.

(4) The circuit breaker is tested one pole at a time.

(5) Trip devices must be allowed to fully reset before performing a check test.

(6) Variations in test results when compared to manufacturers' curves may be caused by non-sinusoidal test currents resulting from low voltage high current test sources.

The recommended tests for a low-voltage power circuit breaker are (1) timing (long- and short-time delay units if the circuit breaker has both type trip units), and (2) instantaneous pickup.

The recommended values of test current for long-time delay is three times the trip unit setting, and for short-time delay it is one and one half times the circuit breaker short-time delay setting; determine the instantaneous pickup. If the circuit breaker does not operate within the tolerances shown by the manufacturers' time—current curves, then suitable adjustments should be made as recommended by the manufacturer.

15.4.6 *Static Trip Devices for Low-Voltage Power Circuit Breakers.* Recently all major United States manufacturers have brought out low-voltage power circuit breakers equipped with a static trip device. These trip devices eliminate the need for a dashpot to obtain a time-delay trip of the circuit breaker (see Chapter 6). Maintenance of the circuit breaker equipped with a static trip device is the same as any other low-voltage power circuit breaker.

15.4.7 *Electrical Tests for Static Trip Devices.* Electrcial tests for a low-voltage power circuit breaker are performed one pole at a time; therefore, for a complete test all three poles must be checked. These tests may utilize a high-current low-voltage source connected to the primary disconnect assemblies of the circuit breaker, namely, primary injection; or a low-current source connected to the static trip device terminals which are normally connected to the secondary terminals of the circuit breaker input signal sensor (current transformer), namely, secondary injection. Regardless of the method used, pickup of long-time, short-time, instantaneous, and ground-fault units should be obtained. In addition, a timing point should be checked on the long-time and short-time characteristics. Suggested values of test current are three times pickup for the long-time characteristic and one and one half times short-time unit pickup for its time characteristic. The test values obtained should be compared to the manufacturer's specifications and suitable adjustments made as recommended by the manufacturer.

The ground trip feature of a static trip device must be disconnected or bypassed when performing tests to determine the pickup and timing of the long-time, short-time, or instantaneous feautures, because the ground-fault function will otherwise incorrectly interpret the single-pole test current as a ground fault and trip the circuit breaker.

Some static trip devices employ a separate power sensor (current transformer) to energize the circuit breaker trip circuit. If precision results are desired when performing electrical tests on this type static device, it is recommended that the power sensor and signal sensor be energized from separate sources. This action should eliminate any discrepancies in the test results introduced by saturation of the power sensor.

15.5 Protective Relays.
The basic types and applications of protective relays are described in Chapter 3.

15.5.1 *Safety.*
Safe work habits and use of good common sense cannot be overemphasized. Metal ladders or step stools should never be used around switchgear. If safety helmets are required, a fiber or plastic helmet should be worn. Loose key chains, tool pouches, or pieces of wire should never be allowed to hang from the body. Rings and metal watchbands should be removed. Whenever possible, power circuits should be de-energized and properly tagged to conform with approved procedures. It must not be assumed that circuits are de-energized.

15.5.2 *Scheduling.*
Operating personnel should be fully informed on the maintenance schedule and the work to be performed. They should be notified in advance, and, if possible, operations should be shifted so that the circuits may be de-energized. However, if circuits cannot be de-energized, the technician should proceed with caution, bearing in mind that all the time a relay is out of the circuit some of that circuit's protection is removed. Therefore, relay work should never be started unless it can be finished before leaving the work area.

15.5.3 *Knowledge of Control Circuit.*
Another point that should be worked out with the operating personnel is the action to be taken in case a circuit is inadvertently tripped. It is the responsibility of the operating personnel to determine whether the circuit should be re-energized.

It is important to know the peculiarities of the control circuit associated with the relay being maintained, such as, the trip circuit being interlocked with a cubicle door, or that removal of potential from the relay operating coil before opening the trip circuit results in tripping the power circuit breaker.

If the relay technician makes a mistake and trips a circuit, he should admit it immediately. This admission will save many hours of work, trouble shooting the control circuit looking for nonexistent problems. It should be remembered that one of the reasons for planned maintenance and testing is to build confidence in the protective system.

The protective relay is the brain of the electric protective circuit. It is the relay that senses an abnormal condition and then sends the message to other devices on the system. Therefore it is imperative that any relay work be done in a very thorough manner. A job should not be rushed or left half-done.

15.5.4 *Identification of Circuits.*
Double check the identification of the relays on the panel. This can be complicated when work on relay terminals must be done behind the board and there are a large number of similar boards. One way to solve this problem is to have an assistant turn the ammeter switch from the front of the panel while you observe which switch arm turns in back of the board.

15.5.5 *Importance of Target or Operation Indicator.* One device often overlooked on a protective relay is the operation indicator or target. When a circuit breaker trips under abnormal conditions, this device indicates the relay element which initiated the tripping action. This information is important, and therefore a target should never be reset under operating conditions without the supervisor's knowledge and permission. In addition, the target unit's operation closes a set of contacts which are in parallel with the relay's trip circuit contacts and spiral spring, thus preventing burnout of the relay's trip circuit due to high tripping current and opening of the circuit breaker trip circuit through opening of the relay induction disk trip circuit contacts.

15.5.6 *Current Transformer.* The current transformer is a vital part of the electric protective system. It is this device that reduces the current in the power circuit to a value that can be handled by the protective relay, and insulates the relay from the power circuit. There is a danger associated with a current transformer. Consider a 600/5 current transformer. As long as 600 A is present in the primary, the secondary will try to put out 5 A. Applying Ohm's law to the secondary circuit, $E=IR$ (voltage equals current multiplied by resistance). Inasmuch as there is current, the higher the resistance of the secondary circuit, the higher the voltage across the secondary. This means that if the secondary circuit opens, a very high voltage is obtained. Therefore under no circumstances should the secondary of a current transformer be opened while the primary circuit of the transformer is energized.

15.5.7 *Inspection.* The manufacturer's instructions for the specific relay should be on hand; they will supply much useful information concerning connections, adjustments, repairs, timing data, etc. The technician should make sure to obtain the proper instruction leaflet.

15.5.8 *Visual Inspection*

(1) Remove cover from relay case.
(a) The trip circuit is a live circuit and on some relays it is possible to cause an instantaneous trip while removing the relay cover.
(b) Inspect the cover gasket.
(c) Check glass for tightness in the frame, cracks, etc.

(d) Clean glass inside and out.

(2) Remove relay from case.
(a) To eliminate uncertainties, short the current transformer secondary by jumpering the relay operating coil terminals on the back of the relay case. This jumper should be clipped on with square jaw clips such as crocodile clips.
(b) Open trip circuit by opening the red-handled switch or removing the test block, depending on the type of relay involved.
(c) Open the rest of the switches, on all black-handled relays.
(d) Open the latches that hold the relay in the case and carefully remove relay from case. Remember that the switch blades attached to the case as well as the bars in the bottom of the cases are still energized. If extreme care is not exercised, the circuit breaker may be tripped. With capacitor trip, voltages as high as several hundred volts are present in the trip circuit.

(3) Foreign material such as dust or metal particles should be removed from the relay case and the relay. This foreign material can cause trouble, particularly in the air gaps between the disk and magnet.

(4) Dust can be removed by blowing air from a small hand syringe.

(5) Remove any rust or metal particles from disk or magnet poles with a magnet cleaner or brush.

(6) Hold relay up to the light to make sure that disk does not rub and has good clearance between magnet poles.

(7) Inspect relay for the presence of moisture. If free moisture is present or rust spots are noted, it may indicate that the relay is in the improper atmosphere and presents a design problem.

(8) Connections, especially taps, should be checked for tightness. Tighten all screws, nuts, and bolts that are not pivotal joints.

(9) Sluggish bearings may be detected by noting smoothness of relay reset. Rotate the disk manually to close the contacts and observe that operation is smooth. Allow the action of the spiral spring to return the disk to its normal de-energized position. If the bearings appear bad or operation of the relay is questionable, the relay should be returned to the manufacturer for overhaul. On an instantaneous plunger-type relay, occasionally a burr or groove may develop on

the plunger which would cause the relay to hang up.

(10) Check mechanical operation of targets by lifting the armature and observe showing of target.

(11) Check for damage to the relay coils caused by prolonged high currents by smelling or squeezing the coil with the fingers. A burned smell, or spongy or brittle insulation indicate thermal damage.

(12) Components of the relay that touch when the relay is in a "normal" position and part as the relay "operates" should be cleaned. These parts may become dirty and prevent the relay from operating properly on relatively low overloads.

(13) Pitted or burned contacts should be cleaned with a burnishing tool. Never use a solvent on these contacts or touch with fingers as the residue left on the contacts may cause improper operation.

15.5.9 *Electrical Testing*

(1) Test methods

(a) Testing with the relay disconnected from the power and trip circuits (most popular for field testing)

(b) Testing across the secondary of the current transformer with primary de-energized (secondary injection)

(i) This method may be used whenever it is possible to de-energize the power circuit being tested.

(ii) This test includes checking the operation of the circuit breaker, the presence of energy to trip the circuit breaker, etc.

(iii) Testing is done by introducing the test current at the secondary terminals of the instrument transformer. This test checks the relay connections as well as the relay.

(c) Primary circuit test; high current, low voltage (primary injection)

(i) Complete system is checked including the current transformers.

(ii) The primary bus must be de-energized and exposed to perform this type of test.

(iii) This method requires that caution and safety practices be closely observed as test connections are made on the primary conductors.

(iv) test simultaneously checks current transformer ratio, secondary wiring, polarity, relay operation, and identity of each phase on the switchboard.

(v) This test is valuable for initially checking bus differentials where many current transformer secondaries are paralleled at the relay.

(2) Test connections

(a) Relays with draw-out construction

(i) Whenever practical, the relay should be tested in place, disconnected from the power and trip circuits. The use of the proper test plug is convenient.

(ii) If the relay cannot be tested in place, carefully remove it from case and proceed to set up at a convenient location. Make sure that the relay is level. If possible, a spare case should be used to hold the relay during the test.

(b) Relays permanently fixed to panel

(i) Usually these relays have test facilities installed to disconnect the relay from the power and trip circuits and to permit access to the relay circuits from the front of the board.

(ii) If such facilities do not exist, it is necessary to test from rear of panel.

(3) Typical test for an overcurrent relay would include

(a) Zero check

(b) Induction disk pickup

(c) Time—current characteristics

(d) Target and seal-in operation

(e) Instantaneous pickup

The importance of these tests and the procedures to be followed in conducting them are given below.

(a) *Zero check.* This test is to determine that the relay contacts close when the time dial is set at zero. For this test a continuity light is used. Consult manufacturer's instruction leaflet or relay schematic to identify relay trip circuit contact terminals.

(i) With the continuity light connected across the relay induction disk trip circuit contact terminals, manually turn time dial until the continuity light glows.

(ii) Note reading of time dial.

(iii) If reading is not zero, consult manufacturer's leaflet for proper adjustment so that the time dial reads "zero" when contacts just make.

(iv) In some relays the position of the back-stop may be changed on the shaft. In others the position of stationary contacts may be changed. Some relays have no adjustment.

(b) *Induction disk pickup test.* This test is to determine the minimum operating current of the relay, that is, the minimum current needed to close the relay induction disk trip circuit contacts for any particular tap setting.

Consult manufacturer's instruction leaflet or relay schematic to identify operating coil terminals and the trip circuit contact terminals.

(i) The pickup value should be equal to tap value ± 5 percent.

(ii) Connect the induction disk operating coil terminals of the relay to the source of test current.

(iii) Connect a continuity light to relay induction disk trip circuit contacts.

(iv) Energize the relay operating coil with 150 to 200 percent of relay tap value so that the relay induction disk trip circuit contacts will close rapidly as indicated by glow of continuity light.

(v) Gradually back off current until contacts open as indicated by extinguishing of the continuity light.

(vi) Alternately increase and decrease current to find the point where continuity light flickers. This indicates that the contacts are just making.

(vii) Record this value of current as "pickup." Usually pickup is adjusted by changing tension of the spiral spring that is employed to return the relay trip circuit contacts to their de-energized position. Increase spring tension to increase pickup. The relay manufacturer's instruction booklet should be consulted to assure the proper procedure.

(c) *Time—current characteristics.* A timing check should be made to see that the relay closes its contacts within a specified time for a given abnormal value of current. Normally this test is run with the relay tap in its designated position.

It is suggested that a test current of four times pickup be used. However, in any specific plant, management personnel will indicate given test points and acceptable test results.

Consult manufacturer's instruction leaflet or relay schematic to identify operating coil terminals and the relay trip circuit contact terminals.

(i) The operating coil of the relay induction disk unit should be connected to the source of test current. Adjust current to test value and then de-energize the relay.

(ii) Connect relay induction disk trip circuit contacts to a timer.

(iii) Energize relay to determine time of operation at the chosen value of test current.

(iv) Record the values of test current and time. Compare with manufacturer's specified values. If the relay is too fast, either increase time dial setting (up to one half division) or adjust damping magnet. If relay is too slow, either decrease time dial setting or adjust damping magnet. Consult manufacturer's leaflet for method of making adjustment.

(d) *Target (operations indicator) and seal-in operation.* The purpose of the target is to indicate which relay tripped the circuit breaker. The seal-in provides a low-resistance trip circuit in parallel with the relay's induction disk trip circuit contacts and spiral spring.

There are three types of target units in common use:

(i) Combined target and seal-in unit

(ii) Separately operated target and separately operated seal-in

(iii) Mechanical target

(iv) Relay seal-in (contactor switch)

Induction disk relays have the combined target (operation indicator) and seal-in unit. Relays of older design may have a separate target. Some relays, such as the instantaneous overcurrent, have mechanical indicators.

(i) Combined target and seal-in test procedures.

(α) Close relay induction disk unit trip circuit contacts.

(β) Connect a direct-current supply to the relay induction disk trip circuit contact terminals. Note the tap setting of the target unit.

(γ) Gradually increase the magnitude of the direct-current supply until the target shows.

(δ) Open the relay induction disk trip circuit contacts. The current through the direct-current circuit should remain unchanged if the seal-in unit is operating properly.

(ϵ) When installing the relay cover, make sure the target reset operates correctly.

(ii) Separately operated target. The same test procedure as the first three steps for combined target and seal-in apply.

(iii) Mechanical-operation targets. These targets show as soon as the relay trip circuit contacts close. No separate electrical check is required.

(iv) Relay seal-in (contactor switch). Some relays are equipped with a seal-in (contactor switch) independent of the target. Note seal-in rating from the relay nameplate and proceed with testing as for combined target and seal-in. Be sure locknut on seal-in unit is tight.

(e) *Instantaneous-unit pickup.* Consult manufacturer's instruction leaflet or relay schematic to identify instantaneous unit operating coil terminals and instantaneous unit trip circuit contact terminals.

(i) Connect relay instantaneous unit operating coil terminals to a source of test current.

(ii) Connect relay instantaneous unit trip circuit contact terminal to a continuity light.

(iii) Gradually increase magnitude of test current until the continuity light glows.

(iv) Note ammeter reading as instantaneous unit pickup. If pickup value differs from specified value, adjust and repeat test. If target operation is not mechanical, test operations indicator as described above.

The preceding test procedures are given only as an example. There are many hundreds of different kinds of protective relays, and this chapter cannot go into detailed test procedures for each and every one. However, the relay manufacturer's instruction books normally contain data pertaining to maintenance and calibration tests.

15.5.10 *Solid-State Relays.* Testing and maintenance of solid-state protective relays differs substantially from the established practices for electromechanical types. Many of the reasons for maintenance (Section 15.4.2) are eliminated. Since many solid-state relays have built-in operational test provisions, these tests are simple to make. Calibration tests are made in the conventional manner.

15.6 Medium-Voltage Circuit Breakers Using Series Trip Coils. The maintenance and testing of these devices follow the same format as for low-voltage power circuit breakers, except that the trip current is properly introduced through the primary of the external current transformer. Tripping by injecting current through the secondary current transformer terminals is an acceptable alternate if isolation of the current transformer primary is not possible.

15.7 Fuses. Maintenance of fuses is limited by nature of the device to an inspection to ensure that the proper size fuse is installed, that it shows no signs of deterioration, and that the enclosure is clean and the connections are tight. The size and type of fuses should comply with those specified by the engineering department.

15.8 Auxiliary Devices. Periodically the substation battery should be checked for proper water level, voltage, and specific gravity of each cell. Terminals should be cleaned and connections tightened. The battery charger should be checked frequently to make sure that the charging rate is correct and that the charger is actually in operation. Control circuitry and auxiliary relays can be checked periodically through an operations test. The operations test consists of closing the protective relay trip circuit contacts manually and assuring that the proper circuit breaker will open. This test will check continuity of all control circuitry and the ability of the battery to trip the circuit breaker. It will also check operation of the circuit breaker. Since this test requires that the power circuit be de-energized, it must be scheduled at the convenience of operating personnel.

15.9 References

[1] Morrow, L.C., Ed. *Maintenance Engineering Handbook*, 2nd ed. New York: McGraw-Hill, 1966, sec 7, chap 11.

[2] "Procedures for Verifying Performance of Molded-Case Circuit Breakers," NEMA, Jun 30, 1971.

Index